…

OBSERVING LAND FROM SPACE:
SCIENCE, CUSTOMERS AND TECHNOLOGY

ADVANCES IN GLOBAL CHANGE RESEARCH

VOLUME 4

Editor-in-Chief

Martin Beniston, *Institute of Geography, University of Fribourg, Perolles, Switzerland*

Editorial Advisory Board

B. Allen-Diaz, *Department ESPM-Ecosystem Sciences, University of California, Berkeley, CA, U.S.A.*

R.S. Bradley, *Department of Geosciences, University of Massachusetts, Amherst, MA, U.S.A.*

W. Cramer, *Department of Global Change and Natural Systems, Potsdam Institute for Climate Impact Research, Potsdam, Germany.*

H.F. Diaz, *NOAA/ERL/CDC, Boulder, CO, U.S.A.*

S. Erkman, *Institute for Communication and Analysis of Science and Technology – ICAST, Geneva, Switzerland.*

M. Lal, *Centre for Atmospheric Sciences, Indian Institute of Technology, New Delhi, India.*

M.M. Verstraete, *Space Applications Institute, EC Joint Research Centre, Ispra (VA), Italy.*

The titles in this series are listed at the end of this volume.

OBSERVING LAND FROM SPACE: SCIENCE, CUSTOMERS AND TECHNOLOGY

Edited by

Michel M. Verstraete
*Space Applications Institute,
Ispra, Italy*

Massimo Menenti
*Laboratoire des Sciences de l'Image,
de l'Informatique et de la Télédétection,
Université Louis Pasteur,
Illkirch, France*

and

Jouni Peltoniemi
*Finnish Geodetic Institute,
Masala, Finland*

KLUWER ACADEMIC PUBLISHERS
DORDRECHT / BOSTON / LONDON

A C.I.P. Catalogue record for this book is available from the Library of Congress.

ISBN 0-7923-6503-8

Published by Kluwer Academic Publishers,
P.O. Box 17, 3300 AA Dordrecht, The Netherlands.

Sold and distributed in North, Central and South America
by Kluwer Academic Publishers,
101 Philip Drive, Norwell, MA 02061, U.S.A.

In all other countries, sold and distributed
by Kluwer Academic Publishers,
P.O. Box 322, 3300 AH Dordrecht, The Netherlands.

Cover: Information Delivery System
(figure by J.S. MacDonald)

Printed on acid-free paper

All Rights Reserved
© 2000 Kluwer Academic Publishers
No part of the material protected by this copyright notice may be reproduced or
utilized in any form or by any means, electronic or mechanical,
including photocopying, recording or by any information storage and
retrieval system, without written permission from the copyright owner.

Printed in the Netherlands.

TABLE OF CONTENTS

Preface
Michel M. Verstraete and Massimo Menenti   1

1 Welcome address
Kalevi Hemilä   3

2 Activities of the European Commission in Earth Observation: Part of a common European strategy
Alan Cross   7

3 The CEO initiatives to collect and analyze user requirements for Earth Observation
Jean Verdebout   15

4 National strategy for developing Earth Observation applications in Finland
Kari Tilli   29

5 Operationalization of Earth Observation: Its prospects and requirements
John S. MacDonald   35

6 Remote sensing of forest fires
Emilio Chuvieco   47

7 Using current and future remote sensing systems in natural hazards management
Jesus San Miguel-Ayanz, Michel M. Verstraete, Bernard Pinty, Jean Meyer-Roux and Guido Schmuck   53

8 Examples of the use of satellite data in numerical weather prediction models
Joel Noilhan, E. Bazile, J.-L. Champeaux and D. Giard   61

9 The contribution of remote sensing technologies and algorithms to land surface processes studies
Bernard Pinty and Michel M. Verstraete   71

10 Exploitation and evaluation of retrieval algorithms for geostationary satellite data processing: Current and future systems
Yves Govaerts   77

11 The role of remote sensing in land surface experiments within BAHC and ISLSCP
Anne Jochum, Pavel Kabat and Ronald Hutjes   91

12 Earth Observation demands for improved water resources management
Wim G. Bastiaanssen and C. J. Perry 105

13 A biophysical process-based estimate of global land surface evaporation using satellite and ancillary data
Bhaskar Choudhury 119

14 Remote sensing of land cover and land cover change
Barry K. Wyatt 127

15 Land-cover categories versus biophysical attributes to monitor land-cover change by remote sensing
Eric Lambin 137

16 A new approach to characterize global land surfaces: Preliminary results from AVHRR data
Nadine Gobron, Bernard Pinty and Michel M. Verstraete 143

17 Remote sensing requirements to support forest inventories
Erkki Tomppo 151

18 Some research and applications in the CSIRO (Australia) Earth Observation Centre on scene brightness due to BRDF
David L. Jupp 161

19 Remote sensing of albedo using the BRDF in relation to land surface properties
Wolfgang Lucht, Crystal Schaaf, Alan Strahler and Robert E. d'Entremont .. 175

20 Experimental study of statistical characteristics of plant canopy radiation regime
Madis Sulev and Juhan Ross 187

21 Light scattering models and reflectance measurements in remote sensing of snow
Risto Kuittinen 201

22 Ray optics approximation for random clusters of Gaussian spheres
Karri Muinonen 209

23 Backscattering of light by snow: Field measurements
Jukka Piironen, Karri Muinonen, S. Keranen, H. Karttunen and Jouni Peltoniemi 219

24 Australian sites for the validation of satellite retrievals of the radiative properties of land surfaces
Ian F. Grant, A. J. Prata, G. Rondeaux and M. D. Steven 229

25 Instruments and methods for the ground-level reference measurement of solar radiation, albedo and net radiation
Madis Sulev, Juhan Ross and Enn-Märt Maasik   243

26 The NASA Earth System Science program in 21st century
Ghassem Asrar   257

27 ESA's plans and strategy for optical remote sensing of terrestrial surfaces in the next decade
Michael Rast   269

28 Early results from ADEOS and future Earth Observation missions
Tamotsu Igarashi   279

29 VEGETATION: An Earth Observation system to monitor the biosphere
Gilbert Saint   291

30 Hyperspectral imager survey and developments for scientific and operational land processes monitoring applications
Bernd P. Kunkel, J. Harms, U. Kummer, E. Schmidt, Umberto del Bello, B. Harnisch and Roland Meynart   303

31 Summary and conclusions
Michel M. Verstraete and Massimo Menenti   329

Appendix
List of participants   333

PREFACE

M. Verstraete (1) and M. Menenti (2, 3)
(1) Space Applications Institute, Ispra, Italy, (2) The Winand Staring Centre for Integrated Land, Soil and Water Research, Wageningen, The Netherlands and (3) Université Louis Pasteur, Illkirch, France.

The European Network for the development of Advanced Models to interpret Optical Remote Sensing data over terrestrial environments (ENAMORS) is a consortium of academic and research institutions involved in methodological research and in applications of remote sensing techniques for Earth Observation. It was supported initially through a Concerted Action from the Environment and Climate Research and Technology Development Program in the 4th Framework Program of the European Commission. Its activities include the organization of international scientific conferences, the first of which took place in Tuusula, Finland, from September 17 to 19, 1997. This book contains the proceedings of that conference and effectively summarizes the discussions and conclusions reached by the participants.

The title of this meeting was 'Optical Remote Sensing of Terrestrial Surfaces: New Sensors, Advanced Algorithms, and the Opportunity for Novel Applications'. It aimed at assembling representatives from the policy maker, remote sensing research and end-user communities, as well as from national and international space agencies and aerospace industries. Together, they discussed the need for R&D support, as well as the contents and priorities of such a program in this economic sector during the period covered by the 5th Framework Program (1999--2002). Information on this conference as well as on other activities of the ENAMORS network are available on the World Wide Web (http://www.enamors.org/).

Over 50 participants from 14 countries (Australia, Belgium, Canada, Estonia, Finland, France, Germany, Israel, Italy, Japan, Spain, Sri Lanka,

Sweden, The Netherlands, United Kingdom, and United States of America) participated in the meeting and discussions. A list of participants is enclosed in the Appendix. Their participation in this endeavor is testimony of their confidence and expectations in the European Commission to play an important role in Earth Observation activities in the future. The conference was organized around five sessions, whose contents are reflected in the various parts of this book.

The papers included in this book provide a useful overview of the state of the art in the interpretation of Earth Observation data acquired in space, for a range of thematic applications. They discuss some of the policy implications of recent research and technological advances, as well as the degree to which existing and future instruments and methods address specific user needs. They certainly reflect the broad range of issues that were addressed during that meeting, and provide food for thought on the opportunities and challenges before us in the coming years.

Chapter 1

WELCOME ADDRESS

K. Hemilä
Minister of Agriculture and Forestry, Helsinki, Finland.

Dear participants in the ENAMORS-workshop,

The topic of your meeting is optical remote sensing and its development and exploitation for the observing, inventorying and monitoring of changes taking place on terrestrial surfaces. This is an important topic, because most human activities—starting from food and raw materials production and ending in processing the waste produced by society—take place on the terrestrial surface. Up-to-date information concerning these matters forms the basis for systematic decision making on both the national and the international levels.

On the European scale of things, Finland is territorially a relatively large country. Because our country is also sparsely populated, favorable conditions exist for using remote sensing to inventory our natural resources and monitor the state of our environment. Therefore the development of remote sensing has been viewed in Finland as the most important field of development of space technology. Remote sensing is being developed in three areas:
- Finland as a member of the European Space Agency participates in its Earth Observation program,
- Finland as a member of the European Union participates in its Environment and Climate Program, collaborates with its Joint Research Center and carries out operative remote sensing in agriculture, and
- Finland develops its national remote sensing capability.

The present meeting is a part of our collaboration with the European Union.

The Ministry of Agriculture and Forestry plays an important role representing both developers and users of remote sensing in Finland, as the field of competence of the Ministry includes agriculture, forestry, mapping activities, game and fisheries as well as an important part of environmental protection.

Finland has for natural reasons a long tradition in the remote mapping of its forests, which started in the sixties using aerial photography. Main objects of interest are forest inventory and planning, as well as the monitoring of forest ecosystems. The work comprises the use of both satellite and airborne imagery in operational activities. In this meeting, you will also hear about the results of these activities.

When Finland joined the European Union, the application of remote sensing to agriculture received a new impetus. Remote sensing is now being used for controlling farmers' declarations for EU support, and methods are being developed for obtaining yearly crop yield estimates. In both cases the work is being done in collaboration with the EU Commission and the Joint Research Center.

The updating of small scale topographic maps using new high-resolution satellite imagery is currently starting up in Finland, made possible by progress in satellite technology. The use of satellites plays nowadays a very central role in the geodetic measurements needed for mapping. During the past ten years, Finland has started to very extensively use satellite positioning (GPS) in mapping work. Significant resources are being expended yearly on the development of mapping techniques and satellite geodesy, which will ensure that our country has a modern map production based on digital data. This again will offer good possibilities for Finnish participation in the European harmonization of mapping and co-ordinate systems in years to come.

Geographic information systems clearly offer the best possibility to manage information on parts of the Earth's surface. Both national and international systems are needed, and these should be compatible. The CORINE (Co-ordination of Information on the Environment) Land cover program, being prepared in the European Union, and the Finnish national land use maps serve as examples. In both cases, remote sensing is used to enter up-to-date information into the system, which would be very hard to obtain by any other means. Thus, remote sensing is becoming an important technique for obtaining input data for geographic information systems as well as for various geophysical models. However, one should be aware that GIS-systems are useful only if the data in it is accurate both in location and in content. You, in your research work, will strive to make these items of information as accurate as possible.

1. WELCOME ADDRESS

Finland in its science and technology policy is consistently increasing its funding of research and product development. Currently, 2.6% of our country's GNP is being allocated for research and product development, and this fraction is intended to grow still. Remote sensing constitutes one part if this development activity. Because of the increasing demand for information within the field of competence of my Ministry, I view the meeting starting today as important for Finland. This meeting is also important because, through European collaboration, we can achieve results that will further the rapid collection of up-to-date, reliable information concerning our natural resources and environment.

I hope, in particular, that efforts will be expended on the development of operational procedures and systems, because there is a great need to take new methods into use in a society in which the rate of change is increasing all the time. One should aim for these operational methods to produce cost savings in data collection and to increase the effectiveness of information management.

Your work in developing remote sensing methods is demanding and fascinating. I wish success to your work and to this meeting.

Chapter 2

ACTIVITIES OF THE EUROPEAN COMMISSION IN EARTH OBSERVATION
Part of a common European strategy

Alan Cross[1]
European Commission, Brussels, Belgium.

1. INTRODUCTION

1.1 A European success story

The European Earth Observation (EO) sector has reached a pivotal stage. Over the past two decades, Europe has developed an outstanding technological base for EO through both the co-operative programs of Member States in ESA, and through the various national space programs. EUMETSAT successfully operates weather satellites for European meteorological services and contributes to related activities worldwide. The European Commission has made a substantial effort in stimulating EO applications through its Research and Technology Development (RTD) programs, including the major contribution of the Center for Earth Observation. As a result of these and other initiatives, European industry and research bodies have developed and demonstrated a wide range of EO applications. EO is increasingly seen as a useful and sometimes essential component of the technological infrastructure of the 'Information Society'. Synergies with other space technologies such as telecommunications and navigation can only enhance this role. The potential social, economic and strategic importance of EO over the coming decade is considerable:

---

[1] The views expressed in this paper are those of the author and they should not be regarded as an official position of the Commission.

- Global scale measurements of parameters such as ocean color, sea surface temperature, continental land cover and biomass burning, vital for initiating and validating models of global change, can only be measured with satellite EO. Earth system science entails a global partnership in which Europe has clear obligations.
- EO will also form a key input to new environmental monitoring requirements in the context of international agreements, such as the Montreal Protocol, the Biodiversity Convention and the Framework Convention on Climate Change. This last agreement has given rise to the Kyoto Protocol, which recognizes forest cover change as a component to be taken into account in future international audits of carbon sources and sinks.
- EO can be a cost-effective tool for the implementation of environmental and civil protection policies. EO is already used pre-operationally for monitoring changes in land cover for the Natura 2000 network as part of the Habitats Directive, and for detecting oil spills in the context of the Bonn Convention. The information offered from space can help in the management and even prediction of natural hazards such as floods, storms and forest fires. Any contribution of EO to their mitigation could have an enormous economic impact.
- EO is already used to check claims for agricultural subsidy, and for gathering statistics. EO will also play a role in the context of an evolving Common Agricultural Policy, particularly with an increasing emphasis on environmental impact.
- The anticipated enlargement of the EU to the East will bring the need for harmonized statistical information over very large areas—a task to which EO can make a unique contribution. An independent European EO capability will provide a secure source of information as we move towards an increasingly global economy.
- Dual use aspects of EO are becoming increasingly important as the distinctions between military and civil technologies become blurred. Civil data has been widely used by NATO and UN forces for operations in the Gulf, Bosnia and Central Africa. Formerly classified US and Russian high resolution data collected during the cold war era is now available for civil use, providing a valuable information source, but potentially disturbing the market for commercial systems. Sensor technologies developed for the US Department of Defense (DoD) will form the basis of new American commercial high-resolution systems.

1.2 An uncertain future

Nevertheless, despite its past successes and clear future potential, the European EO capability is fragile.

A long period of application development and demonstration has been sustained by publicly subsidized space and ground infrastructures. Most European governments now envisage a shrinking of this subsidy, and instead foresee the introduction of private capital into the funding of new missions. However, the customer base for EO, both public and private sector, remains dispersed and underdeveloped. Many potential users[2] are unaware of the possibilities of this data source. Others are deterred by the lack of long-term continuity in data supply caused by the absence of an operational framework. The weak customer-demand deters industry from making the major investments necessary to initiate private space ventures, and financial institutions will not risk loans whilst the commercial viability of EO is considered unproven.

Meanwhile, Europe faces intense and increasing international competition. Private operators in the US are on the verge of providing potentially lucrative end-to-end commercial services based on data from very high-resolution optical systems. Europe is still a leader in Synthetic Aperture Radar (SAR) technology, but Canada is rapidly gaining experience in commercial SAR, and NASA is working with the private sector to promote a lightweight mission using advanced radar technology. Although not yet so clearly market-driven, operators from Russia, India, China and Brazil will play an increasingly active role in the global arena.

Decisive action is needed to capitalize on the past investments in EO, to ensure that the social, economic and strategic benefits accrue to Europe in the coming decade, and, in so doing, to create a European industrial sector competitive in world markets.

1.3 The way ahead

The main market for EO will continue to be the public sector, stemming from requirements arising from scientific programs and from the demand for information in public policy implementation.

The challenge over the coming decade will be to change the role of government from that of source of subsidy, to that of customer. This will create a climate of 'user-pull', allowing industry increasingly to finance and deploy services meeting real demand, taking account of tradeoffs between

---

[2] The term 'user' refers here to any person or body deriving value from information obtained from EO for a given application.

cost and performance. Full advantage will be made of small satellites developed quickly, optimized for specific applications, and lofted using cheap launch opportunities. A market-driven approach will lead to a more efficient allocation of public resources, and should eventually reduce the cost of operational services to the public authorities. In such an environment European EO industry (ranging from aerospace manufacturers to value adding services) should undergo whatever restructuring, consolidation, and integration may be necessary to maintain and strengthen competitiveness in the global arena. A robust EO industry will ensure the necessary continuity of service.

The role of the space agencies in EO will be to ensure that the underpinning essential technologies and scientific data are available, to help industry mount user-led missions during a transitional phase, and to assist other public bodies in procurement. This is implicit in the proposed Earth Explorer and Earth Watch initiatives of ESA.

In certain circumstances it may be appropriate to create new public operators, drawing perhaps on the experience of EUMETSAT, or to establish novel public/private partnerships. One key obstacle is the widespread under-valuation of EO in the eyes of users. This is partly due to the past availability of free or cheap data, and partly to a disappointment in current systems and services which have often failed to meet requirements because of an inadequate space segment for operational needs.

2. A COMMON EUROPEAN STRATEGY

The European Commission does not act alone, and it does not intend to replace or unnecessarily duplicate other initiatives. A tripartite framework for European co-operation in the field of EO has been proposed jointly by the main institutions concerned: ESA, EUMETSAT and the Commission. This proposal was tabled at Ministerial level at the ESA Council of 1995. The Commission then set out its broad intentions for action in its third Communication on space, which was published in December 1996. The Communication has been largely endorsed by the Council of Ministers, the European Parliament, and the Economic and Social Committee.

The European Space Agency, in consultation with its European partners, has undertaken a profound review of its strategy for EO in the coming 15 years, focusing on the post-ENVISAT era. The new strategy is laid out in a proposal for a 'Living Planet' program, built around the concept of Earth Explorer Missions for scientific goals, and operational Earth Watch Missions for eventual operational applications.

2. ACTIVITIES OF THE EUROPEAN COMMISSION IN EARTH OBSERVATION

The ESA Council and the EU Council of Research Ministers both adopted a joint resolution in early 1998 which called for continued co-operation between the two bodies in the space domain, and looked forward to the agreement of an action plan in the field of Earth Observation.

The convergence of thinking in Europe was confirmed in May 1998, on the occasion of the JRC Annual Users Seminar in Baveno, Italy. Representatives of key European organizations (ESA, EUMETSAT, CNES, DLR, BNSC, EARSC, EC) recognized the impetus that could be given to the EO industry by new demands from public policy, particularly linked to environmental security.

The Commission must now carry forward the ideas set out in earlier policy documents. The strategic goal itself flows from the earlier work: to establish a sustainable European capability in the provision of key operational services for monitoring the Earth from space, in response to user-demand.

- A service is sustainable when the full economic cost is met by those demanding it, and when it is reasonable to suppose that such demand will continue, or increase, in the foreseeable future.
- A European capability in EO ensures a measure of independence and helps bring about the double prize of user benefit and industrial growth.
- The key operational services will be lead to improvements in our understanding of the Earth system, more effective conservation of our natural heritage, enhanced management of resources, and mitigation of major hazards.

Within the broad European effort, the Commission's competence complements those of its partners. It bases its actions on four aspects of its mandate:

- as a manager of European-level RTD programs;
- as a user, and as the initiator of policies that entail the use of EO;
- as a proposer of policies bearing upon the political, economic and regulatory framework within which the EO sector operates (including a commitment to boost the competitiveness of European industry);
- as a Community institution representing the EU in certain international forums.

Given the above, the Commission can pursue the strategic goal over the coming five years by taking steps in line with four specific objectives.

- *OBJECTIVE 1: Expand the effective use of EO in the implementation of EU policies*

There is untapped potential for operational EO in the context of many EU policies. The use of EO by the Directorate Generals (DGs) of the Commission, whilst always likely to be limited, will continue to provide highly visible evidence of the effectiveness of this information source. Perhaps more significantly, EU policies can be framed in such a way to allow national administrations, program participants, and other actors such as the European Environment Agency to take full advantage of the benefits of EO.

- *OBJECTIVE 2: Increase the value of EO services*

Demand from users will increase with the perceived value of the information delivered. This perception will depend on the provision of a quality service, linked to a willingness to integrate new types of information into existing management structures. Efforts to increase user-demand will first boost the value-adding sector of the EO industry. These may lead to requirements for new space missions, to the advantage of the upstream aerospace industry. The Fifth Framework Program, including both direct and indirect actions[3], will build on work initiated in the Fourth Program by improving, demonstrating and facilitating access to EO. The emphasis will be on helping industry and users to identify opportunities jointly, both sides becoming stakeholders in eventual exploitation. The Framework Programs will also be used to stimulate use in other areas of EU policy.

- *OBJECTIVE 3: Prepare users and industry for future sustainable services*

The Commission does not intend to intervene directly in the space segment. This is the domain of the space agencies and industry. Nevertheless, the breadth of the Commission's mandate, together with the market links between ground and space activities, means that it cannot solely restrict its focus to applications. The Commission will therefore act to help bridge the gap between the users and the space segment. The emphasis will be on defining and marshalling requirements, and on policy initiatives designed to move towards sustainability. Against this background, one key focus will be the Commission's input to the Earth watch initiative of ESA.

- *OBJECTIVE 4: Strengthen the impact of European action*

The Commission will ensure that its actions are co-ordinated with its partners in Europe. As an actor in European space policy with a competence in affairs beyond Europe, and as a funding agent for global scale research, the Commission will also co-operate with non-European bodies, whilst safeguarding the competitiveness of the European sector. It will ensure that Europe plays a leading role in international forums such as the G-7 body, Committee on Earth Observation Satellites (CEOS) and related initiatives, necessitating a close co-ordination between EU Member States.

---

[3] 'Direct action' refers to research carried out by the Joint Research Centre; 'indirect action' refers to research programmes managed by the Commission but undertaken by consortia, principally through calls for proposals and shared-cost mechanisms.

3. CONCLUSIONS

This paper describes the rationale for a continued involvement by the European Commission in the field of Earth Observation over the next five years, responding to new challenges and opportunities, as part of a wider European effort involving ESA, EUMETSAT, national space bodies, industry and other European actors.

It aims to secure the user benefits of EO, particularly those linked to priority socio-economic issues. The approach taken is to foster market-driven solutions as far as possible, leading eventually to a self-sustaining and competitive European EO sector.

The RTD Framework Programs will be used to mobilize much of the necessary effort, drawing on both the direct and indirect actions. A coherent action on Earth Observation in the Fifth Framework Program, including activities under the key action on Global change, climate and biodiversity, will play a key role during the period covered by this plan.

Both the current and future Framework Programs will be used as vehicles to stimulate operational EO applications in other EU policy areas. Other Community competencies, for example, those linked to the functioning of the single market, will be invoked if needed.

4. REFERENCES

Proposal for a European Policy for Earth Observation form Space, ESA/PB-EO (95) 7 rev. 2, presented to the ESA Ministerial Council in Toulouse on 20 October 1995.

The European Union and Space: Fostering applications, markets and industrial competitiveness; Communication form the Commission to the Council and the European Parliament, 4 December 1996 (COM (96) 617 final).

Chapter 3

THE CEO INITIATIVES TO COLLECT AND ANALYZE USER REQUIREMENTS FOR EARTH OBSERVATION

J. Verdebout
Space Applications Institute, Ispra, Italy.

1. INTRODUCTION TO THE CEO PROGRAMME

1.1 Objectives

The mission of the CEO (Centre for Earth Observation) Program is to develop and promote the use of Earth Observation data from space. To achieve this goal, the CEO Program seeks to increase the number of customers of EO derived information (in research, governmental and commercial fields), to encourage and stimulate the EO related industry (research and commercial) and to improve access to and availability of EO and related data, information and services.

The CEO program is funded both by the European Commission (EC) via the Directorate General Joint Research Centre in Ispra, Italy and by the Space Unit of Directorate General XII in Brussels. By funding such a program the EC hopes that it will re-balance the effort between the space segment of Earth Observation, and the ground segment (i.e., the use of these data), and in so doing help to develop an operational Earth Observation industry.

1.2 The CEO program in a global context

The CEO complements developments elsewhere in the world. In the United States of America (USA) the National Aeronautics and Space Administration (NASA) has begun to develop and implement an Earth Observation System Data and Information System (EOSDIS) for its planned Earth Observing System, supplemented by the Global Change Data and Information system (GCDIS). Japan has proposed an international Earth Observation Information System (EOIS) for its Advanced Earth Observation Satellite (ADEOS). Canada and Australia are planning similar programs.

Earth Observation activities are international in nature. Many users in Europe require access to data and information that are held elsewhere. Organizations and companies based in Europe wish to advertise their products and services to an international audience. International programs on climate and environmental changes are key users of Earth Observation information, and require worldwide collaboration. Hence, in this context a major goal of the CEO is to help European users to have access to such international links, which include interoperability of catalogues, agreement of common standards and formats and better access to non-European sources of data or information.

At the same time the CEO has become a major part of the activities currently being undertaken in Europe. CEO works closely with the European Space Agency (ESA), EUMETSAT and national agencies in an effort to ensure a coherent approach to the use and availability of Earth Observation data in Europe.

The actions implemented in the CEO program have been defined as a result of a pathfinder phase, which lasted from mid 1993 to end 1995. The design and implementation phase concluded at the end of 1998, together with the 4th Framework Program for Research and Development of the EC. In the future, the CEO will evolve to a more operational phase.

1.3 Concept for the CEO design and implementation

The CEO program seeks to build upon the core relationship between customers and service providers. Service providers are encouraged to develop products, which are tailored to the needs of customers. New customers are attracted by demonstrating how EO derived products can effectively be used in their particular professional tasks.

Four interrelated components have been identified as the basis for actions to meet these objectives, namely Application Support, Enabling Services, User Support and Monitoring & Co-ordination. Many initiatives have been

3. THE CEO INITIATIVES TO COLLECT AND ANALYZE USER REQUIREMENTS FOR EARTH OBSERVATION

launched within these components and are ongoing. These include the following:

1.3.1 Application support

The CEO Program is encouraging the development of EO applications by shared-cost funding of appropriate projects. These include the Calls for Proposals issued by DGXII of the European Commission, under Area 3.3 (Space techniques applied to environmental monitoring and research) of the Environment and Climate program, and a number of projects in support of the Services of the Commission. Actions to make ECMWF and EUMETSAT data sets accessible have also started.

1.3.2 User support

User Support implements initiatives on metadata standards, on a long-term European Digital Data Archive (EDDA) for Earth Observation, on an inventory of EO algorithms, models and software, on EO education resources and on market development for new customer segments of EO. An operational Helpdesk service was launched in 1996.

1.3.3 Enabling services

The EWSE (European Wide Service Exchange), which represents the testbed for the CEO's electronic data and information exchange system, was launched in September 1995. Since then, over 1500 users and over 800 resources (encompassing 341 organizations, 331 EO products and 130 case studies) have registered with this service. Every week the EWSE received in excess of 7000 inquiries.

The next generation system developed by the Enabling Services is now called INFEO. The corresponding URL is `http://infeo.ceo.org/`. The vision is to provide European EO Users with a 'one-stop shopping' facility that will be an easy-to-use system for querying data repositories on geographic, thematic, sensor or temporal subjects. INFEO builds on the performance and interoperability of existing information systems rather than replacing them. It is based on the EWSE prototype and designed such that, although the databases are located in various places across Europe, it is perceived by its users as a single integrated system. A number of Middleware Nodes (MWND) allow customers to search for the data, information and services they need, and allow catalogue owners, service providers, commercial, public and research organizations to announce and advertise their data, information and services.

1.3.4 Monitoring and co-ordination

The CEO is currently in discussion with large data providers (such as ESA, NOAA, NASA, EUMETSAT and CCRS, as well as national initiatives within Europe, such as NEONET in the Netherlands and ISIS in Germany) regarding collaboration and joint working agreements. The program has also made significant contributions to the Committee on Earth Observation Satellites (CEOS) and G7 initiatives on the accessibility and availability of data and information, as well as progress towards an agreed international catalogue interoperability protocol (CIP).

2. CAPTURE OF USER REQUIREMENTS IN THE PATHFINDER PHASE

2.1 User requirement studies

Although the goal was to identify the requirements of the current and potential users of Earth Observation data for the CEO program, many requirements for EO can be found in the results of this action. The following groups of EO users were addressed:
- professional groups, through a series of workshops organized on a national basis,
- remote sensing Institutes in Europe, through four workshops organized by groups of countries,
- experimental user communities four (studies were conducted addressing marine altimetry, coastal zones, agriculture, urban development, and global change/atmosphere),
- services of the European Commission, and
- commercial agencies.

From the 11 studies, 59 organizations were contracted to collect user requirements. In total, 407 different organizations were contacted, providing 1745 "requirement statements" for the CEO program. It was clear from an early stage that these requirement statements could not be used on their own; the method of expression varies, they overlap and duplicate in many areas and there are too many of them to handle in any sensible manner. A method of synthesis was developed to convert these requirement statements into requirements for the CEO. This involved a process of sorting and evaluation

3. THE CEO INITIATIVES TO COLLECT AND ANALYZE USER
REQUIREMENTS FOR EARTH OBSERVATION

to achieve a consistent method of grouping and of expression. A principal task was to be faithful to what users need, which in many cases differed in certain ways from how the requirement statements had been written. This work has resulted in various documents, which can be retrieved from the following URL: http://www.ceo.org/.

We believe that these documents can also be of benefit to other activities concerning EO in Europe. Care should be taken, however, in their use. These user requirements were collected for the CEO; they will not be directly applicable to all other areas of Earth Observation.

2.2 Proof of concept studies

The requirement capture was also supported in a different way, by conducting "Proof of Concept Studies". These aimed at prototyping potential CEO applications and communication techniques. The Application Proof of Concept Studies aimed at promoting and fostering the use and dissemination of data, products and services developed by application-oriented projects. The Communication Technologies Proof of Concept Studies tested critical data and information management capabilities, especially with regard to information exchange with the user and thus the applications.

The common link of both areas is the utilization of information exchange via established means (e.g., World Wide Web) and by the utilization of already existing technologies and approaches. This is directly reflected in the contributions that the Proof of Concept Studies have provided to the development of the European Wide Service Exchange testbed which is in turn a proof-of-concept study used to test the CEO concept of Enabling Services.

The 'Proof-of-Concept' Studies were mainly undertaken via contracts to European organizations. Institutes of the JRC also conducted a limited number of activities. Including sub-contractors, 64 European organizations participated in these studies.

Some preliminary lessons learned from the projects were:
- High level information (in simple formats, for non-expert users) is needed to promote EO to a wider community. Several ideas for addressing customers and demonstrating the applicability of EO to their needs were implemented by the projects. In a few cases, this resulted in successfully acquiring new customers, sometimes with sufficient interest to be potential new funding sources for the projects concerned.
- Non-space data have been demonstrated to be of vital importance to complement space data when these are either spatially/temporally

inadequate or because they are not sufficiently accurate for use on their own in many applications.

Information on the individual studies can be found in the INFEO servers of the CEO.

3. DESIGN AND IMPLEMENTATION PHASE ACTIVITIES

3.1 Area 3.3 (CEO) Shared Cost Actions

The CEO follows on the collection of requirements for EO by supporting pre-operational projects aimed at satisfying the needs of well identified customers. The main mechanism is through Shared Cost Actions (SCA). Concerted actions (CA), aimed at coordinating research for developing applications are also supported. All these actions form the area 3.3 of the Environment and Climate Program.

The objectives of area 3.3 are to:
- stimulate the production of information and related services from Earth Observation data in response to customer needs,
- encourage the provision of such services, and
- help increase the quantity and quality of available data, products and services.

Projects should help to develop pre-operational applications and place service providers in a good position to develop the market further. Shared-cost application projects will:
- help existing service providers to find customers,
- help existing users of Earth Observation to find service providers,
- encourage new customers to use Earth Observation,
- encourage existing customers to extend their use of Earth Observation into new fields,
- draw more participants into the CEO program, and
- populate and use the data bases maintained by the Enabling Services.

The CEO SCAs are intended to introduce the use of Earth Observation to as diverse a set of potential participants as possible, to ensure that workers in many disciplines become informed of the ways in which Earth Observation might be useful to them. For this reason the CEO does not prioritize any particular theme for the application of Earth Observation information in Area 3.3, and gives equal weight to any thematic area relevant to European Union policies. Normally these policies should be related to the environment, but projects in thematic areas that are indirectly

3. THE CEO INITIATIVES TO COLLECT AND ANALYZE USER REQUIREMENTS FOR EARTH OBSERVATION

related to environmental policies are considered provided that the activities of the project would be of considerable potential benefit to the CEO program and its participants.

The CEO SCAs contribute significantly to the aims of the CEO, and to the aims or interests of the Earth Observation users for which they are targeted. Where appropriate, they build on and exploit the achievements and experience of existing national or European programs or initiatives in Earth Observation. Concerted and shared-cost actions will make recommendations for further actions, including those for possible implementation in future Framework programs.

Regarding specifically how the SCAs will contribute to the documentation of requirements for EO, each project shall deliver a User Requirement Document, an Implementation Report and a document stating how well the user requirements were fulfilled by the developed products and services.

Following a first call, 19 Shared-Cost actions and 2 Concerted Actions are presently supported. The projects cover a wide range of thematic areas, from crop yield forecast at regional level to general atmospheric circulation. A summary of all the projects can be found in the CEO homepage (URL: http://www.ceo.org/projects.html). Projects submitted following a second call are presently under evaluation.

3.2 Projects in support of services of the Commission

These demonstration/pilot projects are very similar to the Shared Cost Actions in their overall objectives. Their specificity is to address needs and requirements of the European Commission itself; the customers are here Services of the Commission. In order to better monitor the progresses of these projects and to assist the customer DGs in the definition of the requirements, the thematic Units of the Space Applications Institute are providing support in the conduct of the projects and are in charge of the management. The projects are summarized in Table 1 below.

A short description of each project and its current status and future activities is given below.

3.2.1 LACOAST project

The aim of this project is to estimate quantitative changes of land cover and/or land use in European coastal zones especially due to human activities. The second part of the project consists in identifying and interpreting the factors responsible for the change itself. The customers for project

LACOAST are DG XI and the EEA who require specific environmental change information to contribute to major reports on the coastal zones of Europe (e.g., Dobris+3 and 'State of Environment' reports) at the end of 1997 and the end of 1998. If EO can successfully fulfil the requirements of this customer, then there is significant potential for repeated use of EO within coastal zone management by the Commission, and to expand this use to national governments and regional organizations.

Table 1. The projects in support of services of the Commission

Project Acronym	Thematic Area	Customer EC Service
LACOAST	Land cover changes in European coastal areas	DGXI (Environment, Nuclear Safety and Civil protection), DGXVI (Regional Policies), EEA
SEARRI	Monitoring of agriculture (wetland rice paddies) in SE Asia	DGIB (External Economic Relations), DGVI (Agriculture), DGXI
FMERS	Forest monitoring in Europe with remote sensing	DGVI
ATLAS	Statistical analysis on urban agglomerations in Europe	EUROSTAT
DESIMA	Decision support system for coastal areas	DGIB, DGIII

The LACOAST Project is composed of three main components:
1. The European wide analysis: This consists of a quantitative estimation of Land use/Land cover changes that have occurred over a period of about 15 years (1975 and 1990) on the entire European coastal zones. The classification for 1975 uses Landsat/MSS data, whilst for 1990 the Corine classification is already completed (Landsat/TM and SPOT data).
2. The quantitative statistical analysis: This consists of estimations of changes on samples of European coastal zones on 3 to 4 dates (1955, 1975, 1985 and 1995). The classification will make use of SPOT/HRV, Landsat/MSS and aerial photographs.
3. The case study analysis: Three selected zones (along the Belgian coast, along the border between Spain and Portugal, and in France) will be the subjects of a more detailed analysis. This will be used as prospective work for implementation of other coastal zones studies.

The Agricultural Information Systems (AIS) Unit of SAI co-ordinates this project.

3. THE CEO INITIATIVES TO COLLECT AND ANALYZE USER REQUIREMENTS FOR EARTH OBSERVATION

3.2.2 ATLAS project

The project has been scoped to fully satisfy EUROSTAT requirements. Given that EUROSTAT is already quite knowledgeable regarding EO capabilities, the interest focuses on the investigation of the potentiality of high-resolution satellite for urban land use maps. Two cities (Athens and Berlin) have been selected to provide maps of urban land-cover, as well as land-cover changes, that have occurred within the last few years, at a nominal scale of 1:25,000. Data from IRS-1C, KVR-1000 will be used in addition to coherence maps generated by ERS-1 and ERS-2.

The Environmental Mapping and Modeling (EMAP) Unit of SAI coordinates this project.

3.2.3 DESIMA project

The project aims at developing a Decision Support System for Coastal Zones (CZ-DSS). A generic system will be developed which is customizable according to specific requirements related to the monitoring and management of coastal zones. In particular, the system focuses on four initial themes:
1. aquaculture management,
2. development of fisheries information systems,
3. coastal ecosystem management, and
4. sea surface level estimation.

Regarding theme 1, aquaculture management, the activity foresees the design and planning of a prototype DSS for coastal areas in Thailand, responding to a specific request of DG-I B/C (External Economic and Development Co-operation: South- and Southeast Asia), which provides counterpart funding to this project. A detailed work plan has been developed for theme 1, including an analysis of user requirements, a conceptual system design and a prototype CZ-DSS demonstrator. Theme 1 acts as a model to develop the generic CZ-DSS. This will be adapted according to themes 2-4, following the prototype development for theme 1.

The Marine Environment Unit of SAI coordinates the project.

3.2.4 SEARRI project

The SEARRI project aims at developing a system to monitor wetland rice areas in Southeast Asia and to provide this information to two major user groups: agro-economists (foreign agricultural crop production) and the global change community (methane emission). The customer DGs are DG IB (External Economic and Development Co-operation), DG VI (Agriculture) and DG XI (Environment). The project will mainly use radar images of ERS-1/2, complemented by Radarsat and JERS data for selected test sites. A work-plan has been produced and discussed with the customer DGs. A science report has been produced to validate the use of radar images for rice paddy detection. An ERS data acquisition plan has been developed, which indicate that approx. 300 ERS SAR scenes are required.

The Global Vegetation Monitoring (GVM) Unit of SAI coordinates this project.

3.2.5 FMERS project

The FMERS project will map the European forest areas with remote sensing data and investigate the usefulness of remote sensing for the retrieval of forest biomass parameters. Various sensors and analysis techniques will be used, including standard optical high-resolution optical (5 m) and radar data. The project is building on the experience of past work at the Environmental Mapping and Modeling Unit of JRC-SAI and will integrate the available experience.

The work is divided into two parts:
- The first part of the work will focus on forest area and forest composition. More precisely, the three main aims are: a) evaluate and quantify forest composition and species in two study regions; b) compare satellite data usage vs. traditional methods; c) investigate portability of methods to the entire continent of Europe on a 3/5 year basis.
- The second part will focus on forest volume by providing basis for the evaluation of potential use of existing methods and techniques for mapping forest volume.

The Environmental Mapping and Modeling (EMAP) Unit of SAI coordinates this project.

3.3 Demonstration case studies

The CEO Program has put together a collection of Demonstration Case Studies, which serve to illustrate the broad range of application areas of Earth Observation (EO) information to existing and potential customers of EO. Since September 97, users can consult two CEO databases:
1. The Demonstration Case study database is moderated by CEO, it provides practical examples of the operational use of Earth Observation data and techniques with reference to costs and specific customers. The CEO

3. THE CEO INITIATIVES TO COLLECT AND ANALYZE USER REQUIREMENTS FOR EARTH OBSERVATION

program financially supported the population of the Demo case database in 1996. Two "Requests for Proposals" resulted in 136 cases. Since April 97, the population of the database is not funded anymore. Thirty-nine new cases have been submitted, some of which are still under review.
2. A general EO Demonstration non-moderated database complements the previous database, to give a more general picture. It has been obtained through a search on the Internet and it provides examples of Earth Observation applications currently available on other servers. INFEO currently provides a wide variety of links to EO examples in remote sensing. This database is also structured by keyword.

3.4 Additional assessment of the requirements from the Directorates General of the EC

This action was initiated in June 1997 to rationalize and co-ordinate the increasing volume of information that relate to the usage of EO data and derived products in policies and programs of interest for the European Commission. In particular, it is the intention of this action to synthesize the requirements for information deduced from policy and legislature documents generated in the frame of the EC competencies and subsequently to understand how these requirements (or part of them) could be fulfilled by Earth Observation. The final step is to propose and possibly implement a methodology to satisfy the EC services.

The following themes related to EU policies are being analyzed:
- the European Spatial Development Perspective (ESDP) which deals about the long-term trend of the economic, social and spatial developments of the EU which have a territorial impact;
- the EC regulation on European Regional Development Fund (ERDF);
- the Spatial/Ecological Assessment of the Trans European Network (SEA of the TEN) in the transport field aims at assessing the impact of multi-modal transport;
- the cooperation programs with South-Asia Countries promoted by DG IB External Relations to promote sustainable development;
- the Dobris report on the state of the Environment in Europe;
- the EC fishery policies (DG XIV);
- the MEDA program promoted by DG I and DG III for the Mediterranean region.

The requirements raising from the above mentioned issues are currently being assessed and the synthesis will be completed by end December 1997. In few cases, such as for the ESDP and ERDF, the interactions with the relevant DG is resulting extremely positive and constructive and could possibly lead to the definition of a new methodology, based on EO techniques, to satisfy part of the information requirement of the ESDP.

3.5 Customer segments workshop studies

A principal aim of the CEO program is to increase the number of customers for EO data and information. An initiative to help achieve this aim is to undertake a series of workshop discussions with specific customer segments in governmental or commercial organizations, which have been identified as significant potential customers for EO derived information.

The nature of the requirement analysis is as follows:
1. to conduct a series of workshops with representatives from the customer segment;
2. to identify in detail which of their professional tasks can potentially benefit from EO information;
3. to derive a specification of EO products that are currently feasible, to address these professional tasks;
4. to provide a comprehensive report, characterizing the customer segment, its potential usage of EO, and in particular, identifying the value (benefits against costs) that EO information can provide; and
5. to provide an "information paper", to brief other individuals and organizations in the customer segment of the benefits of using EO information.

The goals of the selection were to identify professional groups, who have the potential to be major users of EO information, and to become recurring customers. Mature customer segments (such as oil exploration companies) were excluded. It is inevitable that certain organizations within the selected groups would already have been contacted, or become customers of the VA (value adding) industry. However, where it could be seen that the group as a whole was still largely unaware of the value of EO, or their requirements for using EO information (although having potential) were still unclear, then these were still considered as target groups.

Each study has produced an information paper destined to the potential customers and a detailed final report. These documents are available on-line from the CEO WWW site mentioned earlier. Four additional studies will soon be undertaken, addressing agro-business, shipping, land navigation/ digital mapping and water companies.

Brief descriptions of the 5 segments for which the studies are completed follow.

3. THE CEO INITIATIVES TO COLLECT AND ANALYZE USER REQUIREMENTS FOR EARTH OBSERVATION

3.5.1 Insurance companies

This group has a wide range of land, sea and atmosphere information requirements. Here, the overriding need is to understand in some detail exactly where EO can help. At present, two characteristic activities have been identified. Risk assessment (done in advance of any event) is likely to have a recurring need of EO information. Damage assessment after the effects of an event (storms, floods, etc.), will have one-off requirements for information, but may have substantial funds available.

3.5.2 Travel/Tourism industry

With increasing leisure time and income levels, this is an industry that has grown enormously in recent years. The use of EO information at present is low, confined largely to products like posters and images. The potential for gaining new customers includes real time services (e.g., what is the weather like in Majorca now?), historical records (e.g., snow cover, climate statistics, archive images such as 3-D views) and products to holiday makers (e.g., posters, CD-ROMs, fly-through, etc.). For the latter, the growing popularity of large area adventure travel (e.g., Himalayas, river tours, cruises, etc.) make EO products more technically possible, whilst average disposable income of these types of holiday makers is high.

3.5.3 Environmental protection organizations of land resources

Here a range of regional, national and pan-European (e.g., Alpine group) organizations can be identified. They are often governmental, but other groups (local, environmental, etc.) are also be included since they have a high profile and can be a catalyst for the adoption of new techniques. The benefits of EO for mapping and monitoring natural land resources were evaluated as the largest potential profitable area; hence sea and atmosphere problems are not addressed by the study.

3.5.4 Town/City local government departments

The two main contributions that EO information can make are in the process of planning and in the detection of changes. The potential customers are public bodies, with a local interest for a particular town, city or urban area, and include architects, road and infrastructure planners, environmental control and health, parks and woods, etc. The expected availability of higher resolution satellite data (but confined to small image sizes), is seen as a technical stimulus to this customer area.

3.5.5 Civil engineering companies

Here the main customers are commercial engineering companies, typically with a large size and an international client base. The use of EO is currently happening in some areas, but the potential is not fully realized. Areas include route planning, environmental impact assessments of construction projects, landscape architecture, risk analysis and infrastructure planning. However, civil engineering projects and their use of EO information tend to be one-off and site specific in nature; hence the use of EO is not often recurrent for a particular area.

4. CONCLUSION

The CEO program has conducted and is conducting number of actions to collect and analyze requirements to space EO. It has put the emphasis to sectors which are potential major users of the information derived from space sensors (e.g., governmental bodies at all levels and industry) but which are less naturally attracted to space EO than, for instance, the scientific programs dealing with global issues.

The CEO program follows on the assessment of the requirements by supporting a large number of demonstration/pilot projects aimed at establishing space EO as a routine source of information, contributing to help well identified customers in performing their professional tasks.

These actions have generated and will continue to generate a substantial documentation, which the CEO makes available to the Remote Sensing Community, in particular via its information servers, such as INFEO.

Chapter 4

NATIONAL STRATEGY FOR DEVELOPING EARTH OBSERVATION APPLICATIONS IN FINLAND

K. Tilli
Director (Space Technology), Technology Development Centre (TEKES), Finland.

1. INTRODUCTION

Earth Observation and satellite geodesy are used extensively in Finland. Utilization of these technologies has been studied in Finland since the 1960's. Today about 6 universities, 11 research centers, 15 companies and 12 end-user companies do research or use these methods in practical applications. The most important application areas are forestry, sea ice and snow, agriculture, meteorology and atmosphere, environment, geodesy and mapping. The quality of the research done and of the expertise acquired in this field both meet international standards. Since becoming first an Associate Member and later a full Member of the European Space Agency and a member of the European Union, Finland has systematically increased participation in international programs and correspondingly broadened the scope of national research in the field of Earth Observation (EO).

This paper describes the framework for Finnish research and development in the EO area. The objectives and strategies for developing EO applications will be presented together with results and experience gained from the national technology program, which plays an important role in the implementation of strategies.

2. CHARACTERISTICS OF THE R&D FRAMEWORK

Finnish EO activities are part of national science and technology policy and research and development (R&D) framework. Space activities are closely integrated to other basic and applied research sectors. So the results can easily diffuse to all potential user sectors.

Research areas are connected to national needs. The most important criterion in directing R&D is the potential financial benefit from the use of space-related methods. The work is, therefore, focused on applications and utilization of basic research.

Financing and research activities have been distributed among several authorities and government agencies. Research institutes and government agencies using EO technologies are also responsible for financing development expenses. The distributed organization is, however, supported by advisory boards, of which the most important is the Finnish Space Committee.

3. RECOMMENDATIONS BY THE SPACE COMMITTEE

The Finnish Space Committee co-ordinates R&D work and Finnish international space co-operation at the Government level. The Committee works in conjunction with the Ministry of Trade and Industry as an advisory body. Its members represent all applications fields and parties involved in space research and development.

The Committee recommends the following objectives for the Finnish Earth Observation activities:
- R&D should focus on economically important applications.
- Development of applications should be customer-oriented.
- The growth of industrial applications should be promoted.
- National and international co-operation should be increased.
- Acquisition of data should be ensured by selective participation in international programs.

The administration, finance and implementation of the above objectives have been distributed among several government agencies and research institutes. Finland does not have a centralized space agency. Therefore, a clear-cut division of tasks in the decentralized organization is needed. The main players are the Academy of Finland, TEKES (Technology Development Centre) and the research institutes.

4. NATIONAL STRATEGY FOR DEVELOPING EARTH OBSERVATION APPLICATIONS IN FINLAND

The research institutes are responsible for developing new applications. In the EO area, 11 research centers are carrying out research and developing new applications. Almost all of these centers are also end-users, which utilize remote sensing technologies in their public sector services. The research centers are responsible for self-financing research and development, but the Academy of Finland and TEKES as well as other financing agencies can support some of the project expenses.

The Academy of Finland and TEKES are financing and coordinating research and development. The Academy of Finland is financing basic research and supports mainly scientists in universities. TEKES is responsible for supporting industrial activities as well as applied technical research and demonstration projects, and finances this development of applications together with research institutes.

4. THE NATIONAL TECHNOLOGY PROGRAMME

Because Finnish research activities are scattered, TEKES established in 1996 a national technology program for Earth Observation activities. The goal of the program is to increase co-operation at national level and to support Finnish participation in international programs. The duration of the program is 5 years and its budget is about 6 M Euros (FIM 33 million). Projects are financed together with companies, research centers and TEKES. Financing is shared and companies and research centers are taking care of about 40–50 % of the project expenses.

The recommendations of the Finnish Space Committee have been taken into account when defining specific objectives of the Technology Program. The objectives are to develop new EO businesses, to utilize remote sensing technologies for raising productivity in public sector services and for monitoring the environment, to support innovative research, and to increase international co-operation.

In 1997, in the second year of the program, the program consisted of 25 projects, which can be classified as follows:

Table 1.

Classification	Number of projects	Proportion (%)
Basic research	8	32
Pilot projects	3	12
New applications	6	24
New businesses	8	32
Total	25	100

From the total volume of the program, 6 M Euros, about 5 M Euros are allocated to public sector projects and 1 M Euros to industrial projects. Over half of the projects are dealing with practical applications.

Today, 8 research centers and universities, and 7 companies are participating in the program. Co-operation between researchers is intense. From the 25 projects started so far, 6 projects include co-operation between research groups and 14 projects co-operation between research centers and companies.

5. EXPERIENCE GAINED FROM THE PROGRAM

To fulfil the first objective, to develop new businesses, companies are actively taking part in the program and are participating in 14 projects. Co-operation between research centers and companies is intense, as can be seen from the high number of projects started in co-operation with research centers and companies. From this point of view, the program will fulfil its expectations.

Although the role of companies is increasing, the work is mainly done by scientists in research centers. In projects lead by the industry, satellite data are used together with other observations and satellite technologies, consequently, don't play a critical role. This shows that the market for industrial applications is still small and in some cases it does not encourage companies to develop commercial applications. Consequently, companies do not have enough experienced personnel, which further limits their possibilities to develop new businesses.

Although almost all Finnish research institutes active in EO are participating in the program, the number of proposals for feasibility studies and application projects is lower than expected and the role of the basic research projects is prominent. There are a number of reasons for this situation; one of them being the fact that end-users will be willing to finance 50% of the development expenses only if they get enough benefit from using satellite data in their information systems. Therefore, the feasibility of new applications should be demonstrated before starting the development phase. In addition, data continuation should be guaranteed more effectively than today, the price of satellite data should be attractive for users and the needs of users should be taken into account in the planning of missions.

6. CONCLUDING REMARKS

Finnish experience in developing both commercial applications and public sector services reveals that Earth Observation methods can be utilized in a large number of potential application areas. In Finland this can be seen from the large number of research units active in this field. Co-operation between these scientists is intensive, which makes the diffusion of results effective and also helps bring together the experience needed in a new development project.

Development of applications, however, suffers from technical and commercial drawbacks. From the end-users point of view, EO policy should support the development of new applications. Space programs should be designed to take into account the whole value-added chain. Commercial market for EO products and services is currently small, and companies are, therefore, entering this field slowly. To promote commercial activities, it is important that national research centers support the growth of EO companies and that activities focus on the creation of new markets for EO products.

Chapter 5

OPERATIONALIZATION OF EARTH OBSERVATION
Its prospects and requirements

J. S. MacDonald
MacDonald Dettwiler, Richmond, B.C., Canada.

1. INTRODUCTION

The concept of "operationalization" of Earth Observation refers to the use of Earth Observation data in an "operational" way, which we will define as the use of such data as a *routine* source of *information*, which has economic, social, strategic, political or environmental *value* to a segment of society. The implication of this definition is that operational users have, as their objective, the generation of *wealth* through the application of derived information in a continuous and routine way to the problems and issues of our society. It also emphasizes the fact that the business of Earth Observation is an *INFORMATION* business. The ultimate output product of each acquisition by a space-borne (or airborne) instrument is *information* about the surface and/or atmosphere of the Earth at a particular point in time. The tangible result of all the effort that goes into building and operating an Earth Observation spacecraft and its associated infrastructure is a set of measurements of the Earth system from space from which we can derive information of economic, social, strategic, political or environmental value. It is this concept of the value of information which is key to understanding the economics of Earth Observation and the issue of gaining a return on investments made in such systems be they financed by public or private means.

The Earth Observation field has reached a critical point in its development. The environment in which we all operate is, at the present time, undergoing rapid and fundamental change. Government funding is decreasing in most of the world and there are several initiatives by the private sector to get into the business of Earth Observation on a commercial basis. As with any endeavor where financial resources are in short supply, the key issue faced by decision-makers is that of benefits realized for the investments made, i.e., the return on the investment. Earth Observation in the public and private sectors is no exception.

Historically, civilian remote sensing of the Earth's surface from space at moderate to fine spatial resolution (with a ground spatial sampling resolution < 100 meters) began in 1972 with the launch of Landsat 1. Since that time, many successor spacecraft have been launched, billions of dollars have been spent and thousands of trillions of bytes of data have been downlinked and stored away. Aside from its positive impact on the fortunes of the aerospace industry and the employment of a few thousand scientists, it is often difficult to attribute significant economic or social impact to this endeavor. However, if Earth Observation is to continue to flourish in the long run, whether it is financed by commercial or public means, the question of its economic and social impacts must be addressed, for it is these impacts which provide the returns on the investments made. This is certainly true in the purely commercial case. If there is sufficient economic demand for the products and services that spaceborne Earth Observation produces, then a commercial industry will materialize and develop. If there is insufficient demand for its output to cover the costs and produce a profit, such an industry will not materialize, or if it does it will not last long. This is also true, in a different way, for publicly funded Earth Observation programs. In this case the ultimate investor, the taxpayer, must perceive a benefit in terms of direct economic, social, strategic, political or environmental "good" sufficient to justify public investment in the system and its operation. Failing this, public support for such programs will weaken and potentially disappear. Thus, as we approach the mature phase of the remote sensing business, those of us who are in the business are well advised to consider, in critical terms, the impact of our business on the societies and economies of which we are a part, and to address how the economic and social benefits which we believe are possible can be realized through observation of the Earth from space.

It is worth pointing out here that the criteria which determine whether or not an Earth Observation enterprise flourishes or dies, are essentially the same in both the public and private sectors. What differs is the mechanism by which judgements are made. In the private sector success depends on the ability of the enterprise to sell sufficient product to cover the costs of acquiring the data/information, archiving and distributing it and, at the same time, generate a profit thus providing a return to the investors. This, in turn, requires that the enterprise satisfy the needs of a sufficient number of customers (information consumers) to generate the necessary revenues. In

5. OPERATIONALIZATION OF EARTH OBSERVATION

the public sector case, the decision process happens at a slower pace, the taxpayer is both customer and investor, but the essential criterion is the same. The public must, in the long run, perceive value commensurate with the cost in order for the endeavor to continue to be supported. Thus, in both cases, the key is to be able to satisfy a need for information at a cost perceived by the consumer to be commensurate with the value of the product delivered.

Looked at from a business viewpoint, the operational user is the true *information consumer* and hence the *customer* upon whom the business must focus, if it is to be successful from an economic standpoint and generate a return. Heretofore, aside from those involved in weather forecasting using spaceborne sensors, there have not been many truly operational users of Earth Observation data. In fact most of the users of data from such spacecraft as Landsat, SPOT and ERS-1 have, up to now, been from the scientific community. Scientific users have a vital but different role to play than operational users. Whereas the objective of the latter group of users is to generate wealth by extracting information of value to society, the objective of the former group is to generate the knowledge necessary to ultimately derive useful information from the data. This role is critical to the overall success and when one views the system in this way, the scientific user plays the role of being part of the system which delivers information rather than a customer who consumes the products generated by the system.

In order to realize the return on investment required to make Earth Observation self-sustaining, many factors, economic, technological, political and organizational come into play. The following sections examine these factors.

2. AN INFORMATION DELIVERY SYSTEM

The process of sensing the Earth from space begins when radiant energy reflected by or emitted from the Earth strikes a spaceborne sensor. The sensor converts that radiant energy into data describing its intensity, its wavelength(s), and in some cases its phase and/or polarization. These data are then processed in various ways to extract information, which is of interest to a set of users.

One can think of the process of converting radiant energy into information of value in an operational way as an *information delivery system,* where the data generated by the sensors are processed and combined with other data to ultimately yield information which meets the requirements of a set of operational users. As the data advances through the system its value to society increases, reaching a maximum, and hence an ability to attract the highest prices, when it satisfies all the information requirements of an operational user.

Figure 1 illustrates the Information Delivery System in more detail. On the left side of the diagram raw data is downlinked from the spacecraft and is stored in an archive. The process of conversion of these data into useful information can be divided into two steps, which we will designate here as "Pre-processing" and "Data to Information Conversion". The latter is sometimes referred to as the "value added" step.

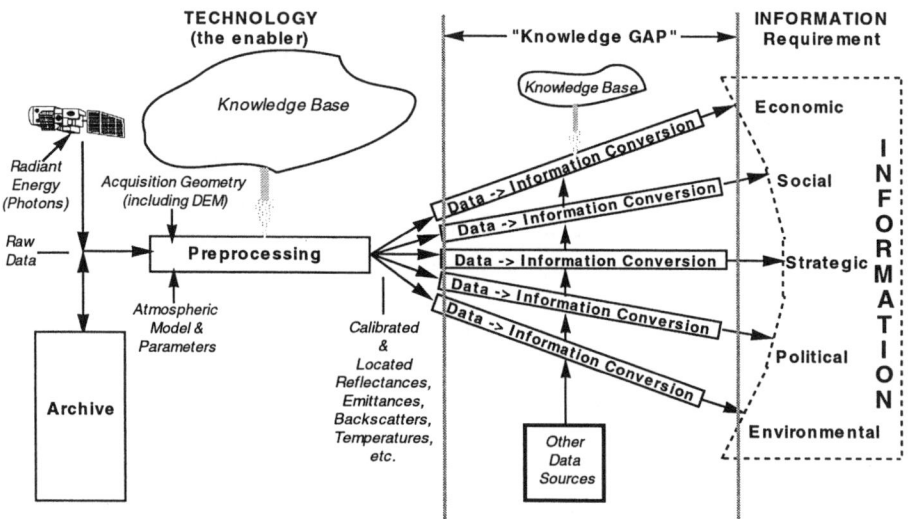

Figure 1. Information Delivery System

2.1 Pre-processing

The pre-processing step is concerned with the conversion of raw data received from a sensor into accurately calibrated, accurately located physical variables such as reflectance, emittance, backscatter, temperature, etc. For a given sensor, the output of this step is a representation of the spatial distribution of a certain physical variable over a segment of the Earth system at a particular point in time. It is important that the pre-processing step be executed in a way which leads to a result which is as *quantitatively accurate as possible,* both in magnitude of the measurement and its location. There are several reasons for this: Firstly, true understanding of physical and biological processes within the Earth system comes from our ability to model these processes and understand the differences and similarities

5. OPERATIONALIZATION OF EARTH OBSERVATION

between what our models predict and what we observe. This process is intrinsically quantitative and therefore demands that its input data be quantitatively expressed in a way that is compatible with the way that these process models express physical and biological phenomena. Secondly, as operational uses of these systems develop, the routine screening and preliminary analysis of the data/information will increasingly be done by machines, owing to the huge volumes of data that it will be necessary to handle. Thus, if we are to be successful in developing the information processing structures to handle information in this way, it is essential that the data be processed in a way that preserves the quantitative accuracy, be as internally consistent as possible, be located as accurately as possible and that the measurements be expressed as physical variables consistent with the ways of expression within the process models.

We have characterized the information delivery process as consisting of two principal steps: pre-processing and data to information conversion. As stated above, the output of the first of these is a set of accurately calibrated, accurately located physical measurements, in essence a set of numbers which represent, to the greatest degree possible, the spatial distribution of the physical variable(s) measured by a particular instrument at a particular time. The case has already been made that in order for these data set to be useful in an operational mode and hence have maximum intrinsic value, they must be calibrated both in magnitude and position in as accurate a manner as possible. This is achieved by modeling the physical process of reflectance or emission of electromagnetic energy as accurately as possible. There are several crucial considerations in being able to achieve maximum accuracy.

- The complete geometry of the acquisition must be known, including:
  - The position of the spacecraft
  - The orientation of the spacecraft
  - The time of the acquisition (hence the position of the Sun)
  - The height and shape of the land (using a digital elevation model)
- The state of the atmosphere at the time of acquisition must be known including:
  - A suitable model describing the absorption and scattering properties of fixed and variable gases
  - The quantity of variable gases present (principally ozone and water vapor) at the time of acquisition
  - The quantity and type of aerosols present at the time of acquisition
  - Intrinsic surface reflectance model(s) for the surface(s) being measured (Bi-directional Reflectance Distribution Function or BRDF).

These considerations affect not only the processing of the data, but the design of the spacecraft instrument complement as well. The necessity to acquire atmospheric information simultaneously with the primary data set is essential to the proper quantitative derivation of physical variables. The requirements are not onerous since the atmospheric data need not be acquired at spatial resolutions finer than 100 meters or so. In practice thus far, however, this data is almost never acquired.

The same is true for the digital elevation model (DEM). Without knowledge of the height and shape of the land surface, it is impossible to adequately model the reflectance process or to correct in any accurate way for atmospheric effects. The height of the land is necessary to derive how much atmosphere the radiation being measured is passing through, and hence how much correction to apply to each pixel once the atmospheric model has been parameterized. The shape of the land is a critical factor in applying corrections for the BRDF since they depend on the slope and aspect of each pixel. The DEM is also critical to the accurate location of measurements from space, in order to correct for relief displacement in the case of measurements made by optical instruments, and layover distortion in the case of measurements made with synthetic aperture radar both of which require the height of the surface to be known.

The point made here is that since quantitative accuracy directly affects the value of the information ultimately derived, the simultaneous acquisition of atmospheric parameters and the availability of an adequate DEM are critical factors influencing this value and hence the return on investment achievable. A more detailed summary of atmospheric and geometric corrections of Earth Observation data is to be found in MacDonald (1993).

What the data produced by the pre-processing step describes is what was measured and the spatial distribution of that measurement. It tells us nothing about what the measurement means. That is the purpose of the second step in the process.

2.2 Data to information conversion

The second step we call Data to Information Conversion. In contrast with the pre-processing step, which is driven entirely by the physics of the measurement process and the technology, this second step is driven by the information requirements of the operational user. In other words, *the processing methodology employed in the data to information conversion step differs depending on the use to which the information is to be put*. A simple example will illustrate this point. Suppose we have measured the distribution of spectral reflectance over a forested area. To the forester, this data can provide input to a model which, when combined with other information, can lead to conclusions about the health of the trees, the vigor of regrowth in recently harvested areas or the onset of disease and things of that nature. To the geologist, on the other hand, precisely the same data speaks of characteristics of the soils in which the trees are growing and, when combined with other knowledge of the geology of the area, can possibly lead to important conclusions about mineralization in the soil etc. What this illustrates is that precisely the same measurement has different meaning, is thought of in entirely different ways and is an input to entirely different models depending on the objectives of the person who needs the information the system provides. This is illustrated in Figure 1 by many *data to information* conversion processes connecting the data delivered by the pre-processing step to the information required. The knowledge base, which supports this second step in the process, is minuscule to non-existent, and this shortage of knowledge is a critical factor inhibiting use of remotely sensed data in an operational way having direct economic and social impacts on our society.

3. THE KNOWLEDGE GAP

The information delivery system described in the previous section has been characterized as being composed of two steps. The methodology of the first step is determined by the physics and geometry of the measurement process together with the technology of the sensor and the satellite, which carries it. The knowledge base behind this process is quite extensive; one could even call it virtually complete. It is the result of over a quarter century of effort by the engineers and scientists who have built the systems and worked to understand their operation.

The methodology of the second step is determined by the information requirements of operational users. These people come from many disciplines, including economists, environmentalists, agronomists, foresters, land managers, lawyers, geologists, military commanders and politicians, to name a few. They are a diverse lot, but in general are not well versed or even knowledgeable in the technologies of measurement of the Earth from space. Furthermore, operational users have little, if any, interest in how the information is derived. They only care that it be accurate, timely and appropriate to their needs. In most cases they already have sources of information which they rely on to do their jobs. If they are to use a new source of information and pay for it, they must be convinced that the new source is in some way better than the old one and that it is worth what they are asked to pay.

Thus we have a knowledge gap. On one side of the gap are the scientists and engineers who know how to measure variables in the Earth system from space, but have little if any idea of what the real information requirements of operational users with economic, social, strategic, political or environmental objectives may be, nor are they likely to understand how those requirements are expressed. On the other side of the gap are the operational users (information consumers), each of whom expresses their information requirements in different ways using the jargon of their particular discipline and has little, if any, knowledge of what can be done to deliver information from spaceborne platforms, how it can benefit them or what it is worth. In addition, the motivation they might have to explore such things is quite weak. This is illustrated in Figure 1 by showing a "Knowledge Gap" separating those with enabling technology from those with a requirement for economic, social, strategic, political or environmental information.

4. BRIDGING THE GAP

When viewed from the perspective developed in the preceding sections, it becomes apparent that for over 25 years those of us in the Earth Observation business have essentially been custodians of a solution in search of problems. While it is true that some examples exist of the operational use of information derived from Earth Observation data, they are rare, and the case for an adequate return on investment has not been an easy one to make. Most of us believe, however, that the potential to achieve such a return exists and continued progress in Earth Observation requires that this vision be fulfilled. Our challenge is to find ways to bridge the knowledge gap.

The key inhibitor to bridging the knowledge gap is the fact that the knowledge base supporting our understanding of how to convert measurements from space into information of economic, social, strategic, political or environmental value is so small and so weak. Bridging the gap involves expanding and strengthening this knowledge base. It also involves understanding and dealing with additional factors, which have inhibited the use of space, based measurements for operational purposes. While it can be credibly argued that there are many of these, three stand out:
1. The difficulty of processing data in a quantitative fashion due of the absence of some of the ancillary data which are necessary.
2. The existence of alternate sources of information which are regarded as adequate to the task.
3. The difficulties of obtaining data in a usable form.

If Earth Observation technology is to expand as a source of operational information, these inhibitors must be recognized and dealt with.

It is this author's view that the knowledge gap, in general, can only be bridged from the technology side. There are two major reasons for this:
1. There is considerable motivation on the technology side of the gap to build a bridge, whereas on the consumer side there is little or no

motivation since the source of data from which to derive information is of marginal or no concern, so long as the information provided is accurate and suitable to the task at hand.
2. To understand how to satisfy an information requirement using remotely sensed data sources requires knowledge of the physics of the sensing process and the technology which supports it. This knowledge, in general, only exists on the technology side of the gap and when one couples this with the differences in motivation on the two sides of the gap, one is soon led to the conclusion that the initiative to bridge the gap must come from the technology side.

In the following sections we explore a process by which the knowledge gap may be bridged.

4.1 The data to information knowledge base

There are two major activities involved in expanding the knowledge base supporting data to information conversion methodologies:
1. Understanding requirements for information. It has already been stated that these requirements differ depending on the use to which the information is to be put. Additionally, the ways in which information consumers express their requirements differs between disciplines and have no explicit relationships with remote sensing technologies. Thus, at this stage of the process, one's technological expertise and knowledge are of little use. One must "get inside the head" of the consumer and understand in the way he/she does, what the information requirements are and how they relate to the task(s) at hand. This is a major undertaking and its difficulty should not be minimized;
2. Having understood an information requirement, one must then determine whether and how this requirement can be satisfied using Earth Observation data sources, the costs of doing so and the quality of the information produced relative to that derived from alternative sources. It is here that technological expertise and knowledge are critical.

Each of these activities is a substantial undertaking. The first involves getting answers to the generic question "When you do your job, what information do you need?" If the field is forestry, for example, the answers will be expressed in terms of such things as biomass, plant nutrition, erosion factors, crown closure, species mix, road access, forest diseases, regrowth rates and the like. Furthermore, information delivered by the system must also be expressed in these terms. Images of forestlands from space are, in general, of passing interest only.

The second task involves applying one's knowledge of the physics of measuring physical quantities from space, the processing of raw data into calibrated physical measurements and of a wealth of data processing methodologies available to derive the information required. Additionally, in addressing solutions to data to information conversion problems, it must always be recognized that, in general, utilization of other sources of data and information will be necessary in order to derive the necessary information from remote sensing sources, and that remote sensing sources alone will not satisfy all the requirements nor will they necessarily be the most cost competitive.

4.2 Quantitative measurement

Once all these things are understood, an interesting question arises: "Are we making the right measurements?" Our knowledge of the physics of the process is now complete enough that a lack of answers to that question no longer needs to be the case. At the present time, with the exception of systems being developed to monitor global change, Earth Observation satellites and sensors are launched (and many millions of dollars are spent) without sufficient attention being paid to this question which is, or at least should be, based on the economic and social impact the use of the systems will have. The relationship between our ability to accurately derive quantitative physical variables from the raw data and the value of the ultimate information supplied has been pointed out in previous sections. At the present time, with few exceptions, simultaneous acquisition of the concentration of variable gases and aerosols in the atmosphere along with the principal data does not occur and adequate DEM's do not exist (or are not available) for most of the land surface of the world. We know how to acquire these data. Ozone and water vapor have well known spectral characteristics and algorithms exist to parameterize atmospheric models such as MODTRAN and 6S using measurements made simultaneously with those of the main instrument. The necessary measurements could be made with a small companion instrument having a spatial resolution in the 100 - 300 meter range and 6 or 7 narrow spectral bands in the VNIR range. A DEM could be acquired with a specialized mission flying a true SAR interferometer such as the TOPSAT mission proposed a number of years ago. There are no technological mysteries here.

4.3 Alternate sources of information

Viewed from the perspective of information requirements, space-based sources of data are simply one source of the data required to produce information. To be utilized in an operational way, these sources must be competitive with alternative sources of data. In other words, they must be more cost effective than the alternative sources either by being less costly or by producing information of higher value for essentially the same cost.

In promoting the operational use of Earth Observation data, one is often in the position of convincing the information consumer to replace, or in some cases augment, a current source of information with a space-based source which, given the conservative nature of most operational users, requires a clear demonstration of increased cost effectiveness and an essentially complete knowledge of the consumer's information requirements and their value. Two instances at least where Earth Observation data have an advantage are those cases where routine coverage of a large geographic area is essential or data are not ordinarily available by other means

4.4 Data availability

Historically, spaceborne Earth Observation data have been poorly archived, difficult to obtain, and in practice accessible only to specialists with the necessary knowledge to be able to acquire it, process it and use it. These are circumstances that operational users will not tolerate. Development of the operational use of remotely sensed data will require that a person be able to request information using nothing more sophisticated than a telephone or a piece of browser software and a modem, request data in any quantity or from any location and expect to receive the information within a time frame that is appropriate to the requester's requirement. Customers should not have to know anything about orbits or processing techniques. They simply want to know if the information they require is available or can be made available and how much it is going to cost. From a technical point of view this is not a difficult thing to achieve, but it is vitally important if the potential economic impact of Earth Observation technology is to be realized.

5. CONCLUSIONS

The economic impact of space-based remote sensing technology is directly related to its use as an *operational source of high value information* relevant to the economic, social, strategic, political and environmental issues which are prevalent in modern societies. Economic impact of a magnitude sufficient to provide an economic return on the investments in such systems will be necessary in order to justify their continued development through either commercial or public investment.

The process of delivering such information can be considered as a two step one where the first step produces descriptions of the distribution of physical variables throughout the Earth system, and the second addresses the conversion of these data into information required by operational consumers. The value of the information ultimately produced is heavily dependent on the quantitative accuracy of the measurements made, therefore future systems should be designed to achieve maximum accuracy of the physical variables measured through accurate modeling of the reflectance or backscattering process. Since the knowledge base supporting the data to information conversion step is virtually non-existent, acting to expand this knowledge base while at the same time enhancing the ability of the information delivery system to deliver information in a convenient and timely way through improvement of its archival and data access capabilities will enhance our ability to deliver information of high economic and social value, thereby increasing the return on the investments made in space-borne Earth Observation systems.

6. REFERENCES

MacDonald, J. S. (1993) *The Status of Remote Sensing Satellite Systems*, Proceedings of the Workshop and Conference "International Mapping from Space", Working Group IV/2 of ISPRS Volume 15, Hanover, Germany, September 27 to October 1, 1993, 7–20.

Chapter 6

REMOTE SENSING OF FOREST FIRES
Current limitations and future prospects

E. Chuvieco
Department of Geography, University of Alcalá, Spain.

1. INTRODUCTION

Wildland fires are a critical environmental process worldwide. They are the main responsible of atmospheric pollution in tropical latitudes (Crutzen et al., 1979), the principal agent of deforestation and land use change in the rain forest (Malingreau, 1990), and one of the most prominent sources of land degradation in Mediterranean areas of Europe.

2. REMOTE SENSING OF FOREST FIRES

The main contributions of remote sensing to forest fires may be grouped in three categories, according to the three phases of fire management: risk estimation (before the fire), detection (during the fire) and assessment (after the fire).

2.1 Fire risk estimation

The estimation of fire risk requires obtaining up to date information on vegetation water status, which is closely related to probability of ignition. Traditionally, this parameter has been estimated from meteorological data. However, there are two problems for the operational use of weather data. Firstly, vegetation species do not respond equally to changes in weather conditions, since physiological mechanisms and soil characteristics greatly affect the water stress resistance of live plants. Secondly, weather stations are frequently located in urban or agricultural areas, often far from forested regions, and therefore the weather measurements may not be sufficiently representatives of stress conditions in fire-prone areas.

Satellite data, on the other hand, are acquired directly from the vegetation canopy, and imply an intensive spatial sampling. On this ground, several studies have shown the usefulness of satellite data to estimate fire danger conditions (Chuvieco and Martín, 1994; Desbois et al., 1997). However, a better understanding on the relations between spectral information and the water content of plants is required. For that purpose, satellite observations at different scales should be coupled with ground data, to better control potential sources of noise (landscape patterns, diversity of species, weather conditions, etc.).

Fire risk estimation requires continuous updating to tackle the temporal and spatial variability of risk factors. Within a short-term perspective, new sensors with improved spatial and spectral resolution are required. Currently, NOAA-AVHRR data are the best choice for fire risk estimation, because this sensor provides daily observations. However, the atmospheric interference and radiometric inconsistency of off-nadir observations and cloud cover reduce the actual temporal frequency of AVHRR images. Future sensors, such as MODIS, may reduce these difficulties. Ideally, fire risk estimation would require a sensor with a spatial resolution in the range of 100 to 1000 meters size, with daily observations in the visible, near infrared, middle infrared and thermal infrared, and with internal calibration sources to assure temporal consistency.

Another approach to fire risk estimation considers only static factors that are not altered daily but in long-term trends, such as topography or vegetation structure. For this latter purpose, remote-sensing systems may provide fuel type maps, which are critical for fuel management and fire behavior prediction (Burgan and Rothermel, 1984). Fuel types are defined according to morphological characteristics of plants (size, volume to height ratio, density, etc.) which are not easily discriminated with current satellite systems (Chuvieco and Salas, 1996). Synergism of optical and microwave data may improve current fuel-type mapping, if understory information may be derived from them.

2.2 Fire detection

Fire detection implies a great sensitivity to middle infrared radiance, since this band is very suitable to detect hot targets. Currently, fire detection from space relies on AVHRR data, which includes a middle infrared band (channel 3). However this sensor is not well adapted to fire detection, because of its low temporal coverage (1 image every 12 hours) and, specially, because of the

6. REMOTE SENSING OF FOREST FIRES

limited radiometric resolution that may lead to confusions with other covers (Kennedy et al., 1994). Several prototypes (such as the Fuego and the Focus programs) are being studied to overcome the deficiencies of AVHRR data, namely to increase the temporal coverage and radiometric sensitivity. However, even with dedicated optical systems, fire detection from space can still be unsuccessful, because of cloud contamination or topographic shadows, which may hide the fire underneath. Little experience is available in using microwave systems have not been tested for fire detection.

2.3   Fire assessment

Burned land inventorying and mapping are very important to assess effects of fire on vegetation and soil. Spectral discrimination of burned areas is critical for fire effects assessment (Pereira et al., 1997). However, this method faces several difficulties, such as the confusion with non-vegetated surfaces or shaded areas. Several single-date and multi-temporal techniques have been proposed to improve burned land mapping. The most successful approaches have relied on the use of analytical techniques (vegetation indices, principal component analysis or spectral mixture analysis), and simple multi-temporal methods (subtraction or ratios). When working at a global scale, AVHRR data need to be corrected from atmospheric and cloud effects, which introduce severe distortions in the original data.

High-resolution data, such as Landsat or SPOT images, are more adequate then the global low resolution sensors for the detail assessment of fire-affected areas. In this case, it has been possible to discriminate burning intensities, by comparing radiances from before and after the fire (Chuvieco and Congalton, 1988). This information may be critical to reduce soil erosion after fire, since rapidly growing species may be planted in those sectors more severely affected by the fire (Isaacson et al., 1982). Monitoring recovery after fire also provides critical information to better understand fire effects on vegetation species. Finally, satellite images may be used to analyze changes in landscape pattern as a result of fire.

3.   THE NEED OF USER-ORIENTED DATA

In all three aspects previously reviewed, the use of satellite data greatly enhances forest fire management. However, limitations of current systems notably reduce the operational application of remote sensing technologies to this field and, to some extent, to other environmental hazards.

These shortcomings are partially due to the lack of specific orientations of remote sensing missions. Most satellite platforms were designed for general-purpose users. Consequently, they may satisfy the majority of customers without completely fulfilling their necessities. For this reason, I believe the future generation of satellite platforms should be oriented toward specific user communities, adapting state of the art technologies to solve particular problems. Obviously, if systems are so clearly oriented, the potential applications will be more restricted, but they will really satisfy to a target user group. For instance, in the field of fire detection, high temporal frequency is required (in the range of 15 to 30 minutes). This frequency may not be useful for other applications (urban management or land cover, for instance), and therefore many images of such a sensor may be unusable for those applications. However, they are absolute critical for forest fire managers who could not afford operational fire observations otherwise. Since the development and technical costs of new sensors have been greatly reduced recently, specially if small platforms are considered, investments of such a dedicated remote sensing system might be cost-efficient, even with a relatively small user community.

Another obstacle for the operational use of remote sensing data concerns the availability of images to the end user. Decentralized facilities and low cost products will be the ideal choice. The experience of free and worldwide reception of AVHRR data to create a solid user community has been excellent.

4. REFERENCES

Burgan, R.E. and Rothermel, R.C. (1984) *BEHAVE: Fire Behavior Prediction and Fuel Modelling System*. Fuel Subsystem, USDA Forest Service, Ogden, Utah.

Chuvieco, E. and Congalton, R.G. (1988) Mapping and inventory of forest fires from digital processing of TM data, *Geocarto International,* **4**, 41–53.

Chuvieco, E. and Martin, M.P. (1994) Global fire mapping and fire danger estimation using AVHRR images, *Photogrammetric Engineering and Remote Sensing,* **60**, 563–570.

Chuvieco, E. and Salas, J. (1996) Mapping the spatial distribution of forest fire danger using G.I.S., *International Journal of Geographical Information Systems,* **10**, 333–345.

Crutzen, P.J., Heidt, L.E., Krasnec, J.P., Pollock, W.H. and Seiler, W. (1979) Biomass burning as a source of atmospheric gases, *Nature,* **282**, 253-256.

Desbois, N., Pereira, J.M.C., Beaudoin, A., Chuvieco, E. and Vidal, A. (1997) Short Term Fire Risk Mapping using Remote Sensing, in Chuvieco, E. (Editor), *A review of remote sensing methods for the study of large wildland fires*, Alcalá de Henares, 29–61.

Isaacson, D.L., Smith, H.G., and Alexander, C.J. (1982) Erosion hazard reduction in a wildfire damaged area, in Johannsen and Sanders (Editors): *Remote Sensing for Resource Management*, Soil Conservation Society of America, Ankeny, 179–190.

Kennedy, P.J., Belward, A.S. and Grégoire, J.-M. (1994) An improved approach to fire monitoring in West Africa using AVHRR data, *International Journal of Remote Sensing,* **15**, 2235–2255.

Malingreau, J.P. (1990) The contribution of remote sensing to the global monitoring of fires in tropical and subtropical ecosystems, in J.G. Goldammer (Editor): *Fire in Tropical Biota*, Springer Verlag, Berlin, 337–370.

Pereira, J.M.C., Chuvieco, E., Beaudoin, A. and Desbois, N. (1997) Remote Sensing of Burned Areas: A Review, in Chuvieco, E. (Editor): *A review of remote sensing methods for the study of large wildland fires*, Alcalá de Henares, 127–185.

Chapter 7

USING CURRENT AND FUTURE REMOTE SENSING SYSTEMS IN NATURAL HAZARDS MANAGEMENT

J. San Miguel-Ayanz, M. M. Verstraete, B. Pinty, J. Meyer-Roux and G. Schmuck
Space Applications Institute, Ispra, Italy.

1. INTRODUCTION

Natural disasters have often occurred in history and will continue to take place. Improved technology or knowledge cannot prevent natural hazards. However, technological developments can lead to a better understanding of the environmental and human causes of these hazards. Furthermore, even though hazards cannot be prevented, disasters can be avoided or diminished by taking precautionary measures, by providing early warnings, by an adequate monitoring of risks and by coordinating relief efforts.

Satellite technology is improving rapidly, especially in terms of sensor and telecommunications performances. However, the actual exploitation of satellites in the management of natural disasters is still in its infancy. The weather forecasting community has a long history of operational use of satellite remote sensing to improve predictions, and in particular the occurrence and evolution of tropical cyclones. By contrast, the management of other natural hazards has so far made relatively little use of satellite technologies. The reasons why satellite derived information may not be used in disaster monitoring and management may be summarized as follows:

- Methods and techniques may be lacking to derive useful information on impending disasters.

- The information may not be sufficiently reliable or timely to manage disasters.
- Disaster management staff may not be aware of the capabilities or convinced of the usefulness of such information in their day-to-day activities.

A partial list of some of the better known examples of natural hazards includes:

- *Asteroid:* a celestial object that may collide with the Earth.
- *Avalanche:* a massive uncontrolled transport of snow and ice down topographical slopes.
- *Crop disease:* a crop health event that leads to losses in agricultural production and sometimes to famine.
- *Drought:* a shortage of water for periods long enough to result in losses in agricultural production, cattle or even human life.
- *Earthquake:* a sudden movement of the Earth surface due to the release of stresses in the lithosphere along faults.
- *Floods:* the overflowing of a river, lake, or ocean.
- *Hurricane:* a violent storm, or a tropical cyclone.
- *Insect infestation:* an invasion by desert locust or other pests.
- *Landslide:* a sudden, uncontrolled transport of soil and rock down a topographical slope.
- *Volcanic eruption:* spouting of molten rock and gas from the magma.
- *Tornado:* a highly destructive wind vortex over land.
- *Tsunami:* a major sea wave generated by an underwater volcanic eruption or earthquake.

Although non natural in their cause, there are also disasters that result from human activities, and directly affect the environment, including acid rain, global warming, soil erosion, uncontrolled fires, deforestation, desertification, the melting of ice caps, ozone depletion, extinction of animal or plant species, etc.

Our awareness of the occurrence and impact of natural disasters has been greatly increased by the mass media. In parallel, our capacity as a society to either prevent or reduce the damages they produce has increased thanks to technological and organizational progress. A number of international initiatives to provide information or to improve disaster management strategies have been proposed, including IDNDR (International Decade for Natural Disaster Reduction), IGOS/DMSP (International Global Observation Strategy/Disaster Management Support Project), STRIM (European program on Space Techniques for Major Hazard Management), Eurimage-ESA Earth Watch project, and the NASA/NDRD (NASA/Natural Disaster Reference Database).

2. DISASTER MANAGEMENT AND INFORMATION REQUIREMENTS

The overall concept of natural hazard encompasses the estimation of the risks a priori, the development and implementation of techniques to mitigate the possible effects, the detection and monitoring of catastrophic events as they unfold, and the management of the disaster after the fact. The initiatives or activities that can take place in each of these phases, and in particular the role of remote sensing during these various phases, are quite different:

- *Increasing preparedness:* the goal is to minimize the effects of a hypothetical hazard by taking appropriate precautionary actions. Remote sensing can be useful to map high-risk areas for different types of disasters and generate information which, in turn, can be used to reinforce infrastructures or target populations with educational programs. Other activities requiring information include the planning for efficient responses, increasing public awareness of danger, drafting evacuation plans, etc.
- *Issuing early warnings:* the objective is to limit the impact of disasters by warning the populations concerned of an impending event. For instance, geosynchronous meteorological satellites can be used to provide warnings for weather related disasters. Such systems could be expanded to deliver flood warnings (possibly in conjunction with rainfall and stream level measurements).
- *Monitoring hazards in real time:* The purpose is to determine the current status of the event and foresee its likely evolution in time and space. Remote sensing can be helpful in managing crises by providing up-to-date reliable information on what is happening.
- *Assessing the damage:* Once the disaster is over, it is necessary to evaluate the damages to the populations and properties, to infrastructures, to various economic sectors, as well as to natural resources. Properly assessing the situation is critical to detect possible unstable conditions that may create further damages. Optical and microwave sensors have been used to assess the damage of forest fires, floods, etc.
- *Providing relief:* Remote sensing can also be very useful in providing information to plan rescue and rehabilitation operations and to bring the situation back to normal as soon as possible. International organizations and non-governmental organizations (NGOs) that provide support to communities affected by disasters would be natural customers for this type of information.

Disaster management is a complex endeavor, which requires the participation of many institutions in its various phases. Decisions have to be made on the basis of information that may be incorrect, unreliable, not delivered fast enough, or not provided to the appropriate user. In emergency situations, the value of information decreases exponentially with time. For instance, recent studies show that the information generated by a forest fire detection system is particularly valuable within the first 15 minutes after fire initiation, and decreases very rapidly to become useless after an hour. The acquisition, processing, analysis and interpretation of critical information, as well as its rapid dissemination are thus essential. Clearly, this type of user requirements has direct implications on the choice of applicable technologies since repeated observations at such high frequency can only be achieved with geostationary satellites or clusters of multiple sensors.

An additional constraint, although not directly related to remote sensing, is that such information must be unambiguous and clearly understandable by the intended recipients in conditions of severe stress.

From the point of view of the disaster managers, the main requirement is that accurate and reliable information be received at appropriate temporal and spatial resolutions. Information on rapidly changing situations must be updated frequently and provide enough detail to allow informed decision-making. Remote sensing can thus be expected to play an important role in disaster management.

3. CONTRIBUTION OF REMOTE SENSING TO THE MANAGEMENT OF NATURAL HAZARDS

Space based sensors can provide useful information to prevent and monitor natural disasters, or to support relief operations. To be effective, however, such information must be reliable and timely. Ideally, an early warning system should be able to acquire data anywhere in the world, to process them very quickly, and to deliver high level products to the authorities concerned, both in the field and at centralized facilities. A well-organized processing center or disaster management group should include personnel, hardware and software capable of integrating not only data streams from various sources, including remote sensing data from various sensors, but also ancillary data such as topography, climate, land cover, etc. It should be able to integrate these data layers in a Geographical Information System (GIS), for instance, and deliver synthetic information to the decision-makers. This may imply the installation and operation of receiving stations, processing chains, and dedicated algorithms for the rapid analysis of large

7. USING CURRENT AND FUTURE REMOTE SENSING SYSTEMS IN NATURAL HAZARDS MANAGEMENT

amounts of data. Deliverables such as thematic maps, statistics, etc. should be generated and distributed within hours of data acquisition by satellites.

In many cases, it may be critical to acquire data in all weather conditions. This may impose particular technological constraints, such as the use of microwave techniques. Table 1 summarizes the types of satellite and ancillary data that may be needed for the management of some natural hazards. Although this table is indicative of the type of information needed, further investigations should be conducted to assess, in each case, the accuracy and spatio-temporal resolution that is required to support the various phases of natural hazard management.

Table 1. Data Requirements for several types of disasters

HAZARD	ANCILLARY DATA	REAL TIME DATA
Drought	Average rainfall Average temperature Potential evapotranspiration Land cover type Soil water holding capacity	Actual rainfall Actual temperature Actual evapotranspiration Vegetation condition Actual soil moisture content
Flood	Digital elevation model Soil type Land use, land cover Average runoff Average temperature Average basin snow	Actual rainfall amount, intensity Flood extent, water depth Stream, river flow rates Actual soil moisture content Actual snow amount, condition
Fires	Digital elevation model Vegetation type Average temperature Average fuel load	Actual fuel amount, moisture Actual surface temperature Location of the fire Actual wind speed, direction Burnt area, vegetation damage
Earthquake	Average tectonic activity Faults lineaments	Actual tectonic stresses Actual ground movement
Hurricane	Hurricane climatology Average wind profile Average pressure profile	Location of the storm Actual wind speed, direction Actual sea surface temperature Actual air temperature, pressure
Crop disease	Average rainfall Vegetation type Average temperature Potential evapotranspiration	Actual rainfall Type of stress, pest or disease Actual temperature Actual evapotranspiration

Table 2 exhibits some of the characteristics of current space platforms and the type of disaster in which they could be used. This table is only presented to show the fairly extensive use of space remote sensing in disaster management, however no details are provided on phases of the disaster addressed by each sensor. A more detailed description of the satellite requirements for measuring different parameters, and the way in which disaster management is addressed by existing sensors can be found in (GEOWARN, 1993), the NASA Natural Disaster Reference Database[1], or the ESA-CEOS Dossier[2].

Most applications are still in an exploratory research phase. Preliminary uses of space remote sensing data have been attempted to provide information on risks and to increase emergency preparedness or to issue early warnings to population and/or disaster management organizations. Such data have also been used to evaluate damages caused by natural hazards in a few specific cases, with different degrees of success. However, the only family of applications that systematically takes advantage of satellite remote sensing is weather forecasting and in particular tropical storm warning systems.

Table 2. Current applications of space platforms in disaster management

Platform	Sensor	Spectral range	Resolution (m)	Application (*)
METEOSAT	VIS	VIS/NIR, TH	2500	1, 3, 5, 6, 11, 12
NOAA	AVHRR	VIS, NIR, TH	1100 to 8000	1–3, 9, 10
GOES	IMAGER	VIS, TH	1000	1, 3, 5, 6, 11, 12
LANDSAT	TM, MSS	VIS, NIR; TH	30 to 80; 120	1–3, 5, 7–9, 10
SPOT	HRV	VIS, NIR	20	1–3, 5, 7–9, 10
ERS	SAR	SAR	30	2, 4, 5, 8–10
	ATSR	VIS, NIR, TH	1000	3, 10
RADARSAT	SAR	SAR	10 (variable)	2, 4, 5, 8–10
JERS	SAR	SAR	25	2, 4, 5, 8–10
RESURS-01	MSU-E	VIS, NIR	35	2, 3, 5, 7
	MSU-SK	VIS, NIR, TH	160, 160, 600	2, 5, 7
RESURS-F20	RFA-1000	PAN	5	2, 8, 10
	KATE-200	PAN	30	2, 3, 5, 10
IRS 1-C	WiFS	VIS, NIR	188	2, 3, 5, 7, 8, 10
	LISS-3	VIS, NIR	23	2, 3, 5, 7, 8, 10

(*) Applications: (1) Avalanche, (2) Crop disease, (3) Drought, (4) Earthquake, (5) Floods, (6) Hurricane or tropical storms, (7) Insect infestation, (8) Landslide, (9) Volcanism, (10) Wild fire, (11) Tornado, (12) Tsunami.

[1] http://ltpwww.gsfc.nasa.gov/ndrd/research.html
[2] http://ceos.esrin.esa.it:9000/ceos.html

4. EXPLOITING FORTHCOMING SENSORS IN NATURAL HAZARD MANAGEMENT

A new generation of space-borne sensors is under development and should become operational within the next few years. These advanced instruments will provide data at much higher spatial and spectral resolution, and offer enhanced features such as multi-angular observations, on-board calibration facilities, improved navigational capabilities, etc. Although generally not designed to address natural hazards information requirements in particular, these systems should generate data streams which should prove useful in this context.

Technological advances are occurring along multiple directions. A series of polar-orbiting platforms will progressively replace the NOAA/AVHRR systems. It includes the already flying VEGETATION instrument on board the French SPOT satellite, as well as major Earth Observation platforms such as the EOS AM/1 (recently named Terra) of NASA launched in December 1999 with the MISR, MODIS and ASTER sensors. The European Space Agency (ESA) will similarly launch (in mid 2001) the ENVISAT platform, embarking MERIS and AATSR, while the Japanese Space Agency (NASDA) is planning to start operating the ADEOS-II platform with the GLI instrument, early in this decade.

Advanced geostationary satellites are also being designed and built. For instance, the Meteosat Second Generation (MSG) family of platforms should provide remote sensing data of interest for weather forecasting, as well as climate and atmospheric studies over the next decade or more.

Last but not least, there are a number of public and private projects to launch and operate hyper-spectral and very high spatial resolution instruments in support of specific applications. These include, for instance, imaging spectrometers such as Earlybird (3 m), Quickbird (3.28 m), Orbview (1 and 2 m) and, most recently, Ikonos (1 m). New approaches are also being pursued where the environment is frequently revisited thanks to a constellation of small satellites. The FUEGO program is one such example, which consists of multiple satellites whose radiometric and orbital characteristics are being selected to allow the near real time detection of fires in the Mediterranean region.

Clearly, taking full advantage of these new technological capabilities will require the development, implementation and evaluation of dedicated tools and techniques of data analysis. For simple applications, where even approximate information is required in a very short amount of time, spectral-based methods, such as spectral indices, will continue to play an important role. On the other hand, information with a much better accuracy and reliability will be obtained from these data if and when they will be analyzed with advanced models and inversion procedures. Another promising avenue is to investigate further the feasibility and efficiency of merging two or more data streams from various remote sensing sensors, when they can provide complementary information (synergy).

Last but not least, all applications relying on the analysis of data collected in the past will continue to have to rely on remote sensing data from old platforms such as the NOAA/AVHRR, Landsat, or SPOT. These sensors are quite unsatisfactory compared to today's standards of quality and to the level of expectation of the users, but they have generated the only global data sets available over the last 15 years. A coherent and focused research and development (R&D) program should be implemented to design and use state of the art algorithms and models. In particular, such a program should aim at compensating for the inadequacies of old sensors through the intelligent use of up to date models and environmental information derived from more recent instruments and fundamental research. The future exploitation of remote sensing for disaster management thus largely relies on our capacity to carry out R&D programs to improve existing techniques for processing existing and future remotely sensed data and for retrieving the needed information.

Chapter 8

EXAMPLES OF THE USE OF SATELLITE DATA IN NUMERICAL WEATHER PREDICTION MODELS

J. Noilhan, E. Bazile, J.-L. Champeaux and D. Giard
Météo France/CNRM, Toulouse, France.

1. INTRODUCTION

One of the conclusions of the first phases of the Program for the Intercomparison of Land surface Parameterization Schemes (PILPS) (Henderson-Sellers et al. 1997) is to demonstrate very clearly the positive impact of the representation of vegetation in the simulation of surface fluxes. The simulation errors with advanced surface schemes are significantly lower than the errors with bucket-type surface schemes where vegetation processes are not taken into account. During the last decade, a large number of studies with general circulation models have analyzed the sensitivity of the climate simulation to continental surface processes. Recently, most of the European weather centers adopted advanced surface parameterizations in operational (ECMWF model, Viterbo and Beljaars, 1995) or pre-operational contexts (HIRLAM model, Bringfelt 1996; ARPEGE model, Bazile and Giard, 1996; Giard and Bazile 1999). For high resolution short to medium-range forecast models, two particular problems are related to the implementation of surface schemes:
1. The specification of vegetation and soil characteristics with a spatial resolution of a tens of kilometers in the area of interest, and at least a monthly time resolution to follow the seasonal cycle of the vegetation;

2. The initialization of soil moisture fields because soil wetness can have a strong impact on the forecast of low level atmospheric quantities (Mahfouf 1991, Bouttier et al. 1993).

Satellite remote sensing data in the visible and infrared spectral domains can bring a significant input into operational algorithms. Two examples of exploitation of satellite data in Numerical Weather Prediction models can be summarized as follows:

1. The use of AVHRR/NOAA data for the specification of vegetation properties in relation to the ISBA scheme (Noilhan and Planton, 1989) in the ARPEGE French system, and
2. The use of METEOSAT visible and infrared data in the SEBAL algorithm developed to retrieve the soil wetness in the KNMI modeling system (Hurk et al. 1997).

2. A HIGH RESOLUTION VEGETATION MAP FOR OPERATIONAL PURPOSES

The implementation of a land surface scheme in operational numerical weather prediction model is a long and difficult task. Bazile and Giard (1996) describe the various aspects of the implementation of the ISBA scheme in the ARPEGE computing system. The low complexity of ISBA (only one surface energy balance, and two soil layers for heat and water diffusion) makes it a good candidate for daily forecasts. The scheme has been widely calibrated at the local scale for a large number of land uses, using data collected during numerous large field experiments conducted during the last decade. See Noilhan and Mahfouf (1996) for a review of the calibration of ISBA, and the chapter by Jochum et al. in this book for a description of land-surface experiments. In addition, a sequential assimilation scheme for soil moisture based on optimum interpolation has been developed (Giard and Bazile 1997, 1999). The implementation of ISBA requires the specification of surface parameters prescribed from a vegetation map and a soil map. Up to now, the most widely used data base in the meteorological community is the classification of Wilson and Henderson-Sellers (1985) at the global scale, but with a coarse resolution of one by one degree (latitude and longitude). This resolution is not satisfactory for NWP models operating with a grid box size of a few km. For this reason, a methodology for generating vegetation parameter maps over Europe with a resolution of 1 km has been developed by Champeaux and Le Gléau (1995). The method takes full advantage of the regular remote sensing information provided by the AVHRR/NOAA. Fist of all, an annual archive has been created by Météo-France over Europe since the early 1990s. AVHRR

8. EXAMPLES OF THE USE OF SATELLITE DATA IN NUMERICAL WEATHER PREDICTION MODELS

channels 1 and 2 were processed from April 1992 to September 1994 at a 2-km resolution. After treatment for cloud detection and atmospheric correction, monthly maps of the Normalized Difference Vegetation Index (NDVI) values were assembled. For the purpose of this treatment, the Holben (1986) composite technique was used; it involves the selection of the maximum value of NDVI during the time period considered. Some 29 images have been utilized. Then, an automatic classification is performed on monthly time series of NDVI to classify 11 classes (Figure 1): forests, grasslands, orchards and bare ground, as well as 8 types of crops characterized by distinct radiometric signatures, according to their seasonal variations. In particular, this classification distinguishes very clearly the areas with summer and winter crops. It is important to note that such discrimination is not provided by high-resolution classification like the European Corine database. The second phase of the treatment consists in assigning values to the relevant vegetation parameters (roughness length, leaf area index, minimum stomatal resistance) for each class of vegetation, using look-up tables. As far as possible, the parameter values are derived either from the local calibration of ISBA, or from the literature. In the last step, the parameters are averaged within each model grid box using aggregation rules described in Noilhan and Lacarrère (1995). Recently, the method was improved by Habets et al. (1997) who assumed that the NDVI profiles could be used to infer the seasonal evolution of the leaf area index (LAI), independent of the vegetation class. Thus, Habets et al. (1997) prescribed only the extreme values of LAI for each vegetation type, and the monthly NDVI estimates were used to compute the seasonal variation of the leaf area index between the minimum and maximum values. Such a method has been tested to provide boundary conditions for a hydrological distributed model in the Hapex-Mobilhy area (the Adour river basin). In an operational context, high-resolution maps of vegetation are used over Europe. The impact of either of these vegetation maps on ISBA (compared with the old surface scheme with no vegetation and only one type of soil) has been evaluated by Giard and Bazile (1997b). The new assimilation suite has proved successful in improving the forecast scores against observed air temperature and relative humidity at 2 m for synoptic stations. In the subsequent assimilation process, the soil reservoirs and the soil temperatures are corrected every 6 hours using the forecast errors of screen-level temperature and humidity. The results of 20 consecutive days of assimilation (6 hour cycles) with ISBA and the old surface scheme are shown figure 2 for Spain and France. The statistics represent the mean bias and RMS for temperature (K) and relative humidity (%) at 2 m. For both areas, the forecast is improved, particularly for the daily cycle of humidity, which have been partly corrected with ISBA. Over Spain, the reduction of the temperature bias is important (the error is reduced by a factor 2). Clearly, this impact is related to the improvement of the vegetation map. In the old scheme, the fraction of vegetation over Spain was too high, resulting in a forecast which was too cold and wet.

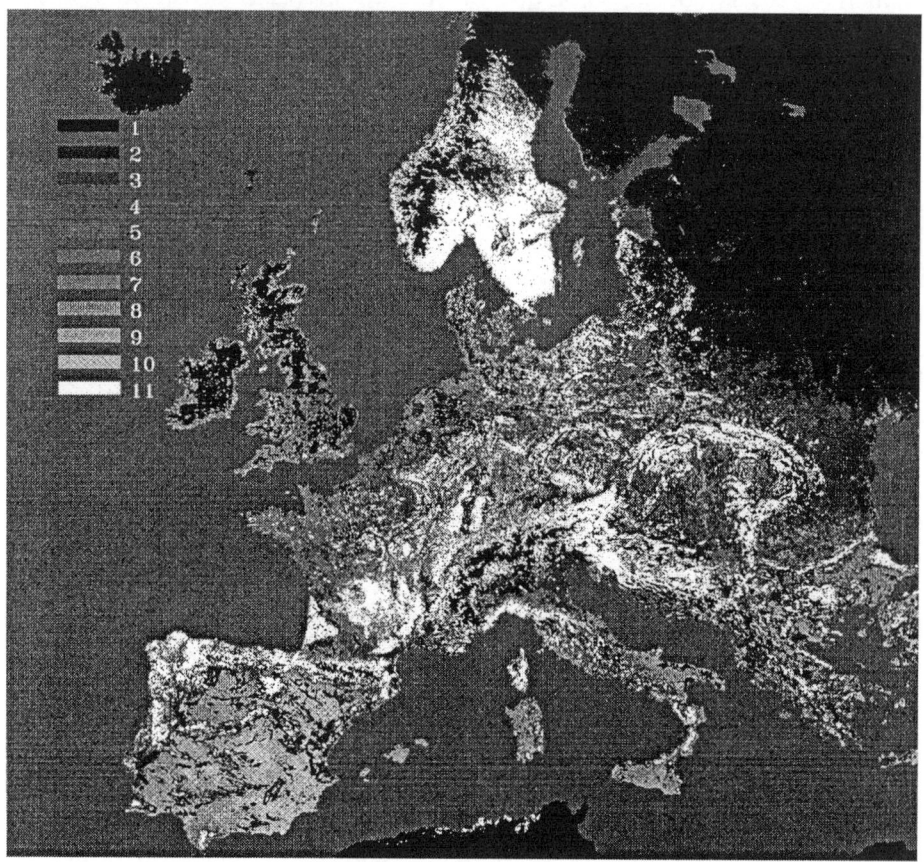

Figure 1. The vegetation classification of Champeaux and Legleau (1995) computed from monthly NDVI/AVHRR information: 1, crop; 2, Mediterranean crop; 3 cereal; 4 crop; 5 cereal; 6 crop and grassland; 7 pasture land; 8 crop; 9 crop; 10 vineyard; 11 forest.

With the AVHRR classification and the calibration of ISBA over Spain (particularly for dry soils, see Giordani et al. 1996), the surface flux treatment has been improved, leading to acceptable score over the Iberian Peninsula. The improvement over France is not as good because the old system was already well calibrated for that domain. Overall, the general quality of the forecast of quantities at 2 m was improved in this experiment for the European domain for which the AVHRR classification was available.

8. EXAMPLES OF THE USE OF SATELLITE DATA IN NUMERICAL WEATHER PREDICTION MODELS

On the other hand, the improvement is not very good in the area where only the low resolution classification of vegetation is available (for instance over the US or over Australia, which are not shown here).

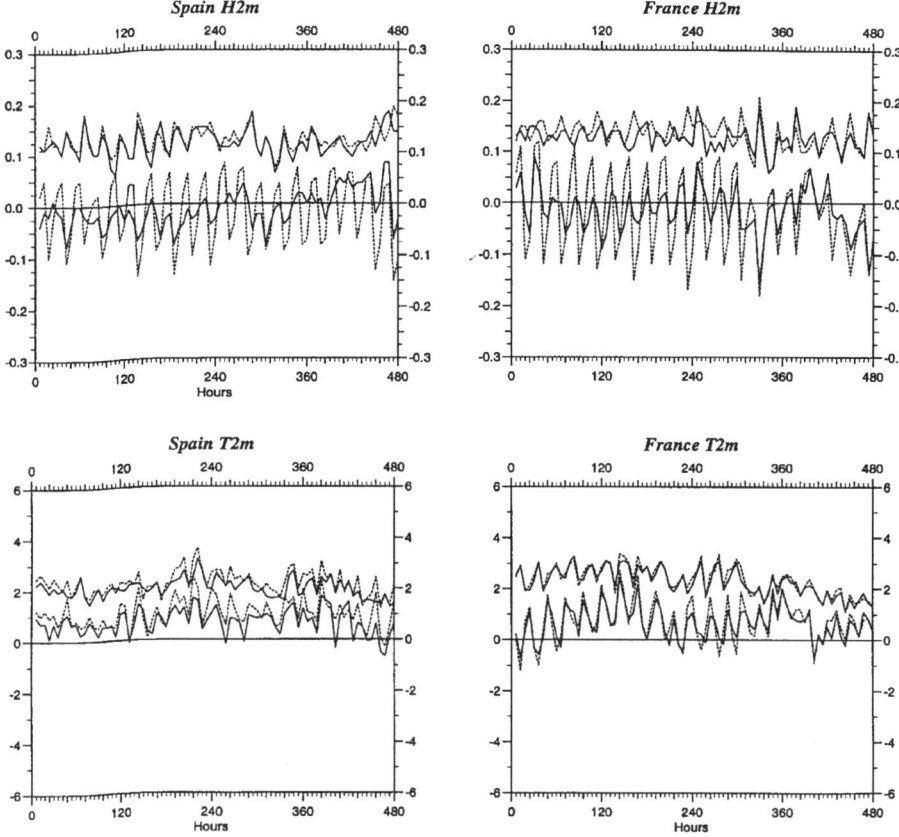

Figure 2. Impact of 20 days (September, 97) assimilation on the forecast errors (mean bias and RMS) of air temperature and relative humidity at 2 m over Spain and France with ISBA (solid lines) and with the old surface scheme (dashed lines) (from Giard and Bazile, 1997).

3. USE OF THE SEBAL ALGORITHM TO RETRIEVE THE SOIL MOISTURE

Recently, Hurk et al. (1997) developed an algorithm for the assimilation of initial soil moisture fields in NWP models using Meteosat and NOAA data. The method is based on the SEBAL algorithm (Bastiaanssen et al. 1997) which determines the surface fluxes from high and low resolution satellite data. The main interest of the method is that information on land use and synoptic data is not required. SEBAL works only during clear sky conditions and with the assumption that the Meteosat image contains a number of wet and dry pixels. A statistical relation between surface temperature and surface albedo is used to identify the dry (high T_s and albedo) and wet (low T_s and albedo) areas. Assuming that sensible heat equals zero for the wet pixels and that evaporation is negligible for the dry pixels, SEBAL computes the sensible H and latent LE heat fluxes for the whole image. The remotely sensed derived input of SEBAL includes surface temperature, albedo and NDVI. The method has been tested over Spain on the basis of EFEDA data. The system proved capable to retrieve the evaporative fraction ($\Delta = LE/(H + LE)$) for soil and sparse vegetation. Hurk et al. (1997) proposed a method to modify the soil moisture from the comparison between the simulated and SEBAL (observed) evaporative fractions. The minimization of forecast errors was achieved by adjusting the soil reservoir. A preliminary study was conducted over the Iberian Peninsula during a 7-day period in the summer of 1994. Figure 3 illustrates the results of the assimilation cycle. In the control run, the initial soil moisture is taken from the climatology and then evolves freely (no correction). In the experimental run, the soil reservoir is corrected every day using the simple assimilation procedure. Figure 3 shows that the assimilation method has a positive impact on the mean bias and root mean square errors of screen level temperature and humidity over Spain. Thus, these results suggest that remotely sensed data could improve the estimation of surface fluxes and help the assimilation of soil moisture in NWP models. However, some important technical points remain to be solved before an operational implementation of the method. For instance, the method has to be used regularly, in particular in cloudy and moderate soil water conditions. The SEBAL algorithm could also benefit from the use of vegetation classification and low level atmospheric information (influence of wind speed, for instance).

8. EXAMPLES OF THE USE OF SATELLITE DATA IN NUMERICAL WEATHER PREDICTION MODELS

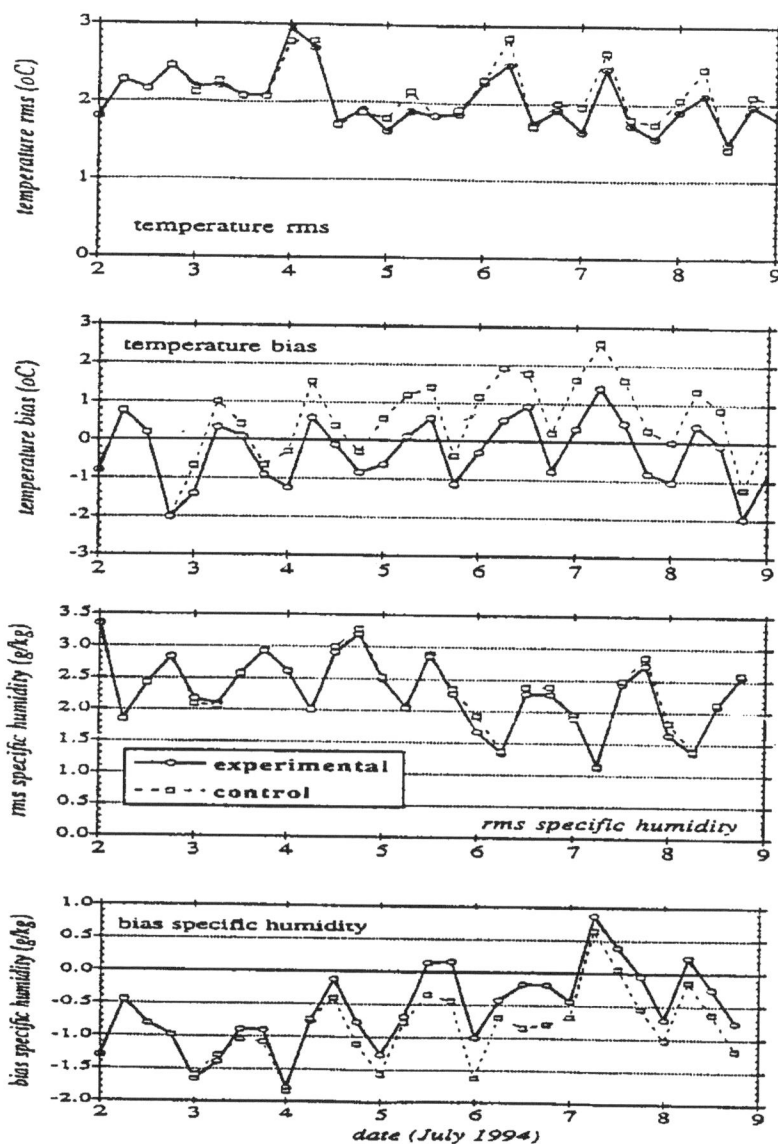

Figure 3. Bias and RMS error of temperature (upper panels) and specific humidity (lower panels) at 2 m. In the control run, soil moisture is not corrected. In the experimental runs, the soil moisture is adjusted daily from the minimization between the observed and forecast evaporative fraction (figure from van den Hurk et al., 1997)

4. CONCLUSION

Most of the operational weather prediction centers are now using advanced land surface schemes, which include the influence of the vegetation on the surface exchanges. Two particular problems are related to the operational implementation of those surface schemes: the characterization of vegetation properties with high enough resolution in space (typically a kilometer) and time (typically a month), and the initialization of the soil water reservoir. Satellite information in the visible and infrared spectral ranges can bring significant input to solve the two problems. Algorithms are now ready to include the satellite information in operational NWP models. Important requirements are the production of high-resolution vegetation maps at the global scale, monthly variations of vegetation properties from vegetation indices and monthly maps for albedo (see Roujean et al. 1997 for a review). Concerning the problem of soil water initialization, remotely-sensed surface fluxes and near-surface soil moisture (Calvet et al. 1997) have a good potential for deep soil reservoir assimilation.

5. REFERENCES

Bastiaanssen, W., M. Menenti, R. Feddes, and A. Holtsag (1997) A remote sensing surface energy balance algorithm for land (SEBAL): Formulation, *Journal of Hydrology*, (in press).

Bazile, E., and D. Giard (1996) Assimilation and sensitivity experiments in the NWP model ARPEGE with ISBA, *Hirlam Newsletter*, **24** (Soil processes and soil surface data assimilation), 73–78.

Bouttier, F., J.F. Mahfouf, and J. Noilhan (1993) Sequential assimilation of soil moisture from atmospheric low level parameters. Part I: Sensitivity and calibration studies, *Journal of Applied Meteorology*, **32**, 1335–1351.

Bringfelt, B. (1996) Hirlam surface parameterization and assimilation: General overview, present status and developments, *Hirlam Newsletter*, **24** (Soil processes and soil surface data assimilation), 26–29.

Calvet, J.-C., J. Noilhan, and P. Bessemoulin (1997) Soil water reservoir retrieval from surface soil moisture and temperature estimates, *Journal of Applied Meteorology*, **37**, 371–386.

Champeaux, J.-L. and H. Legleau (1995) Vegetation mapping over Europe using NOAA/AVHRR, in *The 1995 meteorological satellite data users' conference*, 139–143, Winchester, UK and EUMETSAT.

Giard, D. and E. Bazile (1997) La version pré-opérationelle du schéma de surface ISBA. *Atelier de Modélisation Atmosphérique*, 265–269.

Giard, D. and E. Bazile (1999) Implementation of a new assimilation scheme for soil and surface variables in a global NWP model, *Monthly Weather Review*, (to appear).

8. EXAMPLES OF THE USE OF SATELLITE DATA IN NUMERICAL WEATHER PREDICTION MODELS

Giordani, H., J. Noilhan, P. Lacarrère, P. Bessemoulin, and P. Mascart (1996) Modelling the surface procesess and the atmospheric boundary layer for semi-arid conditions, *Agricultural and Forest Meteorology*, **80**, 263–287.

Habets, F., J. Noilhan, C. Golaz, J.P. Goutorbe, P. Lacarrère, E. Leblois, E. Ledoux, E. Martin, C. Ottle, and D. Vidal-Madjar (1999) Implementation of the ISBA surface scheme in a distributed hydrological model, applied to the HAPEX-MOBILHY area. Part 1: The modelling strategy and the data base, *Journal of Hydrology*, (to appear).

Habets, F., J. Noilhan, C. Golaz, J.P. Goutorbe, P. Lacarrère, E. Leblois, E. Ledoux, E. Martin, C. Ottle, and D. Vidal-Madjar (1999) Implementation of the ISBA surface scheme in a distributed hydrological model, applied to the HAPEX-MOBILHY area. Part 2: Simulation of streamflows and of the annual water budget, *Journal of Hydrology*, (to appear).

Henderson-Sellers, A. et al. (1997) The Project for Intercomparison of Land-Surface Parameterization Schemes (PILPS): Phases 2 and 3, *Bulletin of the American Meteorological Society*, **76**, 489–503.

Holben, B.N. (1986) Characteristics of maximum-value composite images from temporal AVHRR data, *International Journal of Remote Sensing*, **7**, 1417–1437.

Mahfouf, J.F. (1991) Analysis of soil moisture from near-surface parameters: A feasibility study, *Journal of Applied Meteorology*, **30**, 1534–1547.

Noilhan, J. and P. Lacarrère (1995) GCM gridscale evaporation from mesoscale modelling, *Journal of Climate*, **8**, 206–223.

Noilhan, J., and J.-F. Mahfouf (1996) The ISBA land surface parameterization scheme, *Global and Planetary Change*, **13**, 145–159.

Noilhan, J. and S. Planton (1989) A simple parameterization of land surface processes for meteorological models, *Monthly Weather Review*, **117**, 536–549.

Roujean, J.L., D. Tanré, F.M. Bréon, and J.L. Deuze (1997) Retrieval of land surface parameters from airborne POLDER bidirectional reflectance distribution function during HAPEX-Sahel, *Journal of Geophysical Research*, D10.

VandenHurk. B., W. Bastiaanssen, H. Pelgrum, and E. Meijgaard (1997) A new methodology for assimilation of initial soil moisture fields in weather prediction models using Meteosat and NOAA data, *Journal of Applied Meteorology*, **36**, 1271–1283.

Viterbo, P., and A.C.M. Beljaars (1995) An improved land surface parameterization scheme in the ECMWF model and its validation, *Journal of Climate*, **8**, 2716–2748.

Wilson, M. and A. Henderson-Sellers (1985) A global archive of land cover and soils data for use in general circulation climate models, *Journal of Climatology*, **5**, 119–143.

Chapter 9

THE CONTRIBUTION OF REMOTE SENSING TECHNOLOGIES AND ALGORITHMS TO LAND SURFACE PROCESSES STUDIES

B. Pinty and M. M. Verstraete
Space Applications Institute, Ispra, Italy.

1. INTRODUCTION

Solar radiation is the primary source of energy driving the atmosphere, the hydrosphere and the biosphere. The models needed to estimate or predict the amount of radiative energy available as a function of space and time must therefore be based on a thorough understanding of the processes controlling the absorption and scattering of light in these geophysical environments. The bulk of the solar energy available to drive the global Earth system is in fact absorbed at the lower boundary of the atmosphere (oceans and terrestrial surfaces). Surface-atmosphere interactions play an important role at a variety of spatial and temporal scales, including those relevant to the phenomena that control the possible occurrence of convective precipitation. This is especially the case in the context of Soil-Vegetation-Atmosphere Transfer (SVAT) models that simulate the environment with a time step of a few minutes to an hour. These processes have been extensively studied at the local scale (e.g., Geiger 1965, Oke 1978, Monteith and Unsworth 1990).

At a minimum, atmospheric General Circulation Models (GCM) require a specification of the mass, energy and momentum fluxes at their lower boundary. These boundary conditions may take the form of expressions similar to that for the well-known radiation balance at the Earth's surface. Stand-alone SVAT schemes and advanced land surface processes codes, such as those pioneered by Prof. Dickinson, are designed in large part to be used in GCMs. Computer codes such as the Biosphere-Atmosphere Transfer Scheme (BATS) and the Simple Biosphere (SiB) require a detailed description of the transfer of radiation within these surface layers (vegetation, soil). This is particularly so if they try to represent physiological processes within the plant canopy (stomatal control of the water and CO_2 fluxes) or terrestrial elements of biogeochemical cycles (photosynthesis, carbon assimilation, and respiration). SVAT models have been developed by and for atmospheric scientists to describe essential physical processes (namely the mass, energy and momentum exchanges) occurring at the interface between the terrestrial surface and the atmosphere.

Land Surface Processes (LSP) models require the specification of spatial distributions and temporal evolutions of a number of surface state variables controlling the initial and boundary conditions for the exchanges of energy, water, carbon, etc. at the surface. One of the major conceptual advances over the last two decades has been to consider land surfaces as 'active boundaries' for the dynamical, radiative and hydrological processes in the atmosphere (e.g., Dickinson 1983, Sellers 1985). The global monitoring of terrestrial surfaces largely relies on appropriate measuring tools and techniques, such as remote sensing. However, the exploitation of these data streams to extract some of relevant variables required by the biogeochemical models and describe the corresponding exchanges must necessarily rely on an analysis of the radiance fields emerging at the top of the atmosphere.

2. CAN REMOTE SENSING HELP LSP STUDIES?

Remote sensing can, in principle, provide invaluable information on two categories of LSP variables:
1. those controlling the radiative fluxes between the surface and the atmosphere above (e.g., surface albedo, surface skin temperature, and soil moisture), and
2. the state variables of the radiative transfer problem of the surface itself (e.g., the Leaf Area Index, the vegetation fractional cover, the plant architecture, etc).

Category 1 corresponds to instantaneous spectral- and angular-dependent properties. They are not intrinsic properties of the surface since, for instance, their values are changing with the state of the atmosphere.

Category 2 corresponds to variables representative of land surfaces. They are intrinsic surface properties controlling not only the radiative exchanges but also the dynamical and hydrological processes.

9. THE CONTRIBUTION OF REMOTE SENSING TECHNOLOGIES AND ALGORITHMS TO LAND SURFACE PROCESSES STUDIES

The ultimate objectives of remote sensing data interpretation are to extract reliable and accurate information on the state of the system being observed from the data, and, simultaneously, to reduce or eliminate all undesirable variations present in the original data. These unwanted variations can generally be attributed to the sensor itself (calibration) or to the observation conditions (anisotropy effects), rather than intrinsic changes in the target of interest (Verstraete et al. 1996). The quality and reliability of this interpretation phase will directly affect the accuracy, sometimes even the feasibility, of all products generated downstream by LSP models from these data. The immediate benefit of a high performance interpretation scheme is access to a large number of highly accurate and reliable end products.

The accurate and reliable estimate of Category 1 variables requires an adequate sampling with respect to the *spectral* (visible, near-infrared and mid-infrared) domain of interest and the *angular* domain of integration (hemisphere). The spectral sampling is needed because the radiative properties of land surfaces are changing significantly from visible (0.4–0.7 µm) to near-infrared (0.7–1.3 µm). The absorption of light in the visible domain by plant leaves provides the energy required by the photosynthetic activity and permits plant growth. On the other hand, light is very efficiently scattered away in the near-infrared region, most probably as a result of adaptation since the amount of energy per photon is not sufficient for the synthesis of organic molecules in that spectral range (Gates 1980). An adequate angular sampling is required to guarantee a sufficiently accurate and reliable atmospheric decontamination (Martonchik et al. 1998a), and is needed anyway to estimate surface albedo values. In fact the latter quantities can be defined in various ways, directional hemispherical reflectance or bihemispherical reflectance (Nicodemus et al. 1977) but they always correspond to quantities integrated angularly over an hemisphere.

Obviously, the poorer the sampling, the more it is necessary to formulate hypotheses and/or acquire data from additional independent sources. For instance, without the angular sampling capability, simplifying assumptions to convert radiances (bi-directional reflectance factors) into fluxes (albedos) cannot be avoided. These assumptions can consist in using a Lambertian hypothesis or in adding some a priori knowledge on the surface under study. However the Lambertian hypothesis is not accurate enough (e.g., Deering et al. 1995); and a priori knowledge on the typical anisotropy of land surfaces is quite limited on a global scale because these features have not been documented yet.

In the case of Category 2 variables, an extensive sampling in both the spectral and angular domains is also a prerequisite to constrain the inversion process and therefore limit the number of possible solutions. Indeed, two or more different geophysical situations may lead to quite similar sets of measurements for a given instrument, but the more complete the sampling, the easier it is to distinguish between these geophysical situations (Gobron et al. 1997 and Martonchik et al. 1998b).

3. WHERE DO WE STAND, AND WHERE DO WE GO FROM HERE?

Few instruments, if any, have been dedicated so far to the acquisition of data relevant for LSP studies on a global basis. Existing sensors suffer from significant drawbacks, including lack of reliable and accurate calibration, low spatial resolution, and very poor or biased sampling. In addition, support for the development of advanced algorithms has been quite limited, partly as a result of the absence of appropriate sensors.

This situation is evolving quite rapidly, in response to both scientific and technological developments. All major Space Agencies have invested sizable resources to design and implement a new generation of Earth Observation platforms and sensors (e.g., MERIS on the ESA ENVISAT, MODIS on the NASA Terra, or GLI on the NASDA ADEOS-II platform.) Some of these new sensors are partly or fully dedicated to the acquisition of data on the state of the Earth surface, either at a medium spatial resolution but on a global scale, or at a high spatial resolution for limited regions. The upcoming availability of such advanced instruments has motivated significant algorithmic developments. For instance, users of information on land surfaces will soon have access to extensive documentation of environmental variables such as albedo, Leaf Area Index (LAI), or the Fraction of Absorbed Photosynthetically Active Radiation (FAPAR). These values will ideally be accompanied by documented accuracy and reliability estimates, directly from the ground segments set up by the Space Agencies or from third party companies. This is in contrast with recent or even current practices, when users were (are) provided with raw data or at best vegetation index values, without clear indications as to their meaning or intrinsic variability.

In the same vein, the acquisition of a more representative sample of the angular and spectral signatures (including observations in the blue band, as well as narrower and well-positioned bands in the red and near-infrared regions) will lead to products of higher quality. This is because these new sensors will be much less sensitive to atmospheric perturbations, for instance. The multi-angular capability of advanced sensors such as MISR and POLDER will further lead to improved products and services. Indeed,

9. THE CONTRIBUTION OF REMOTE SENSING TECHNOLOGIES AND ALGORITHMS TO LAND SURFACE PROCESSES STUDIES

such data are much more apt at effectively constraining the inversion problem and limiting the number of solutions which can acceptably account for the observed variations in spectral and directional reflectance (e.g., Martonchik et al. 1998a, Martonchik et al. 1998b, Knyazikhin et al. 1998).

Thus, technological advances in aerospace and computer technologies will allow the acquisition of data under much better defined observation protocols, and recent theoretical and simulation achievements will permit to take better advantage of these new measurements. For instance, the observation of the reflectance of the Earth in the blue spectral region has already permitted the development and implementation of a new generation of spectral indices optimized to characterize the surface without being too affected by atmospheric composition (e.g., Govaerts et al. 1999, Gobron et al. 1999). More generally, the advent of new sensors and the implementation of advanced algorithms of information extraction will yield new or better characterizations of land surface processes (e.g., Pinty et al. 1999), which, in turn, will lead to more accurate forecasts or descriptions of our environment.

The next few years should thus prove quite exciting, both from the technological and scientific point of view. For instance, the European Space Agency (ESA) is currently planning a new series of Earth Explorer missions, which are small to medium size space platforms with one or a few instruments to support focused scientific applications. One such mission currently under study is the Land Surface Processes and Interactions Mission (LSPIM). It would permit to better understand biosphere-atmosphere interactions and land surface processes, as well as help addressing scaling issues which are so critical to transfer the understanding one derives from local field studies to the regional scales that are relevant to address climate change and environmental degradation problems.

New applications of satellite remote sensing will likely develop in a variety of fields, and these should spawn a number of value-added operational activities (Glackin 1999). This progressive evolution closer to applications and users will probably change in some profound ways the research and development activities in satellite remote sensing. At the same time, the improved capability to address some of the issues raised in LSP studies and the progressive use of information on the state of our environment should justify, a posteriori, the investments made in the recent past as well as those that are still needed to bring these projects to fruition. A clear demonstration of the benefits that can be derived from the analysis of multi-angular and multi-spectral data, such as those that will be generated by MISR, is thus critical for this field.

4. REFERENCES

Deering, D. W., Ahmad, S. P., Eck, T. F. and Banerjee, B. P. (1995) Temporal Attributes of the Bidirectional Reflectance for Three Boreal Forest Canopies, *International Geoscience and Remote Sensing Symposium* (IGARSS'95), Quantitative Remote Sensing for Science and Applications, Florence, Italy, 10–14 July 1995, IEEE Cat. No. 95CH35770, 1239–1241.

Dickinson, R. E. (1983) Land Surface Processes and Climate-Surface Albedos and Energy Balance, *Advances in Geophysics*, **25**, 305–353.

Gates, D. M. (1980) *Biophysical Ecology*, Springer-Verlag, New York, NY.

Geiger, R. (1965) *Climate Near the Ground*, Harvard University Press, Cambridge, MA.

Glackin, D. L. (1999) Earth Remote Sensing Business Goes Public, *The Industrial Physicist*, **5**, 22–25.

Gobron, N., Pinty, B., Verstraete, M. M. and Govaerts, Y. (1997) A Semidiscrete Model for the Scattering of Light by Vegetation, *Journal of Geophysical Research*, **102**, 9431–9446.

Gobron, N., Pinty, B., Verstraete, M. M. and Govaerts, Y. (1999) The MERIS Global Vegetation Index (MGVI): Description and Preliminary Application, *International Journal of Remote Sensing*, **20**, 1917–1927.

Govaerts, Y., Verstraete, M. M., Pinty, B. and Gobron, N. (1999) Designing Optimal Spectral Indices: A Feasibility and Proof of Concept Study, *International Journal of Remote Sensing*, **20**, 1853–1873.

Knyazikhin, Y. V., Martonchik, J. V., Diner, D. J., Myneni, R. B., Verstraete, M. M., Pinty, B. and Gobron, N. (1998) Estimation of Vegetation Canopy Leaf Area Index and Fraction of Absorbed Photosynthetically Active Radiation From Atmosphere-Corrected MISR Data, *Journal of Geophysical Research*, **103**, 32,239–32,256.

Martonchick, J. V., Diner, D. J., Kahn, R. A., Ackerman, T. P., Verstraete, M. M., Pinty, B. and Gordon, H. (1998a) Techniques for the Retrieval of Aerosol Properties Over Land and Ocean Using Multiangle Imaging, *IEEE Transactions on Geoscience and Remote Sensing*, **36**, 1212–1227.

Martonchick, J. V., Diner, D. J., Pinty, B., Verstraete, M. M., Myneni, R. B., Knyazikhin, Y. and Gordon, H. (1998b) Determination of Land and Ocean Reflective, Radiative, and Biophysical Properties using Multi-Angle Imaging, *IEEE Transactions on Geoscience and Remote Sensing*, **36**, 1266–1281.

Monteith, J. L. and Unsworth, M. H. (1990) *Principles of Environmental Physics*, Edward Arnold, London.

Nicodemus, F. E., Richmond, J. C., Hsia, J. J., Ginsberg, I. W. and Limperis, T. (1977) *Geometrical Considerations and Nomenclature for Reflectance*, US Department of Commerce, National Bureau of Standards, NBS Monograph No. **160**, Washington, DC.

Oke, T. R. (1978) *Boundary Layer Climates*, Methuen and Company, Ltd., New York.

Pinty, B., Roveda, F., Verstraete, M. M., Gobron, N. and Govaerts, Y. (1999) Estimating Surface Albedo From the Meteosat Data Archive: A Revisit, *Proceedings of the ALPS 99 Conference*, Meribel, France, 18–22 January 1999, CNES, **O-03**, 1–4.

Sellers, P. J. (1985) Canopy Reflectance, Photosynthesis and Transpiration, *International Journal of Remote Sensing*, **6**, 1335–1372.

Verstraete, M. M., Pinty, B. and Myneni, R. B. (1996) Potential and Limitations of Information Extraction on the Terrestrial Biosphere From Satellite Remote Sensing, *Remote Sensing of Environment*, **58**, 201–214.

Chapter 10

EXPLOITATION AND EVALUATION OF RETRIEVAL ALGORITHMS FOR GEOSTATIONARY SATELLITE DATA PROCESSING
Current and future systems

Y. Govaerts
EUMETSAT, Darmstadt, Germany.

1. INTRODUCTION

Two ENAMORS objectives are especially relevant with respect to meteorological remote sensing activities: the validation of operational remote sensing products and the identification of further actions in methodological research for the development of novel applications, in particular in the framework of new European meteorological geostationary satellites. These two activities are closely related since validation is an important phase of new algorithm design. This paper covers therefore these two different topics. In section 2, the current Meteorological Product Extraction Facility (MPEF) exploited at EUMETSAT is presented and the problem of validation and operational quality control addressed.

The characteristics of the Meteosat Second Generation (MSG), the new European meteorological geostationary satellite to be launched in 2000, are presented in section 3. Its potential for the monitoring of land surfaces is explored. Albeit MSG spectral characteristics are similar to the NOAA/AVHRR instrument, the observation sampling is completely different. The additional information conveyed by the frequent observations is illustrated with a simple example. It turns out that an optimal exploitation of geostationary observations will require dedicated retrieval algorithms.

2. EXTRACTION AND EVALUATION OF METEOROLOGICAL PRODUCTS

The meteorological geostationary satellite currently exploited by EUMETSAT has three channels[1] with a 30 minutes image repeat cycle. Quantitative meteorological products such as cloud motion winds (CMW), sea surface temperatures (SST), clear sky radiances (CSR), cloud cover, cloud top temperature and cloud top height (CTH) are automatically extracted from these images. This operational processing chain, initiated almost 20 years ago (ESA 1987), has reached a very high level of maturity and automation. The main product extraction processing steps are summarized here after:

1. *Data reception.* MPEF real-time input is composed of Meteosat observations, forecast data from the European Centre for Medium-range Weather Forecasts (ECMWF) and finally meteorological observations.
2. *Pre-processing.* The pre-processing phase consists in a series of data manipulations. These include satellite data calibration based on ground observations and radiosondes (Gube et al. 1996; van de Berg et al. 1995), the computation of scene predicted radiances (Schmetz 1986) and finally image analysis. The latter involves cloudy pixel identification and cloud top height estimation, accounting for the presence of semi-transparent clouds (Szejwach 1982).
3. *Product extraction.* Meteorological products are then automatically generated. Examples of the CMW product can be found at the following URL address: http://www.eumetsat.de. Separate wind vectors are extracted from the different channels. Each wind vector is characterized by an altitude (low, medium or high), a direction, a speed and finally a quality value (Schmetz and Holmlund 1990).
4. *Quality control.* The quality control step generates information concerning product accuracy and reliability. In the case of the CMW product, each wind is assigned a quality score based on a series of consistency checks, which are combined to give an overall value. These verifications concern (1) a test on the temporal direction and speed change, (2) a test on the spatial direction and speed change, and (3) a comparison with forecast values (Elliott and Holmlund 1996). Consistent quality flags are of primary importance to ensure the optimal product interpretation.
5. *Product dissemination.* The products are both archived in the Meteorological Archive and Retrieval Facility (MARF) and distributed

---

[1] Visible: 0.4–1.1 µm, water vapor: 5.7–7.1 µm, and thermal infrared: 10.5–12.5 µm.

10. EXPLOITATION AND EVALUATION OF RETRIEVAL ALGORITHMS FOR GEOSTATIONARY SATELLITE DATA PROCESSING

in real-time through the meteorological Global Telecommunication System (GTS).

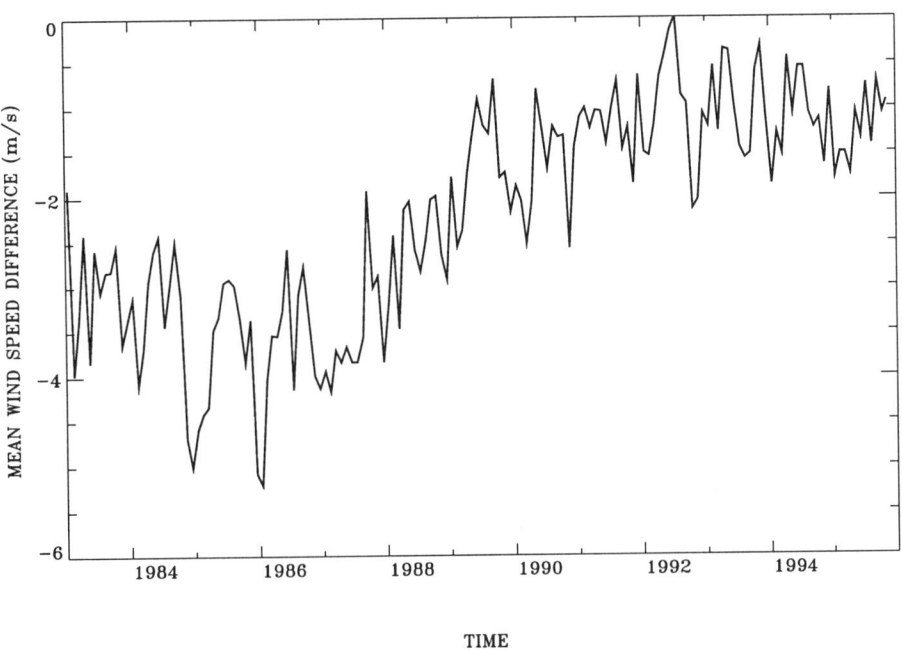

Figure 1. Wind speed difference (m/s) between collocated radiosondes and high infrared cloud motion winds

In addition to this fully automated on-line processing, an off-line product validation is performed. This validation monitors the product's spatial and temporal consistency with respect to the user's accuracy requirements. In the case of the CMW product, the validation is based on three main methods:
1. *Comparison with in-situ observations.* The CMW product is compared with collocated radiosonde wind observations. These comparisons provide relevant information to assess the impact of algorithm modifications and input data quality (such as sensor signal-to-noise ratio) on the quality of the final product. This approach has, however, two major limitations. First, CMWs represent large-scale horizontal tracer (i.e., cloud or water vapor) displacements. Conversely, radiosonde observations result from instantaneous and local wind observations. It is therefore not expected that the difference between the two variables should vanish, as can be seen on Figure 1. Because both quantities result from different processes, the interpretation of these differences in terms

of CMW product accuracy is therefore not straightforward. Second, radiosonde observations are sparse in both space and time. They occur at best twice a day, mainly over Europe. Large areas such as the Atlantic Ocean or the African continent are covered only with very sparse radiosonde observations. Consequently, this method cannot be used to detect the presence of systematic spatial biases in the CMW product.

2. *Product monitoring against assimilated data.* A second type of validation mechanism is performed by ECMWF, comparing the bias and standard deviation between the CMW product and the forecast winds. These comparisons provide meaningful information on the spatial consistency between these two fields. Once again however, the two quantities do no represent exactly the same natural process and cannot be used to estimate the product absolute accuracy.

3. *Impact studies.* The strongest validation test is performed by ECMWF. Numerical weather forecasts are generated with and without the assimilation of the CMW product. The validation is based on the impact of the CMW on the forecast quality. This test verifies whether the product accuracy matches the user's requirement. When the user's requirements are met, the product assimilation has a positive impact on the forecast quality.

Quality control and product validation are two important processing steps of operational meteorological product extraction. Quality control is applied everywhere and all the time, as the products are generated. The quality estimation associated with the products ensures an optimal interpretation and usage of the delivered information. Validation is an off-line activity, which provides an indirect estimation of product accuracy. It is applied in a limited number of cases and permits to monitor the effect of retrieval algorithm modification and input data quality on the product reliability. If necessary, validation results can be used to adjust quality control mechanisms.

It has been seen that quality control is of primary importance to increase the intrinsic value of a product, providing a way to improve its interpretation. Hence, the definition of new products as well as the development of advanced retrieval algorithms should account for the possibility to assign consistent quality scores. These issues will be addressed in the next section in the context of the exploration of the MSG potential for land surface monitoring.

3. POTENTIAL OF GEOSTATIONARY OBSERVATIONS FOR LAND APPLICATIONS

Meteosat Second Generation (MSG) will have enhanced spectral capabilities with respect to the present Meteosat satellite, in addition to an improved image repeat cycle (15 minutes). The spectral characteristics of its main radiometer, called Spinning Enhanced Visible and Infrared Imager (SEVIRI) (Aminou et al. 1997) are presented in Table 1. As can be seen, SEVIRI has a spectral capability similar to the NOAA-AVHRR instrument, (specifically, the RED, NIR, IR3.8, IR10.8 and IR12.0 channels), in addition to the IR1.6 and IR9.7 channels that can also be used to monitor land surface properties. These familiar spectral characteristics have already prompted scientists to investigate the potential of MSG for large-scale biospheric applications. Most of the AVHRR-based applications rely, however, on an empirical or statistical exploitation of the spectral information, using image compositing to deal with a variable cloud coverage, but neglecting the anisotropic effects because of the under-sampling in the angular domain. This type of approaches may introduce biases and prevents consistent quality control to be applied. Conversely, any given location is always observed with the same viewing angle but different Sun angles with a geostationary orbiting satellite. Thus, the effects of surface anisotropy on the observed radiances can be properly sampled and studied.

Table 1. Spectral characteristics of the SEVIRI instrument on MSG. (# Detect. is the number of detectors per channel, Resol. is the sub-satellite sampling distance in km).

Channel	Band (μm)	# Detect.	Resol.	Remark
HRV	0.50-1.00	9	1	Solar
RED	0.56-0.71	3	3	Solar
NIR	0.74-0.88	3	3	Solar
IR1.6	1.50-1.78	3	3	Solar
IR3.8	3.40-4.20	3	3	Window
IR6.2	5.37-7.15	3	3	Water vapor
IR7.3	6.85-7.85	3	3	Water vapor
IRS.7	8.30-9.10	3	3	Window
IR9.7	9.38-9.94	3	3	Ozone
IR10.8	9.80-11.80	3	3	Window
IR12.0	11.00-13.00	3	3	Window
IR13.4	12.40-14.40	3	3	CO_2

Obviously, surface biophysical properties in the solar spectral range will not change tremendously every 15 minutes, so that the advantage frequent observations may be questionable. These frequent observations should permit a better documentation the radiative transfer processes in the atmosphere-vegetation-soil system, and therefore an increase of the retrieved product accuracy and quality control, provided appropriate algorithms are developed. Consequently, instead of providing an exhaustive list of new potential biospheric variables that can be derived from SEVIRI, it is desirable to focus on the implications of exploiting geostationary observation and on the design of appropriate advanced algorithms. For the sake of simplicity, the discussion is limited to the red and near-infrared spectral region.

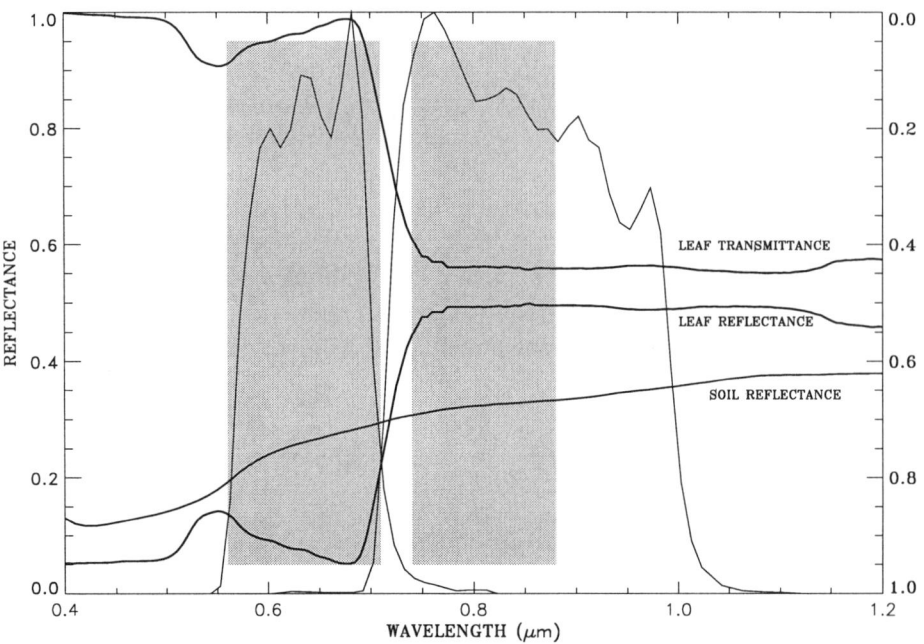

Figure 2. Leaf and soil spectral properties. The shaded areas correspond to the SEVIRI red and near-infrared channels. The corresponding AVHRR channels 1 and 2 spectral responses are also shown.

To illustrate the differences between geostationary and polar orbiting space-borne observations, a simple and limited case study has been carried out on observed radiance sensitivity to Leaf Area Index (LAI) changes. This parameter is particularly difficult to derive because it requires the discrimination of the respective contributions from the soil and the vegetation in the observed radiances. It therefore represents a challenging radiative transfer problem.

A typical Northern Hemisphere mid-latitude pixel has been considered, covered by a homogeneous canopy with a LAI of two. The spectral characteristics of the leaves and the soil are shown in Figure 2. A month (April) of observations has been simulated during which the LAI increases linearly from 2 to 3. The transfer of radiation in the canopy is computed with the model of Gobron et al. (1996). The corresponding top-of-atmosphere reflectance in the AVHRR channels 1 and 2 and SEVIRI RED and NIR channels have been modeled with the 6S code (Vermote et al. 1997), assuming noise-free instruments and perfectly calibrated data. The impact of this LAI increase on radiances, as observed by the two satellites, is then compared.

Figure 3 shows the simulated top-of-atmosphere reflectances. The SEVIRI reflectances are plotted only when the Sun zenith angle is smaller than 70 degrees. The SEVIRI and AVHRR NIR mean level reflectances differ by about 25%. AVHRR channel 2 spectral response extends indeed up to 1 µm and is therefore largely affected by the water vapor absorption band centered around 0.9 µm. This difference illustrates the importance of the water vapor absorption on the observed radiances in AVHRR channel 2.

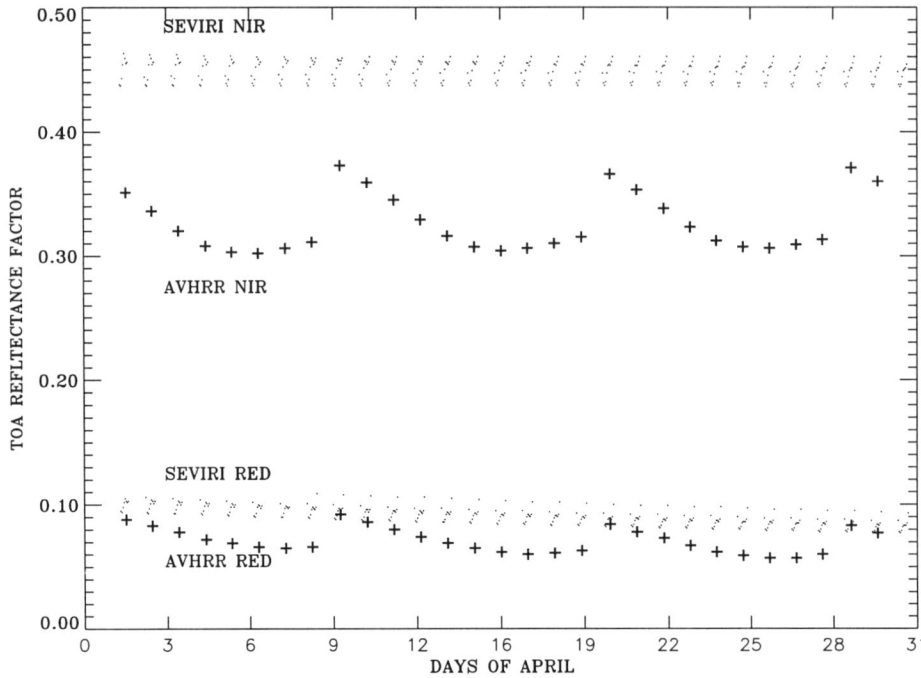

Figure 3. Differences between polar and geostationary satellite observations. The "+" symbol represents AVHRR observations, the "." symbol the MSG ones. The LAI linearly increases from 2 to 3 during the simulation period.

Normally, a LAI increase translates in a decrease of red reflectance and an increase in the NIR, in the absence of atmospheric perturbations. In the present scenario, however, bright soil makes reflectances in NIR less sensitive to LAI changes. AVHRR observations are largely affected by the 9.5 days viewing angle period, giving rise to variations much larger than the variations resulting from the LAI increase. The angular effects are completely different for geostationary observations. Solar angle changes generate regular variations in the SEVIRI observations. Direct comparisons between AVHRR and SEVIRI spectral observations are therefore irrelevant if angular and atmospheric effects are not properly processed. Because of the sampling differences, geostationary and polar observation systems require dedicated retrieval algorithms.

10. EXPLOITATION AND EVALUATION OF RETRIEVAL ALGORITHMS FOR GEOSTATIONARY SATELLITE DATA PROCESSING

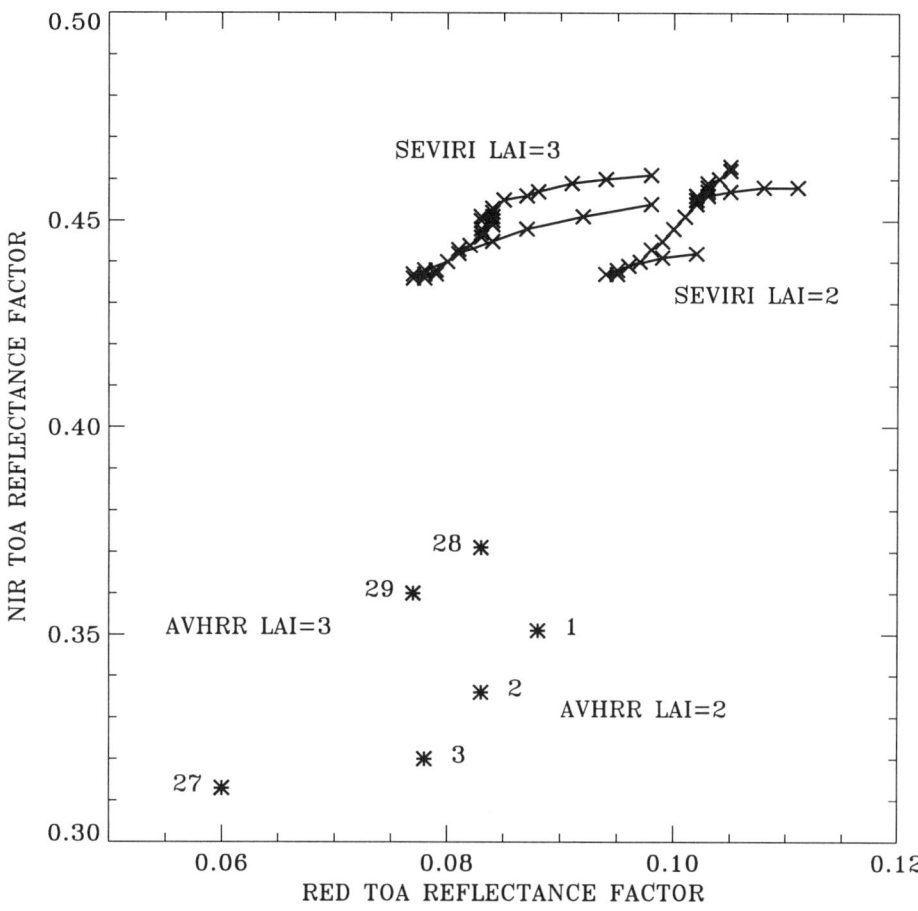

Figure 4. Red versus near-infrared spectral space diagram of the TOA reflectances. The x symbol represents the SEVIRI observations for day one (LAI=2) and for day 30 (LAI=3). The first and last three days of AVHRR observations are indicated with the * symbol.

The differences between AVHRR and SEVIRI observations are summarized in Figure 4. LAI changes generate spectral displacements or "spectral signatures" in the red versus near-infrared spectral space diagram (Verstraete and Pinty 1996). These displacements can be affected by perturbing factors such as the presence of aerosols, clouds, or illumination and observation geometry changes. In particular, AVHRR observations suffer from angular effect under-sampling that may be emphasized by the presence of cloud. To illustrate the consequences of these two effects on the spectral signatures, let's assume that only one out of the first and last three days of the simulation period is cloud free. The resulting spectral displacements may be very different depending on the particular couple of days which is selected. For instance, the displacements corresponding to changes from day 2 to 28 and from day 1 to 27 give rise to perpendicular vectors. A reliable interpretation of these spectral signatures in terms of LAI changes may therefore be quite difficult in the absence of appropriate angular effect processing. By contrast, the hourly observations of SEVIRI show daily trajectories which change in shape as the LAI increases. These daily trajectories are also shifted when LAI goes from two to three. Spectral signatures corresponding to observations acquired at different Sun zenith angles are also quite spread out. A consistent exploitation of these spectral signatures requires accounting for the Sun position. Hence, though angular effects appear as perturbations or "noise" in the case of polar observations, they may convey relevant information in the case of the geostationary observations, if there are properly exploited. These angular variations are now explored in details.

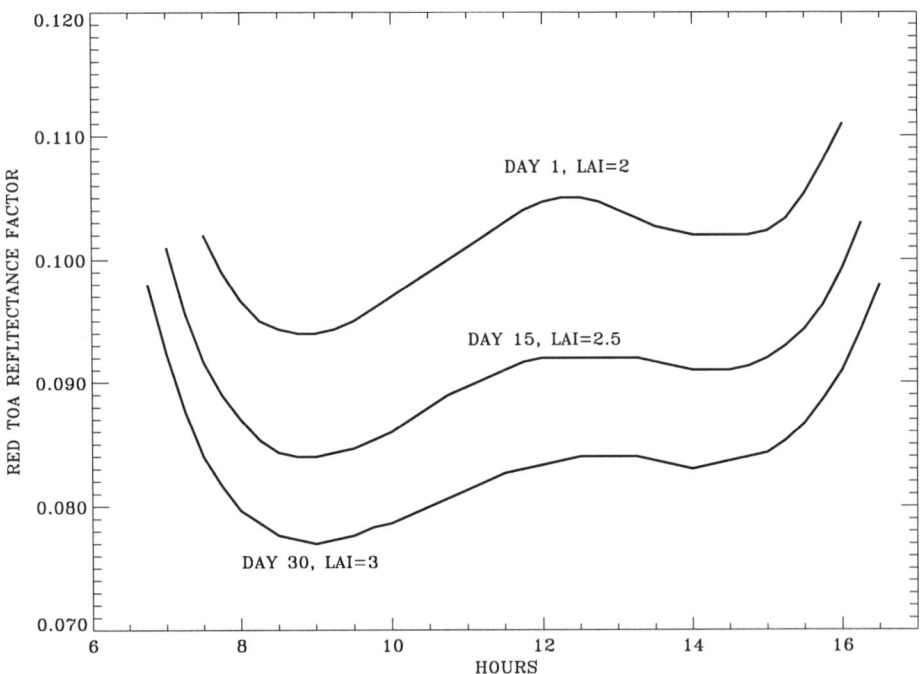

Figure 5. Daily variations of the reflectance in the SEVIRI red band for day 1 (LAI=2), day 15 (LAI=2.5) and day 30 (LAI=3).

Figure 5 shows simulated hourly variations in the reflectance that will be measured the SEVIRI red band for day 1 (LAI=2), day 15 (LAI=2.5) and day 30 (LAI=3). As the LAI increases, the mean reflectance decreases. The observed reflectance around 12 hours, i.e., when the Sun zenith angle is minimum, decreases more rapidly than reflectances observed in the morning or in the afternoon. At noon, the trajectory of photons that reach the soil under the vegetation canopy and are reflected back to the sensor is shortest. It translates into high reflectances, especially when the LAI is small. The probability to be intercepted by the vegetation is indeed minimal. As the LAI increases, the interception probability increases more rapidly for small Sun zenith angles and the "reflectance peak" observed around 12 hours is less pronounced. The high soil brightness is responsible for the same behavior in the NIR band, despite of the importance of multiple scattering (Figure 6).

These frequent observations ease the discrimination between soil and vegetation reflectance contributions for different LAI values. They should also permit to verify that identical LAI values are retrieved over a short time period, under varying illumination conditions. In turn, this constraint can be exploited for the design of relevant quality control mechanism.

4. DISCUSSION

In Section 2, we saw that validation and quality control are key activities of operational remote sensing product extractions, as they ensure a meaningful use of the disseminated information. Hence, the design of new retrieval algorithms should include the associated quality control mechanisms. Current space-borne polar observations of surfaces in the solar spectral region suffer from an under-sampling of the surface anisotropy. This poor angular sampling prevents the use of these observations for quantitative applications with advanced and reliable algorithms. SEVIRI observations will permit a better documentation of the radiation transfer processes in the atmosphere-vegetation-soil system as a function of the Sun's position, and therefore provide improved constraints to solve the inverse problem. In turn, this advantage can be used for the development of relevant quality control mechanisms that generate information on the retrieved product reliability. Through a simple but illustrative case study, we demonstrated that geostationary observations, as will be provided by SEVIRI, will permit the development of such advanced retrieval algorithms, thanks to a proper sampling of the angular effects.

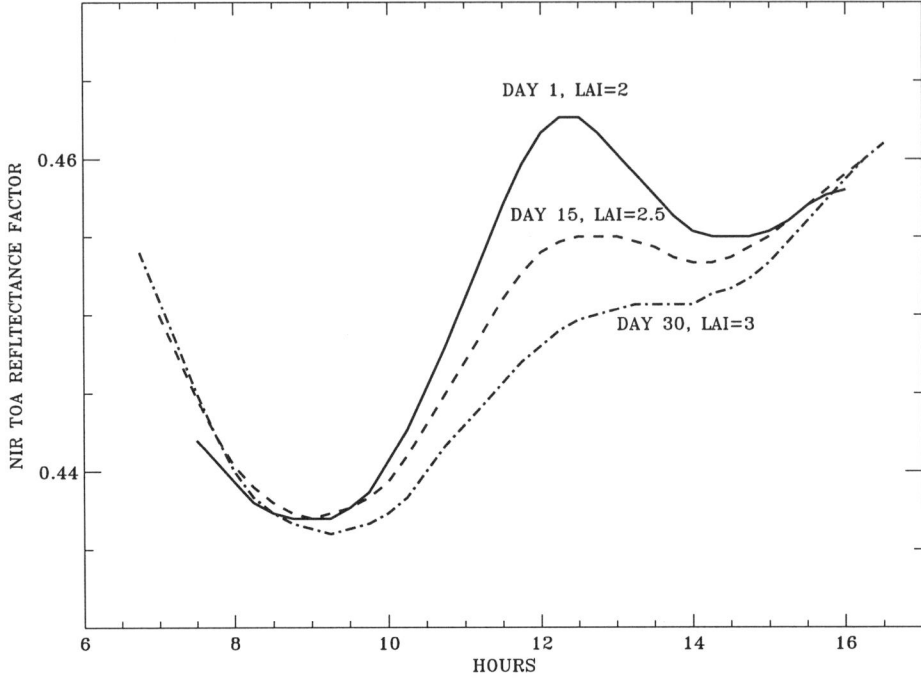

Figure 6. Same as Figure 5, but for the near-infrared spectral band.

5. REFERENCES

Aminou, D. M. A., B. Jacquet, and F. Pasternak (1997) Characteristics of the Meteosat Second Generation (MSG) radiometer/imager: SEVIRI, in EUROPTO (Ed.): *The European Symposium on Aerospace Remote Sensing*, London, United Kingdom.

Elliott, S. S. and K. T. Holmlund (1996) Water vapour winds from cloud free areas, in EUMETSAT (Editor): *1996 Meteorological Satellite Data User's Conference*, Vienna, 327–331.

ESA (1987) *MIEC processing*, Technical Report **STR-224**, ESA.

Gobron, N., B. Pinty, M. M. Verstraete, and Y. M. Govaerts (1996) A semi-discrete model for the scattering of light by vegetation, *Journal of Geophysical Research*, **102**, 9431–9446.

Gube, M., V. Gaertner, and J. Schmetz (1996) Analysis of the operational calibration of the Meteosat infrared-window channel, *Meteorol. Applied*, **3**, 307–316.

Schmetz, J. (1986) An atmospheric-correction scheme for operational application to Meteosat infrared measurements, *ESA Journal*, **10**, 145–159.

Schmetz, J. and K. Holmlund (1990) Operational cloud motion winds from Meteosat and the use of cirrus clouds as tracers, *Advances in Space Research*, **12**, 95–107.

Szejwach, G. (1982) Determination of cirrus cloud temperature from infrared radiances: Application to METEOSAT, *Journal of Applied Meteorology*, **21**, 284–293.

Van de Berg, L. C. V., J. Schmetz, and J. Whitlock (1995) On the calibration of the Meteosat water vapor channel, *Journal of Geophysical Research*, **100**, 21,069–21,076.

10. EXPLOITATION AND EVALUATION OF RETRIEVAL ALGORITHMS FOR GEOSTATIONARY SATELLITE DATA PROCESSING

Vermote, E. F., D. Tanré, J. L. Deuzé, M. Herman, and J. J. Morcrette (1997) Second simulation of the satellite signal in the solar spectrum, 6S: An overview, *IEEE Transactions on Geoscience and Remote Sensing*, **35**, 674–686.

Verstraete, M. M. and B. Pinty (1996) Designing optimal spectral indices for remote sensing applications, *IEEE Transactions on Geoscience and Remote Sensing*, **58**, 201–214.

Chapter 11

THE ROLE OF REMOTE SENSING IN LAND SURFACE EXPERIMENTS WITHIN BAHC AND ISLSCP

A. Jochum (1), P. Kabat (2) and R. Hutjes (2)
(1) Wageningen Agricultural University and (2) DLO Staring Centre, Wageningen, The Netherlands.

1. BACKGROUND

The growing concern, both within the scientific community and governments, about the possibility of climate change induced by the continued expansion of human activities led to the establishment of major international programs devoted to issues of global change.

The World Climate Research Program (WCRP) was established in 1980, jointly supported by the World Meteorological Organization (WMO), the International Council of Scientific Unions (ICSU) and the Intergovernmental Oceanographic Commission (IOC) of UNESCO. Its purpose is *"to develop the fundamental scientific understanding of the climate system and climate processes that is needed to determine to which extent climate can be predicted and the extent of man's influence on climate"*.

WCRP aims both at designing and implementing observational and diagnostic research activities and at developing global models. It is pursuing its activities in the following components:
- ACSYS: Arctic Climate System Study;
- CLIVAR: Climate Variability and Predictability;
- GEWEX: Global energy and Water Cycle Experiment;
- SPARC: Stratospheric Processes and their Role in Climate;
- WOCE: World Ocean Circulation Experiment.

The International Geosphere Biosphere Program (IGBP) was established in 1986 by ICSU with the objective *"to describe and understand the interactive physical, chemical, and biological processes that regulate the total Earth system, the unique environment that it provides for life, the changes that are occurring in this system, and the manner in which they are influenced by human activities"*. For this purpose it has developed science plans in its eleven Program Elements:
1. BAHC: Biospheric Aspects of the Hydrological Cycle;
2. DIS: Data and Information System of the IGBP;
3. GAIM: Global Analysis, Interpretation and Modeling;
4. GCTE: Global Change and Terrestrial Ecosystems;
5. GLOBEC: Global Ocean Ecosystem Dynamics Program;
6. IGAC: International Global Atmospheric Chemistry Project;
7. JGOFS: Joint Global Ocean Flux Studies;
8. LOICZ: Land-Ocean Interaction in the Coastal Zone;
9. LUCC: Land Use/Cover Change;
10. PAGES: Past Global Changes; and
11. START: System for Analysis, Research, and Training (jointly with WCRP and IHDP).

The International Human Dimensions of Global Environmental Research Program (IHDP), sponsored by the International Social Science Council (ISSC) complements IGBP and WCRP by focusing on the socio-economic aspects of global change.

Both WCRP and IGBP include program elements dedicated to the land-surface processes. Land covers a total of 35% of the Earth's surface. Processes at the land-surface represent one of the most important climate forcings. Understanding, modeling and monitoring the physical, biological and chemical processes occurring at the land-atmosphere interface is, therefore, crucial for the analysis and prediction of regional and global climate change.

Internationally coordinated research efforts in this area have focused on extensive field experiments involving considerable resources in terms of ground-, aircraft- and satellite-based observation systems. The driving forces were the global climate modeling (GCM) community on one hand (seeking improved and calibrated land-surface parameterizations) and the remote sensing community on the other (seeking calibration of satellite observing systems).

In the following two sections we describe in some more detail those program elements of IGBP and WCRP which have played a central role in defining and implementing these land-surface experiments (Section 4). These are IGBP-BAHC and WCRP-GEWEX/ISLSCP. They both focus on the interaction between the land-surface and the atmosphere, but each from a different, yet complementary perspective. Several further related program elements are briefly summarized in Section 5, which is followed by a brief review of long-term global aspects (Section 6). An example of land-surface experiments and related data needs and potential is given in Section 7. Remote sensing requirements are summarized in Section 8.

2. IGBP-BAHC (BIOSPHERIC ASPECTS OF THE HYDROLOGICAL CYCLE)

The objective of BAHC is centered on the question *"How do changes in biospheric processes interact with global and regional climates, hydrological processes and water resources, when driven by changes in atmospheric composition and land cover?"*

The basic tasks defined for BAHC are
- to determine the biospheric controls on the hydrological cycle through experiments and modeling of energy and water fluxes in the soil-vegetation-atmosphere system at all spatial and temporal scales;
- to develop appropriate data sets that describe the interactions between the terrestrial biosphere and the physical Earth system, and that can be used for testing and validation of model simulations of these interactions.

Currently a new program structure is being implemented consisting of the following *key themes*
1. Investigate energy, water and carbon fluxes at the patch scale (FLUXNET)
2. Evaluate the role of below-ground processes
3. Parameterizations of land-atmosphere interactions
4. Land use-climate interactions at the regional scale
5. Global terrestrial vegetation-climate interactions
6. Influence of climate change and human activities on mobilization and transport of matter through riverine systems
7. Mountain hydrology and ecology
8. Contribute to development and production of global data sets

and *crosscutting issues*:
A. Integrated terrestrial system experiments–Design, priorities, plans, target areas
B. Analyze impacts of environmental change and climate variability on water and land systems

Although remote sensing data are needed and being used in all of these, there is particular emphasis on two components.

Key Theme 3 explicitly stresses the need to provide the information (input and validation) needed for soil-vegetation-atmosphere transfer (SVAT) models and to "develop an interface between remote sensing and SVATs (jointly with ISLSCP)". A better mutual integration of SVAT models with remote sensing algorithms, and the combined use of multiple sensor types, short-wave, thermal and microwave, seems promising in this respect.

Key Theme 8 places its emphasis on the up-scaling aspects of global data set production, i.e., the aggregation of vegetation and soil characteristics from local scale to the grid-element size of (gridded) global data sets.

3. WCRP-GEWEX/ISLSCP

The goal of the Global Energy and Water Cycle Experiment (GEWEX) Program is *"to reproduce and predict, by means of suitable models, the variations of the global hydrological regime, its impact on atmospheric and surface dynamics, and variations in regional hydrological processes and water resources and their response to changes in the environment, such as the increase in greenhouse gases. GEWEX will provide an order of magnitude improvement in the ability to model global precipitation and evaporation as well as accurate assessment of the sensitivity of atmospheric radiation and clouds to climate change"*.

Among the specific objectives are global measurements of atmospheric and surface properties in order to determine the hydrological cycle and energy fluxes; pilot studies encompassing the full range of experimental scales from local to global; and advancing the development of observing techniques, data management and assimilation systems for operational application to long-range weather forecasts, hydrology, and climate predictions. This latter objective includes making recommendations to space agencies with respect to instruments planned for satellite platforms.

The three GEWEX Research Foci combine modeling and observations and impose challenging observational needs in the fields of radiation (atmospheric and surface radiation fluxes and heating with the precision needed to predict transient climate variations and decadal-to-centennial climate trends), energy and water budget, and soil moisture.

The International Satellite Land Surface Climatology Project (ISLSCP) was established to promote the use of satellite data for the global land-surface data sets needed for climate studies. For more than a decade, ISLSCP has played a key role in addressing land-surface processes, developing climate models, experiment design and implementation, and data set development. In some sense, it can be regarded as an intermediary

11. THE ROLE OF REMOTE SENSING IN LAND SURFACE EXPERIMENTS WITHIN BAHC AND ISLSCP

between the land-surface climate and remote sensing communities. Its objectives are
- to demonstrate the types of surface and near-surface satellite measurements that are relevant to climate and global change studies;
- to develop and improve algorithms for the interpretation of satellite measurements of land-surface features;
- to develop methods to validate area-averaged quantities derived from satellite measurements for climate simulation models;
- to prepare the groundwork for future operational production of land-surface data sets, which can be directly applied to climate problems.

The "ISLSCP Initiative I" data set released on CD-ROM includes global land cover, hydrometeorology, radiation and soils data sets for 1987-88, re-gridded to a common $1° \times 1°$ format. A follow-up is under preparation, extending the time frame (1987-96), increasing the resolution ($0.5°$), and using improved algorithms where appropriate.

4. LAND-SURFACE EXPERIMENTS

It has long been recognized that due to the complex and interacting nature of the exchange processes at the land-atmosphere interface, concentrated interdisciplinary research efforts are needed. For this purpose the concept of land-surface experiments (LSEs) was developed with the objectives
- to document a GCM grid-scale volume subtending the surface;
- to sample surface exchanges, and study their aggregation from sub-grid scale to landscape units;
- to collect and validate satellite data for land-surface models;
- to validate surface-atmosphere models at 10-100 km scales.

Aggregation refers to spatial averaging of surface variables in order to obtain effective values representative for a heterogeneous area.

Initially motivated by the need to better understand the aggregate behavior of the landscape in the formation of weather and climate (HAPEX-MOBILHY), and by the wish to exploit the potential of remote sensing in the same context (FIFE), a series of LSEs was implemented through the Joint IGBP/WCRP Working Group on Land-Surface Experiments, to cover the ecosystem and climate zones of most interest for global change. Table 1 shows the timing, location and spatial scale of these LSEs.

Each LSE involves detailed ground-based measurements (micrometeorology, soils, and vegetation) at selected sites, research aircraft with in-situ atmospheric and remote sensing instrumentation, aerological soundings, and a variety of satellite-based observations. Figure 1 below shows an example of the FIFE experiment design.

Table 1. Major land-surface experiments

Name	Spatial scale	Location	Year	Reference
HAPEX-MOBILHY	100 km	France	1986	André et al. (1988)
FIFE	10 km	Kansas	1987, 1989	Sellers et al. (1992)
EFEDA	100 km	Spain	1991, 1994	Bolle et al. (1993)
HAPEX-Sahel	100 km	Niger	1992	Goutorbe et al. (1996)
BOREAS	1000 km	Canada	1993-1994	Sellers et al. (1995)
NOPEX	100 km	Sweden, Finland	1994-1998	Halldin et al. (1998)
LBA	4000 km	Amazon basin	1998-2001	Nobre et al. (1996)

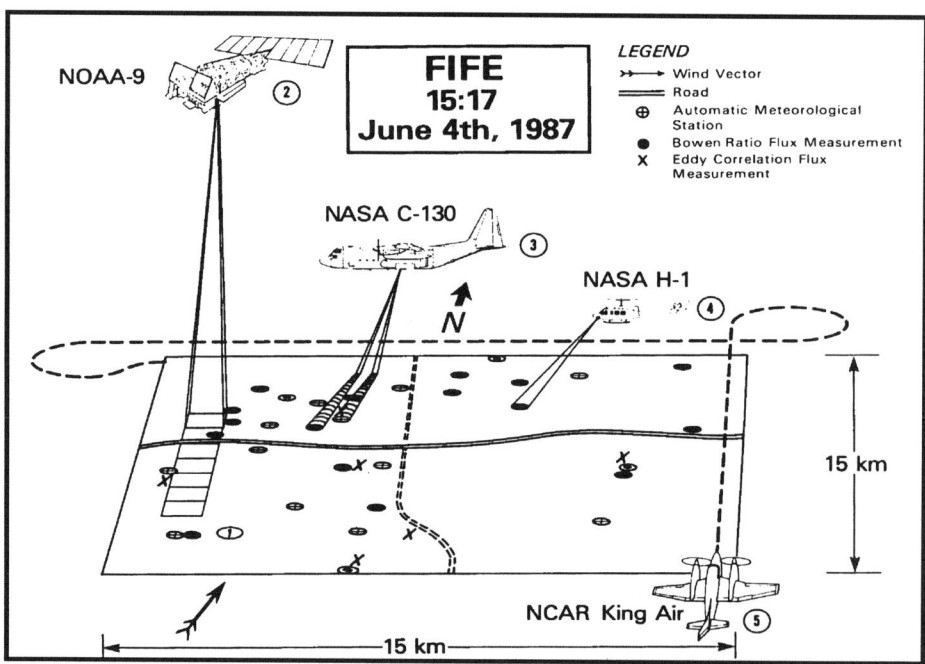

Figure 1. FIFE experiment design (from Sellers et al., 1987)

The earlier LSEs were conducted in campaign mode, covering at best a full growing season, and focused on hydro-meteorological issues in the land surface atmosphere interactions. The more recent (NOPEX, BOREAS)

11. THE ROLE OF REMOTE SENSING IN LAND SURFACE EXPERIMENTS WITHIN BAHC AND ISLSCP

already extended their time frame as also more ecological aspects grew in importance. Currently in its implementation phase, the Large Scale Biosphere Experiment in the Amazon basin (LBA) is the example of the culmination of this trend. It will study the integrated functioning of Amazonia as a regional entity. It specifically addresses the effects of changes in land-use and climate on the biological, chemical and physical functions of Amazonia, including the sustainability of development in the region and the influence of Amazonia on global climate. This involves studies of the multi-annual dynamics of the tropical forest biome and its links with the global energy, water, carbon and other biogeochemical cycles (Nobre et al., 1996, see Section 7).

In 1994, the "Tucson Aggregation Workshop" was held to assess the state of the art in land-surface aggregation research (Michaud and Shuttleworth, 1997). Among the conclusions relevant for remote sensing we find that

- field-based modeling shows for both full and sparse canopies, the measured area-average value of remotely sensed variables is a reasonable estimate of the sub-pixel, linear average value, but their eco-hydrological relevance is only partly explored;
- experimental studies suggest, remotely sensed vegetation information contains valuable information on CO_2 assimilation and stomatal resistance, but the dependence of algorithms on biome and nutrients is not known.

In the area of aggregation of remotely sensed variables, "substantial progress has been made, and it is now recognized that surface temperature and spectral vegetation indices can be aggregated easily at scales ranging from meters to kilometers. However, it is also recognized that there can be significant differences with respect to aggregation strategies for sparse and dense canopies for variables derived from remotely sensed data. We do, however, have some knowledge of the factors that may introduce troublesome non-linearities into the remotely sensed measurements. In order of decreasing importance, these are soil and soil moisture, vegetation, and topography" (Michaud and Shuttleworth, 1997).

In general the LSEs have greatly improved our understanding of the role of the land surface and of changes therein on weather and climate (Feddes et al., 1998). The importance of this role is now well recognized by major operational weather forecasting services and climate prediction groups. Several climate and weather modeling centers are or will soon be ready to assimilate different kind of land surface data for model-initialization and operational purposes. These data could include remotely sensed 'real time' data, e.g., on vegetation and or soil moisture dynamics. This is a new, extremely promising avenue for both land surface climatology research, and for improvements of operational weather prediction.

5. RELATED PROJECTS

The role of IGBP-DIS (Data and Information System) is to assist, as needed, IGBP Core Projects in the development of their individual data system plans; to help provide an overall data system plan for IGBP; to carry out activities leading directly to the generation of data sets; to ensure the development of effective data management systems and act, where appropriate, to ensure the meeting of the data and information needs of IGBP through international and national organizations and agencies. Thus, it addresses the more technical aspects of data, such as the provision of efficient data exchange mechanisms, meta-data and catalogues; the very large volumes and novel character of many new data sets expected to be available from space platforms in the next decade, with distinctive problems of information manipulation and extraction; and the need to ensure the provision of long-term data sets to detect significant global changes in environmentally important parameters.

The IGBP Core Project Global Change and Terrestrial Ecosystems (GCTE) requires many large-scale data sets to develop a global vegetation model, and for other modeling activities related to finer scale ecological changes and impacts on agriculture and forestry. Highest priority is given to improving the acquisition and availability of data sets on land cover (vegetation type), human use and disturbance, soil parameters, topography and climate.

The Land Use and Land Cover Change (LUCC) Project is a joint Program Element of the IGBP and the IHDP. It was established recognizing the fact that over the coming decades, the global effects of land use and cover change may be as significant, or more so, than those associated with potential climate change. Consequently this Core Project is aimed *"to improve understanding of the land use and land cover change dynamics and their relationships with the global environmental change"*. A study of land-cover dynamics involves, amongst others, the regional assessment of land-cover change as determined from direct observation (e.g., satellite imagery and field studies) and models built from these observations.

6. LONG-TERM GLOBAL PERSPECTIVES

Three closely linked global observational systems are being established to provide long-term, interdisciplinary monitoring of global change: the Global Climate Observing System (GCOS; under the aegis of ICSU, IOC/UNESCO, UNEP and WMO), the Global Ocean Observing System (GOOS; under the aegis of ICSU, IOC/UNESCO and WMO), and the Global Terrestrial Observing System (GTOS, under the aegis of ICSU, FAO, UNEP, UNESCO and WMO). These operational programs build on existing monitoring and research activities, with IGBP (primarily through IGBP-DIS) and WCRP playing a significant role in developing their scientific specifications and strategy for use.

A primary requirement for achieving the goals of WCRP is global observations of climate parameters that are needed to improve the understanding of climate-forming mechanisms, to provide a description of the present state of climate, to monitor climate variability, and to serve as a basis for initiating climate predictions.

Existing observing programs implemented through the World Weather Watch (WWW) and other WMO activities provide for the acquisition and processing of important climate information. GCOS is expected to include operational or quasi-operational observing systems that could be used to support well-established climate applications, and to characterize climate impacts on the environment.

The needs of the climate research community, however, include data not usually available from operational centers. WCRP therefore plans to continue observational projects as part of its overall strategy in order to satisfy specific and evolving scientific needs. Examples of such projects include the International Satellite Cloud Climatology Project (ISCCP), ISLSCP and the Global Precipitation Climatology Project (GPCP). Many of these projects include the development of techniques to solve particular observing problems. Large-scale regional data sets such as those from the GEWEX Continental-Scale Experiments are also being integrated to solve global climate problems.

7. LBA (LARGE SCALE BIOSPHERE-ATMOSPHERE EXPERIMENT IN AMAZONIA)

In order to illustrate the Earth Observation data requirements of land-surface experiments (LSEs), we have chosen the LSE currently under progress, viz. the Large Scale Biosphere-Atmosphere Experiment in

Amazonia (LBA; Nobre et al., 1996). This example also shows the process of transition from the classical mesoscale LSEs towards larger scales in space and time (see Section 5).

The Large Scale Biosphere-Atmosphere Experiment in Amazonia (LBA) is an international research initiative lead by Brazil. LBA is designed to create the new knowledge needed to understand the climatological, ecological, biogeochemical, and hydrological functioning of Amazonia, the impact of land use changes on these functions, and the interactions between Amazonia and the Earth system. LBA is centered around two key questions that will be addressed through multi-disciplinary research, integrating studies in the physical, chemical, biological, and human sciences:

- How does Amazonia currently function as a regional entity?
- How will changes in land use and climate affect the biological, chemical and physical functions of Amazonia, including the sustainability of development in the region and the influence of Amazonia on global climate?

Remote sensing contributes to these goals in several important ways. One of the key issues still is the proper and timely delineation of the spatial extent and function of ecosystems with significant impact on the water, carbon and nutrient exchange processes. The most obvious in this context are flooded forest areas (methane and water budget) and re-growth stages (carbon, biomass). Equally important, but much less studied, are selectively logged areas and forests with apparently homogeneous closed canopy top surfaces but exhibiting significant differences in biophysical function. Improved fine-scale classification of these biophysical and hydrological properties (and thus the possibility of its quantification) will provide crucial information for the understanding of the Amazonian ecosystems.

Remote sensing data are also needed to integrate information and processes pertinent to ecosystem-atmosphere exchanges of carbon, trace gases, water, and energy across a broad range of geographic scales. Links between remote sensing data and key variables and parameters of atmosphere and land surface can thus be established and validated at local scales, where extensive ground observations are practical. A combination of remote sensing, mesoscale modeling and other spatial integration techniques will permit an extension of this knowledge to other geographic scales.

Furthermore satellite data have the stability needed for long-term monitoring of these variables at a wide range of temporal frequencies. This temporal and spatial scope of satellite data provides a unique tool that can be used to study the dynamics of vegetation communities (disturbance, succession, fire, etc.) over a wide range of scales.

Remote sensing also helps to place the intensive study sites in their correct ecoclimatological and geographic context, by providing basin-wide maps of topography, land-cover and soil properties. This is important to enable optimization of the field sampling design at basin-wide and more local scales and will be necessary for correct interpretation of research results.

Finally, because biomass burning is an extremely important issue in Amazonia, remote sensing is being used to monitor the frequency of occurrence and extent of fires and the subsequent distribution of atmospheric aerosols.

Data needs for remote sensing in LBA have been identified along with candidate data sources and methodologies to derive the required biogeophysical parameters (Hall et al., 1996; see also list in Section 8). A significant part of these data needs are rather challenging in terms of both sensors and algorithms.

A broad constellation of satellites may add to existing archives during LBA, including Landsat 7, SPOT, ERS, JERS, RADARSAT, ADEOS, ENVISAT and CBERS, as well as the EOS platform and TRMM. Remote sensing technique development can contribute to the scientific aims of LBA to allow LBA to benefit from the enhanced capabilities of new sensors and algorithms while, at the same time, validating the new sensor products and algorithms. Aircraft continue to be a key complementary platform to test and validate new sensors and algorithms and to acquire higher spatial resolution data.

Remote sensing algorithms need extensive development and validation for conditions specific to Amazonia. This particularly applies to radiative transfer modeling, and to algorithms to identify certain types of land cover unique to Amazonia, including forest re-growth stages. Further development is also needed for algorithms to process active and passive microwave sensor images that will be required to deal with the ubiquitous cloud cover and seasonal smoke from biomass burning, and to retrieve atmospheric profiles of temperature and water vapor, cloud properties and precipitation.

8. SUMMARY, CONCLUSIONS, OUTLOOK

The objectives of land-surface related components of WCRP and IGBP have been reviewed. Their common denominator is the focus on the interactive physical and biological processes at the land-atmosphere interface. Earth Observation data play an important role. Two kinds of data requirements (linked by feedback processes in both directions) result from these objectives:

Research data are needed both for process studies and for the improvement and validation of parameterizations in regional and global forecast and climate models. The typical scales in both cases are local to regional, the latter usually with imbedded intensive field sites.

Operational data are needed in the context of global observing systems, for the purpose of serving as operational input for forecast and climate models on one hand, and as long-term data bases for change analysis on the other.

The information that should be derived from remote sensing data includes

- a quantitative description of the general characteristics of the land-surface (topography, soil type and properties, land-use and vegetation type) their spatial extent and changes therein;
- biophysical characteristics of the land-cover (e.g., biomass, leaf area index, photosynthetically active radiation, spectral radiances, roughness length);
- biogeophysical parameters, such as surface temperature, components of the surface radiation and energy budgets, soil moisture, runoff, precipitation; and
- meteorological parameters, such as temperature, wind, humidity, atmospheric composition (near-surface values, vertical profiles, total column values for the latter two), cloud cover.

Current and future Earth Observation data can satisfy many of these requirements, provided adequate algorithms are (further) developed and validated. There will be a continuing need, however, for complementary ancillary data to be supplied by conventional observing systems and numerical models. Increasing emphasis is also on the development of adequate assimilation schemes which are required to blend information from many data sources of very different characteristics into (globally homogeneous for years to decades) gridded datasets.

It is important to note here that many of the methods developed in the research mode described above can, at a certain stage of maturity, be transferred to be used in operational environments for the management of natural resources (land, forest, water). Examples are refined land-use and land-cover classifications or evapotranspiration maps. This kind of spin-off needs to be taken into account when assessing the market potential of EO data. Recent land-surface experiments like LBA have this goal clearly in mind. The outcome of research into the land-surface processes under BAHC, ISLSCP and related projects will, in the long run, increase the usefulness of EO data for customers in sustainable land and water management.

9. ACKNOWLEDGMENTS

Part of this text draws on material published by WCRP, IGBP, and their Program Elements in Science Plans and Newsletters.

10. REFERENCES AND RELATED MATERIAL

André, J.-C. et al. (1988) HAPEX-MOBILHY: First results from the special observing period, *Annals of Geophysics*, **6**, 477–492.

Bolle, H.-J. et al. (1993) EFEDA: European Field Experiment in a Desertification-threatened Area, *Annals of Geophysics*, **11**, 173–189.

Feddes, R.A., P. Kabat, A.J. Dolman, R.W.A. Hutjes and M.J. Waterloo (1998) Large scale field experiments to improve land surface parameterizations, in Dooge J.H. et al. (Editors): *Climate and Water - A 1998 Perspective*, (in press).

Goutorbe, J.P. et al. (1997) An overview of HAPEX-Sahel: A study in climate and desertification, *Journal of Hydrology*, **188–189**, 4–17.

Hall, F.G. et al. (1996) *Role of Remote Sensing in LBA*, NASA, 31 p.

Halldin, S., L. Gottschalk, A.A. van de Griend, S.-E. Gryning, M. Heikinheimo, U. Högström, A.M. Jochum, and L.-C. Lundin (1998) NOPEX: A northern hemisphere climate processes land-surface experiment, *Journal of Hydrology*, (in press).

Michaud, J.D., and W.J. Shuttleworth (1997) Executive summary of the Tucson Aggregation Workshop, *Journal of Hydrology*, **190**, 176–181.

Nobre, C.A. et al. (1996) The Large Scale Biosphere-Atmosphere Experiment in Amazonia (LBA). Concise Experiment Plan. LBA Project Office, CPTEC/INPE.

Sellers, P.J., F.G. Hall, G. Asrar, D.E. Strebel, and R.E. Murphy (1992) An overview of the First ISLSCP Field Experiment, *Journal of Geophysical Research*, **97**, 18,345–18,372.

Sellers, P.J. et al. (1995) The Boreal Ecosystem-Atmosphere Study (BOREAS): An overview and early results from the 1994 field year, *Bulletin of the American Meteorological Society*, **76**, 1549–1577.

Chapter 12

EARTH OBSERVATION DEMANDS FOR IMPROVED WATER RESOURCES MANAGEMENT

W. G. M. Bastiaanssen and C. J. Perry
International Water Management Institute (IWMI), Colombo, Sri Lanka.

1. INTRODUCTION

In recent years, there has been an increasing awareness that water will be the critical natural resource issue within, and between, many countries in the next century (e.g., Serageldin, 1995; Biswas, 1995; Postel et al., 1996; Gleick, 1997). Frederiksen (1996) mentions that a world population growth of 1 billion in the next decade and 2 billion in the next two decades is forecasted and that this growth will place immense demands on the water resources of developing countries. In many developing countries, the scarcity of water is becoming a significant constraint to food production, and shortage of water and food may threaten socio-economic stability. Food projections for 2040 reveal a need for a two to three-fold increase in food productivity, as compared to 1990 (e.g., Penning de Vries et al., 1995). By 2025, as many as fifty-two countries inhabited by some 3 billion people will be plagued by water stress or chronic water scarcity. Falkenmark (1989) computed a water scarcity index for African countries to create awareness among African leaders to reserve water for recurrent drought years. Seckler et al. (1998) have evaluated the available water resources in relation to population development, industrial demands and irrigated areas country wise by means of a water scarcity index to indicate the countries being most vulnerable to proper water resources management.

Worldwide, more water is diverted to and consumed by irrigated agriculture than any other economic sector. Agriculture accounts for some 63% of the world's use of fresh water (70–90% in developing countries). However, increasing water demands for industry, municipalities and natural ecosystems mean that less water will be available to sustain food production while one third of the world's food crops are currently produced by irrigated agriculture, and irrigated agriculture has been the primary source of incremental food production over the last several decades. This alarming situation requires a full understanding of the water cycle at the regional, river basin, and sub-continental scale. The need for better, more scale-appropriate approaches to the measurement of water utilization in agriculture—to improve the positive impacts of water resources development—is complementary to the requirements for reducing negative impacts on other ecosystems in the context of river basins.

The aim of this paper is to discuss the progress achieved in the establishment of international data bases of information on land and water resources, to highlight the information gaps in irrigation management, and to show how remote sensing technologies can potentially contribute to that information gap. The conversion between spectral and angular satellite measurements on one side, and irrigation conditions on the other side, lies outside the scope of this paper.

2. AVAILABLE LAND AND WATER RESOURCES DATABASES

Many databases at watershed, continental and global scale have been developed in the past five years and are being continuously updated (see Appendix 1 for selected examples). Aspects of the global water cycle are essential for diagnosing regional scale water availability for irrigated crops. Soil maps have been digitized and stored in electronic databases (e.g., FAO, UNEP). River flow records have been assembled for a number of major river basins (e.g., Global Runoff Data Center). Precipitation records have been collected and analyzed by Legates and Willmott (1990). Rudolf et al. (1996) characterized precipitations, using data from rain gauges, satellite measurements and numerical models. Mintz and Serafini (1981, 1992) worked on continental scale water balances and assimilated global soil moisture fields from evaporation. Eswaran et al. (1995) used model simulations with rainfall records collected from 27,000 climatic stations around the world to infer soil moisture. In an attempt to identify the regional and sub-continental water resources, Baumgartner and Reichel (1975), Budyko (1986), Henning (1989), Chahine (1992) and Hartmann (1994) undertook efforts to integrate hydrological records to obtain continental and global scale water balances. Besides databases on hydrology records, numerical models can provide essential information on water balance terms. Numerical hydrological models compute water fluxes at a range of space and time scales and are therefore important tools to quantify water resources and evaluate water availability. Land surface models coupled with atmospheric models represent the exchanges of heat and energy between land and atmosphere at the continental and global scale. Stendel and Arke (1997) critically examined the output data of several coupled land surface models against observations of rainfall, evaporation and surface runoff. Their study represents the state of the art on large scale distributed water flow analysis.

Another example of bringing together valuable historical records is the ongoing compilation of the IIMI's World Water and Climate Atlas by Utah State University (Hargreaves and Jensen, 1997) which is based on rainfall and temperature data gathered from 56,000 stations world-wide. The reference evapotranspiration for unstressed grass has been estimated using the approach proposed by Hargreaves (1994). This methodology is based on minimum and maximum near surface air temperatures and extraterrestrial radiation, and gives acceptable (standard error of estimate of 0.88 mm day^{-1}; Allen et al., 1995) to accurate results (standard error of estimate 0.44 mm day^{-1} in non-extreme circumstances). Moisture adequacy—indicating the need for irrigation and drainage—can be derived from this rainfall and reference evapotranspiration information at 2.5 km grids. Choudhury (1997) has published a worldwide assessment of reference evapotranspiration based on the equation of Penman-Monteith (Monteith, 1965) using satellite and assimilated data for a 24 month period (January 1987 to December 1988).

Because vegetation affects the radiation, energy and water balances, the presence of vegetation communities and agricultural areas play a major role in the establishment of water fluxes. Vegetation Indices determined from NOAA satellites are available with a resolution of 1.1 km (LAC) and degraded resolution (GAC), and stored in freely available databases such as 'Pathfinder' (e.g., Defries et al., 1995). Land cover data sets have also been compiled, for example under the auspices of the International Geosphere Biosphere Program (IGBP) in their working group Data and Information System (DIS) using vegetation measurements of the 1.1 km LAC data ('DISCover', Belward, 1996). Classes such as croplands and cropland/natural can be derived from DISCover, though irrigated agriculture does not appear in the legend of this particular database. A similar situation can be found on the 10 Minutes Pan-European Land Use Database which recognizes only arable land. The improved European based databases Co-ORdination of Information oN the Environment (CORINE; Anon, 1992) and

Pan-European Land Cover Monitoring (PELCOM) distinguishes typically non-irrigated arable land, irrigated arable land, permanent crops (e.g., vineyards, orchards), winter crops and complex cultivation practices. These databases are both essential pieces of information for global terrestrial eco-hydrological descriptions, and critical for the description of agro-ecological zones found within river basins. Multi-temporal and multi-spectral classifications have also been worked out for the delineation of agro-eco-climatic zones (e.g., Brown et al., 1993; Menenti et al., 1993; Lambin and Ehrlich, 1996; Maselli et al., 1996) and are essential for the distribution of water among for instance irrigated agriculture, wetlands and forests.

Traditional inventories of the world's irrigated areas are almost exclusively based on information provided by governments, and the procedures followed in the data gathering process is inherently non-systematic. For example, official irrigation statistics often reflect design areas rather than actually irrigated areas. Although FAO promulgates the AQUASTAT database, its accuracy, in the absence of independent estimates, is difficult to assess. The number of irrigation seasons can vary from once in a few years up to three times per year for paddy crops cultivated in the humid tropics. Furthermore, irrigated areas expand and shrink continuously due to market prices, land degradation and unreliability in the irrigation service.

Until recently, the ability to delineate irrigated areas systematically and globally did not exist. A test with NOAA satellite data in Argentina using the Fourier analysis methodology proposed by Azzali and Menenti (1996), revealed that the satellite interpreted permanently irrigated area was 843,300 ha whereas the FAO statistics gave 1,632,000 ha—hence a 94% difference! A comparison for the Iberian Peninsula using FAO figures and data from the CORINE land cover database revealed larger deviations. Hence, our knowledge on the extent of irrigated areas is poor by inadequate conventional methods, and the need exists to get more accurate statistics at the regional scale.

3. INFORMATION NEEDS FOR THE ANALYSIS OF IRRIGATED AGRICULTURE

In spite of the rapid growth of the aforementioned databases on global and regional scale, information to improve the water resources management in large-scale irrigation systems is seriously inadequate. River basins are faced with water competition among users and the quantification of water flowing to the sea, wetlands, flushing, navigation, in-stream recreational purposes and so on is gaining importance. A well-balanced partitioning of

12. EARTH OBSERVATION DEMANDS FOR IMPROVED WATER RESOURCES MANAGEMENT

fresh water resources into agriculture and natural ecosystems is a primary concern for all. Misuse and misallocation damage arable land resources, through salinization or waterlogging, and can threaten the consumptive needs of other eco-systems and wetlands in the downstream end of a river basin. Information on agro-ecological zones, land cover, land use and their changes in time are managerial key issues.

Another need for information relates to the analysis of the performance of irrigation projects. In physical terms, according to FAO, only 45% of irrigation water being diverted to irrigated areas are actually consumed by crops. The remaining 55% are termed as 'losses', although the definition of losses is, hydrologically, not clear. Where water is lost to aquifers, it may not only be recoverable, but also recoverable at times and locations where it is more valuable than at the time it was 'lost'. Water management in, for instance, the Indo-Gangetic plains depends heavily on groundwater availability during the dry season. Under such circumstances, and in simpler cases, where downstream farmers pump water from drains carrying runoff from upstream farmers, it is clearly misleading to consider all such flows as 'losses' without a clear definition of the term. The IWMI paradigm (Perry, 1998) promotes an approach which re-examines losses (quantities of water released from storage that do not reach the crop) and efficiency (ratio between water consumed by the crop and water delivered), noting that the linkages between sources, uses and re-uses must be fully understood to appraise 'losses' and 'efficiency'. River basin models (e.g., Arnold and Williams, 1987) can help in quantifying the space and time components of these water flows. Remote sensing information used as input in these models has been shown to improve the accuracy of the simulation process (e.g., Kite et al., 1994) and there is a trend to replace model parameter estimation procedures by indirect Earth Observation measurements.

In productivity terms, as water becomes the binding constraint to agricultural production, attention has to shift from the productivity of land to the productivity of water. The concept is well known to many farmers in arid areas, who practice deficit irrigation, mulch the soil, and apply many other techniques to conserve water for productive use, thus maximizing tons of crop per cubic meter of water consumed. But this indicator has so far attracted far less attention from policy makers and irrigation managers than the conventional productivity measure—tons per hectare. This is due in part to the technical difficulties of measuring actual evaporation at field and basin scales. The complexity of the physical system is a key problem in shifting our frame of reference from productivity of land to productivity of water. While measures of water delivered to and within irrigation systems are sometimes relatively well documented, return flow to drains, percolation to groundwater, and consumption—in time and space—by crops is far less well known. Further, the spatial and temporal complexity of irrigation systems themselves makes the analysis of their operational performance extremely challenging. In a typical medium sized scheme in a developing country, tens of thousands of farmers each receive perhaps five irrigation events in a season, with each event having varying characteristics of flow rate and duration. Establishing a measuring system capable of observing such a process has challenged a large number of irrigation researchers over the last decade or more.

4. EARTH OBSERVATION DEMANDS FOR IRRIGATED AGRICULTURE

The capacity of remote sensing to identify and monitor vegetation parameters has undergone impressive improvements during the last 10 years. Significant research efforts have been undertaken to better explore the potential that satellite measurements offer: objective and spatial data with a dynamic dimension. The fleet of satellite platforms has been expanded and the availability of spectrally, spatially and temporally different radiative properties of the land surface has grown. A major new breakthrough is expected to occur at the beginning of the next century with the establishment of space platforms. Although the applications of remote sensing were long delayed after the launch of satellites in the seventies, it is generally accepted that the science related to the understanding and interpreting of spectral signatures progressed rapidly. Review papers on the potential of remote sensing for irrigation management have been prepared by Menenti (1990), Vidal and Sagardoy (1995), Thiruvengadachari and Sakthivadivel (1997) and Bastiaanssen (1998). On the basis of these papers, it may be concluded that remote sensing provides an opportunity to obtain much of the data required to improve the management of irrigation and drainage systems.

The current status of research has reached a stage where several interpretation algorithms have proven their power. Visible and near-infrared measurements can be combined to obtain a number of agronomic features that are useful to describe the crop water demands through crop coefficients (e.g., Ahmed and Neale, 1996), net radiation (e.g., Roerink et al., 1997) and transpiration coefficients (e.g., Choudhury et al., 1994). The crop response to water application can be expressed on the basis of fractional vegetation cover and leaf area index using vegetation indices calculated from satellite data (e.g., Choudhury et al., 1994). Complementary information on the hydrological status can be obtained from thermal infrared measurements to arrive at crop water stress (e.g., Moran et al., 1994) and crop consumptive use (e.g., Bastiaanssen et al., 1996). Dry matter accumulation is a direct function of the cumulative amount of solar radiation and water transpired. Remotely sensed estimates of leaf area index (e.g., Gobron et al., 1997), photosynthetic active radiation (e.g., Asrar, 1992) and evapotranspiration can therefore be very helpful in determining the productivity of water at the regional scale, i.e., yield per unit of water consumed. Analysis and management in a variety of ecosystems at the basin level can be assessed by means of remote sensing techniques at a spatial level consistent with our needs—observing with resolutions below 10 m, or truly basin level—is feasible. The data can be collected at a variety of temporal resolutions, from hours to weeks.

To the field researchers on irrigation, this combination of what can be observed, and how it can be observed and interpreted is already a giant leap from farmer interviews and staff gauge readings in irrigation canals. Nevertheless, we see the potential for yet more, including the compilation of objectively measured, remotely sensed data on irrigated river basins as the basis for negotiation among competing sectors—and countries—for scarce and valuable water resources.

Table 1 contains a list of parameters that are relevant to irrigation water management and quantifiable from remotely sensed data, as demonstrated in the international literature. These parameters can further be linked to ancillary data such as precipitation, canal flow rates, river runoff and groundwater table fluctuations, as well as existing global databases such as digital terrain elevation and soils, to evaluate the performance of irrigated areas in the context of inter-sectoral competition for water.

Table 1. Selection of remote sensing determinants necessary for the management of irrigated agriculture at two different scales of water planning: (sub-) river basins and (sub-) continents

Remotely Sensed Parameter	River basin	Continent
Irrigated area	•	•
Agro-ecological zones	•	
Land cover/ land use	•	•
Crop types	•	
Biomass water requirements	•	•
Biomass water consumption	•	•
Bare soil evaporation	•	
Lake evaporation	•	
Crop yields	•	
Soil moisture	•	•
Precipitation	•	•

5. CONCLUSIONS

Ensuring sustainable and productive use of available water in many parts of the semi-arid and arid zones presents critical challenges. Decisions on water allocation, and the impacts of construction, reclamation and rehabilitation projects on the productive use of land and water, while maintaining environment sustainability, can only be addressed in a basin context, but such evaluations are often impeded by incomplete field data. Data availability, and hence the quality of analysis, can be improved by incorporating satellite remotely sensed measurements at the scale of (sub-) river basins and (sub-) continents. Careful validation of the information based on remotely sensed data is necessary, but validation has been demonstrated in the international literature for a number of key parameters. Remote sensing data can further be used to improve the quality of public domain databases for decision making with respect to water and environment at scales ranging from tertiary irrigation units of a few hectares up to heterogeneous river basins containing a variety of ecosystems. It is expected, from the user's side, that monitoring of irrigation processes can be enhanced with Earth Observation data.

Knowledge on irrigated area, ecosystems, land cover, land use, crop acreage, crop growing stages and crop water demands for optimal dry matter production is often incomplete. The performance of water using sectors, especially irrigation, and the productivity of water in terms of yield per unit volume of water consumed needs to be known for making strategic decisions related to the inter-sectoral competition for water. In the absence of adequate spatio-temporal hydrological surveys, water resources management is extremely difficult. Classification of crop types and delineation of agronomic zones can help this process greatly, as can knowledge of crop water stress and relative water availability, especially when monitored on a frequent basis. Estimation of crop yields—measured with respect to both land and water—provides the possibility to evaluate the productivity of land and water and thus to assess allocation policies in relation to demand, supply, and the impact of alternative strategies.

6. REFERENCES

Ahmed, R.H. and C.M.U. Neale (1996) Mapping field crop evapotranspiration using airborne multispectral imagery, *IEEE*, 2369–2371.

Allen, R.G., M. Smith, A. Perrier and L.S. Pereira (1994) An update for the definition of reference evapotranspiration, *ICID Bulletin*, **43**, 1–34.

12. EARTH OBSERVATION DEMANDS FOR IMPROVED WATER RESOURCES MANAGEMENT

Anonymous (1992) CORINE land cover: A European Community project presented in the framework of the International Space Year, in *Proceedings of 1992 European Conference of the International Space Year*, Commission of the European Communities, Brussels.

Arnold, J.G. and J.R. Williams (1987) Validation of SWRRB simulator for water resources in rural basins, *Journal of Water Research Planning and Management*, **113**, 243–256.

Asrar, G., R.B. Myneni and B.J. Choudhury (1992) Spatial heterogeneity in vegetation canopies and remote sensing of absorbed photosynthetically active radiation: A modeling study, *Remote Sensing of Environment*, **41**, 85–103.

Azzali, S. and M. Menenti (1996) *Fourier analysis of temporal NDVI in Southern African and American continents*, Report **108**, DLO Winand Staring Centre, Wageningen, The Netherlands, 149 pp.

Bastiaanssen, W.G.M., T. Van der Wal and T.N.M. Visser (1996) Diagnosis of regional evaporation by remote sensing to support irrigation performance assessment, *Irrigation and Drainage Systems*, **10**, 1–23.

Bastiaanssen, W.G.M. (1998) *Remote sensing in water resources management: The state of the art*, International Water Management Institute (IWMI), Colombo, Sri Lanka, 118 pp.

Baumgartner, A. and E. Reichel (1975) *The world water balance*, Elsevier, New York.

Belward, A.S. (1996) *The IGBP-DIS global 1 km land cover data set*, Report of the Land Cover Working Group of IGBP-DIS, CNRM, Toulouse, France.

Biswas, A.K. (1995) Water for the developing world in the 21 century: Issues and implications, *ICID Journal*, **45**, 2.

Brown, J.F., T.R. Loveland, J.W. Merchant, B.C. Reed and D.O. Ohlen (1993) Using multisource data in global land-cover characterization: Concepts, requirements and methods, *Photogrammetric Engineering and Remote Sensing*, **59**, 977–987.

Budyko, M.I. (1986) *The evolution of the biosphere*, Elsevier, New York.

Chahine, M.T. (1992) The hydrological cycle and its influence on climate, *Nature*, **359**, 373–380.

Choudhury, B.J., N.U. Ahmed, S.B. Idso, R.J. Reginato and C.S.T. Daughtry (1994) Relations between evaporation coefficients and vegetation indices studied by model simulations, *Remote Sensing of Environment*, **50**, 1–17.

Choudhury, B.J. (1997) Global pattern of potential evaporation calculated from the Penman-Monteith equation using satellite and assimilated data, *Remote Sensing of Environment*, **61**, 64-81.

Defries, R., M. Hansen and J.R.G. Townsend (1995) Global discrimination of land cover types from metrics derived from AVHRR pathfinder data, *Remote Sensing of Environment*, **54**, 209–222.

Eswaran, H., E. Van den Berg, P. Reich, R. Almaraz, B. Smallwood and P. Zdruli (1995) *Global soil moisture and temperature regimes*, World Soil Resources, USDA Natural Resources Conservation Service, Washington.

Falkenmark, M. (1989) The massive water scarcity now threatening Africa - why isn't it being addressed, *Ambio*, **18**, 112–118.

Food and Agricultural Organization of the United Nations (1994) *Water for Life*, FAO, Rome, Italy.

Frederiksen, H.D. (1996) Water crisis in developing world: Misconceptions about solutions, *Journal of Water Resources Planning and Management*, **122**, 79–87.

Gleick, P.H. (1997) Water and conflict in the twenty-first century: The Middle East and California, in D.D. Parker and Y. Tsur (Editors): *Decentralization and coordination of water resources management*, Norwell MA, USA, Kluwer Academic Publishers, 411–428.

Gobron, N., B. Pinty and M.M. Verstraete (1997) Theoretical limits to the estimation of the leaf area index on the basis of visible and near-infrared remote sensing data, *IEEE Transactions on Geoscience and Remote Sensing*, **35**, 1438–1445.

Hargreaves, G.H. (1994) Defining and using reference evapotranspiration, *Journal of Irrigation and Drainage Engineering*, ASCE **120**, 1132–1139.

Hargreaves, G.H. and D.T. Jensen (1997) A water and climate atlas for water resources development, *Proceedings of the San Antonio Workshop*, Texas, October 1996, 23–26.

Hartmann, D.L. (1994) *Global physical climatology*, Academic Press, San Diego, 411 pp.

Henning, D. (1989) *Atlas of the surface heat balance of the continents*, Gebruder Borntraeger, Berlin, Germany.

Janowiak, J.E. and P.A. Arkin (1991) Rainfall variations in the tropics during 1986–1989, as estimated from observations of cloud-top temperatures, *Journal of Geophysical Research*, **96**, 3359–3373.

Lambin, E.F. and D. Ehrlich (1996) The surface temperature-vegetation index space for land cover and land cover change analysis, *International Journal of Remote Sensing*, **17**, 463–487.

Legates, D.R. and C.J. Willmott (1990) Mean seasonal and spatial variability in gauge corrected global precipitation, *Journal of Climate*, **10**, 111–127.

Maselli, F., L. Petkov, G. Maracchi and C. Conese (1996) Eco-climatic classification of Tuscany through NOAA-AVHRR data, *International Journal of Remote Sensing*, **17**, 2369–2384.

Menenti, M. (1990) Remote sensing in evaluation and management of irrigation, *Proceedings of the International Symposium Centro Regional Andino*, Instituto Nacional de Ciencia y Tecnica Hidricas (INCYTH), Argentina, 337 pp.

Menenti, M., S. Azzali, W. Verhoef and R. Van Swol (1993) Mapping agro-ecological zones and time lag in vegetation growth by means of Fourier analysis of time series of NDVI images, *Advances in Space Research*, **13**, 233–237.

Mintz, Y. and Y.V. Serafini (1981) *Global fields of soil moisture and land-surface evaporation*, NASA Goddard Space Flight Center, Technical Memorandum **83907**, Research Review - 1980/81, NASA, no. 178-180.

Mintz, Y. and Y.V. Serafini (1992) A global climatology of soil moisture and water balance, *Climate Dynamics*, **8**, 13–27.

Monteith, J.L. (1965) Evaporation and environment, *Symposium Soc. Exp. Biol.*, **XIX**, 205–234.

Moran, M.S., T.R. Clarke, Y. Inoue and A. Vidal (1994) Estimating crop water deficit using the relation between surface-air temperature and spectral vegetation index, *Remote Sensing of Environment*, **41**, 169–184.

Penning de Vries, F.W.T., H. Van Keulen and R. Rabbinge (1995) Natural resources and limits of food production in 2040, in Bouma et al. (Editors): *Eco-regional approaches for sustainable use and food production*, Kluwer Academic Publishers, Dordrecht, The Netherlands, 64–87.

Perry, C.J. (1999) The IWMI water resources paradigm, definitions and applications, *Agricultural Water Management*, **40**, 45–50.

Postel, S., G.C. Daily and P.R. Ehrlich (1996) Human appropriation of renewable fresh water, *Science*, **271**, 785–788.

Roerink, G.J., W.G.M. Bastiaanssen, J. Chambouleyron and M. Menenti (1997) Relating crop water consumption to irrigation water supply by remote sensing, *Water Resources Management*, **11**, 445–465.

Rudolf, B., W. Hauschild, W. Ruth and U. Schneider (1996) Comparison of raingauge analysis, satellite based precipitation estimates and forecast model results, *Advances in Space Research*, **7**, 53–62.

Seckler, D., U. Amarasinghe, D.J. Molden, R. De Silva and R. Barker (1998) *World water demand and supply, 1990–2025: Scenarios and issues*, Research Report **19**, International Irrigation Management Institute, IWMI, Colombo, Sri Lanka.

Serageldin, I. (1995) Toward sustainable management of water resources, Address to the VII *World Congress on Water Resources of the International Water Resources Association*, Cairo, November 22, 1994, Directions in Development, The World Bank, Washington D.C., USA, 33 pp.

Stendel, M. and K. Arpe (1997) *Evaluation of the hydrologic cycle in reanalysis and observations*, MPI Validation Report, Max Planck Institute for Meteorology, Hamburg, Germany, 50 pp.

Thiruvengadachari, S. and R. Sakthivadivel (1997) *Satellite remote sensing techniques to aid irrigation system performance assessment: A case study in India*, Research Report **9**, International Irrigation Management Institute (IIMI), Colombo, Sri Lanka, 23 pp.

Vidal, A. and J.A. Sagardoy (1995) (Editors), *Use of remote sensing techniques in irrigation and drainage*, Water Reports No. **4**, FAO, Rome, Italy, 202 pp.

Weng, F. and N.C. Grody (1994) Retrieval of cloud liquid water using the SSM/I, *Journal of Geophysical Research*, **99**, 25,535–25,551.

Appendix 1

Important Global Centers, Databases and Research Programs related to the assessment of water resources.

Global Center	*Topic*
Population	
UN-New York	World population prospects
Population Reference Bureau	Population growth
University of California-Santa Barbara	Population data for Asia
Agriculture and natural vegetation	
FAO-Rome	Irrigated areas Africa
FAO-Rome	Actual food production
IFPRI-Washington	Global food production
AB-DLO-Wageningen	Global food production
World Bank-Washington	Global major cereal areas
World Resources Institute-Washington	Geography Africa
Space Applications Institute (EC/JRC)-Ispra	European food production
International Geosphere Biosphere Program (IGBP)	Land cover and land use
IIASA-Laxenburg	Global land cover
USGS-Sioux Falls, Space Applications Institute-Ispra	Global land cover
World Conservation Monitoring Centre (WCMC)/IFOR	Tropical moist forests and protected areas

Topography and soils

FAO-UNESCO	Global soil map
FAO-Rome	Global soil map
UNEP-ISRIC-Wagenigen	Global soil and soil degradation maps
NASA	Global soil map
US Geological Survey-Sioux Falls	Digital terrain elevation

Climate

UNEP-Intergovernmental Panel Climate and Climate Change (IPCC)	Climate change map
Committee on Global Change	Climate change map
University of Trier	Global climate data
Oak Ridge National Laboratory (ORNL)-Oak Ridge	Global precipitation, temperature and pressure
International Irrigation Management Institute (IIMI)-Colombo	Global precipitation and temperature data
International Satellite Cloud Climatology Project (ISCCP)	Global cloud behavior
International Satellite Land Surface Climatology Project (ISLSCP)	Radiation and energy balances
Earth Radiation Budget Experiment (ERBE)	Net radiation land surface
World Climate Research Program (WCRP)	Global Short-wave Surface Radiation Budget Database
GEWEX-Water Vapor Project (GVaP)-Langley	Global water vapor
GEWEX-Cloud System Study (GCSS)	Boundary layer and cirrus clouds
Global Precipitation Climatology Project (GPCP)	Global precipitation
European Space Agency (ESA)	International Cirrus Experiment
Australian National University-Canberra	DEM/climate Africa

Hydrology

Baumgartner and Reichel	Global water balance
UNESCO-Paris	Global water balance
National Center for Atmospheric Research (NCAR)-Boulder	Global hydrological cycle
European Centre for Medium-term Weather Forecast (ECMWF)- Reading	Global hydrological cycle
NASA-New York	Global hydrological cycle
Max Planck Institute-Mainz	Global hydrological cycle
National Meteorological Centre (NMC)-Paris	Global hydrological cycle
Utah State University-Logan	World water for agriculture
International Irrigation Management Institute (IIMI)-Colombo	Climatological irrigation water needs
UNEP-World Resources Institute-New York	Water balance of water using countries
Delft Hydraulics-Delft	Water availability for irrigation

12. EARTH OBSERVATION DEMANDS FOR IMPROVED WATER RESOURCES MANAGEMENT

ICID-New Delhi	Irrigation and drainage in the world
University of Texas-Austin	World water balance
NASA Goddard Space Flight Center-Greenbelt	Global monthly soil moisture fields
Global Energy and Water Experiment (GEWEX)	Global wetness fields
USDA Natural Resources Conservation Service	Global soil moisture and temperature regimes
WMO/UNESCO	Global hydrometeorological data
World Meteorological Organization (WMO)	World hydrological cycle observing system
World Bank	World hydrological cycle observing system
Global Precipitation Climatology Centre (GPCC)-Offenbach	Global rainfall
Global Runoff Data Centre-Koblenz	Global runoff data
Max Planck Institute for Meteorology-Hamburg	Global runoff data

Chapter 13

A BIOPHYSICAL PROCESS-BASED ESTIMATE OF GLOBAL LAND SURFACE EVAPORATION USING SATELLITE AND ANCILLARY DATA

B. Choudhury
NASA Goddard Space Flight Center, Greenbelt, MD, USA.

1. INTRODUCTION

The rich history and many fundamental investigations of the global land surface evaporation can be found in Korzun (1978), and Korzun's monograph can also provide an appreciation of the care and the effort involved in such investigations. Recent contributions include those by Willmott et al. (1985), Henning (1989), Mintz and Walker (1993) and Oki et al. (1995). These investigations, except for Oki et al. (1995), are based upon long-term average surface meteorological observations and, in some cases, river runoff data.

Fig. 1 shows zonal average values of annual total evaporation calculated by Baumgartner and Reichel (1975) and Henning (1989) in 5° latitude bands and by Budyko (1978) in 10° latitude bands. One could point to both remarkable agreements and noticeable differences seen in this figure. These calculations provide references for climatological evaporation.

Significant changes of land surface characteristics have occurred and are still continuing, and field observations show that these changes affect the partitioning of available energy and precipitation at the land surface. The representativeness of the previous evaporation values for the present state of the land surface has not yet been assessed. A biophysical process-based model is needed to account for the impact of land surface change on evaporation.

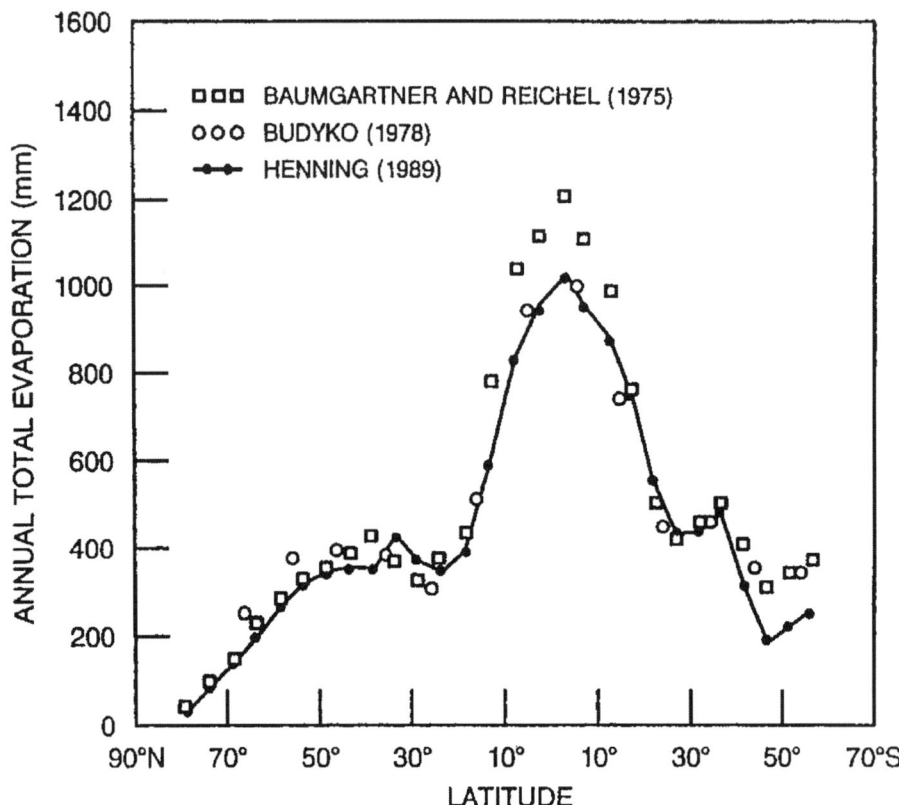

Figure 1. Zonal average values of annual total evaporation calculated by Baumgartner and Reichel (1975), Budyko (1986) and Henning (1989).

The objective of this paper is to present seasonal and annual values of the components of total evaporation obtained from a biophysical process-based model using satellite and assimilated data. Important distinctions between the present results and previous estimates of global land surface evaporation are:
1. Previous studies did not quantify the components of total evaporation,
2. Biological control on evaporation through leaf stomata was not considered in previous studies,
3. Present calculations have been done for each months of two specific years resulting in a 24 months long time series rather than average of many years,
4. Except for precipitation, present calculations have been done using spatially representative meteorological and surface data rather than interpolated values, and

13. A BIOPHYSICAL PROCESS-BASED ESTIMATE OF GLOBAL LAND SURFACE EVAPORATION USING SATELLITE AND ANCILLARY DATA

5. Quantitative comparisons with micro-meteorological, catchment water balance, soil moisture, and atmospheric water budget data have provided a measure of uncertainty for the calculated results at a range of spatial and temporal scales, which is not evident in previous estimates.

2. DESCRIPTION OF THE MODEL AND DATA

A brief description of the model and the data is presented here due to space limitation; details can be found in Choudhury and DiGirolamo (1998). Validation of the model against observations (micrometeorological and soil moisture measurements, catchment water balance, and atmospheric water budget analyses) is also given in Choudhury and DiGirolamo (1998).

Transpiration has been calculated using the Penman-Monteith equation, with the rate of carbon assimilation determining the canopy stomatal resistance. The rate of carbon assimilation by the canopy depends upon interception of photosynthetically active radiation, the maximum rate of assimilation by the leaves, and the efficiency of quantum absorption as determined by temperature. The actual transpiration depends upon the available soil moisture and the fractional duration when the foliage surface is dry. Soil evaporation is considered to occur in two stages; the Priestley-Taylor equation adjusted for fractional vegetation cover was used for the first stage (i.e., the energy-limited rate), while Philip's equation is used for the second stage (i.e., exfiltration-limited rate). The exfiltration-limited rate is not allowed to exceed the energy-limited rate. Interception has been calculated using the Horton's equation, adjusted for fractional vegetation cover. A daily water balance model, which includes surface runoff and drainage, together with snow accumulation, evaporation and melt, was run at $0.25° \times 0.25°$ spatial resolution over the global land surface using satellite and ancillary data. Net radiation and sensible heat flux are obtained by solving the energy balance equation.

Satellite observations are used to obtain the fractional vegetation cover, albedo, photosynthetically active and solar radiation, air temperature and vapor pressure. The friction velocity is derived from a four-dimensional data assimilation procedure, while precipitation values are derived from gauge and satellite measurements. The spatial resolution and sensors (or sources) of these data are given in Table 1, where correspondence with the future sensors is also noted. Choudhury (1997) has assessed the accuracy of some of these data. The monthly total precipitation data has been disaggregated to obtain the daily values. The spatial distribution of biophysical parameters of the model has been prescribed using a land use and land cover data, which include the intensity of agriculture.

Table 1. Spatial resolution of some of the key data sets used in the present study, the present source of these data and likely correspondence with the source during the EOS era.

Data	Resolution	Source[*]	EOS Era[#]
Meteorological:			
Albedo	2.5° × 2.5°	ISCCP	CERES
Solar Radiation	2.5° × 2.5°	ISCCP	CERES
PAR	2.5° × 2.5°	ISCCP	CERES
Fractional Cloud Cover	2.5° × 2.5°	ISCCP	CERES
Air Temperature	1° × 1°	TOVS	AIRS
Vapor Pressure	2.5° × 2.5°	TOVS	AIRS
Friction Velocity	2° × 2.5°	4DDA	4DDA
Precipitation	2.5° × 2.5°	GPCP	GPCP
Surface:			
Fractional Vegetation Cover	0.25° × 0.25°	AVHRR	MODIS
Land Use / Cover	1° × 1°	Prescribed	MODIS
Roughness Height	2° × 2.5°	Prescribed	VCL

[*] ISCCP (International Satellite Cloud Climatology Project), TOVS (Tiros Operational Vertical Sounder), 4DDA (Four-Dimensional Data Assimilation), AVHRR (Advanced Very High Resolution Radiometer), GPCP (Global Precipitation Climatology Project), Prescribed (Various sources, e.g., maps, satellite data, "guesstimate").

[#] CERES (Cloud and Earth's Radiation Energy System), AIRS (Atmospheric Infrared Sounder), MODIS (Moderate-Resolution Imaging Spectrometer), VCL (Vegetation Canopy Lidar).

3. RESULTS AND DISCUSSION

Zonal average (5° latitude bands) values of annual evaporation (1987-1988 average) are shown in Fig. 2 (a, b), and compared with previous estimates. Although the pattern of zonal variation agree well with the previous calculations, our values are generally closer to Budyko than Baumgartner and Reichel, Mintz and Walker or Henning. For latitudes around the equator, our evaporation values are higher than those of Budyko and Henning by about 100 mm, but lower than those of Baumgartner and Reichel and Mintz and Walker by about 50 and 100 mm, respectively. Both Henning and Baumgartner and Reichel show minimum evaporation in the latitude band 45°-50°S, which does not appear in our calculations, although a weak minimum appears in the latitude band 50°-55°S. Evaporation values of Mintz and Walker are 150-200 mm higher than ours for the latitude band 45°-55°S, although the difference is less than 20 mm for the latitude band 35°-40°S. Henning, and Mintz and Walker show a small peak in the latitude band 30°-35°N, while Baumgartner and Reichel show a similar peak in the latitude band 35°-40°N. Such peaks do not appear in our calculations.

13. A BIOPHYSICAL PROCESS-BASED ESTIMATE OF GLOBAL LAND SURFACE EVAPORATION USING SATELLITE AND ANCILLARY DATA

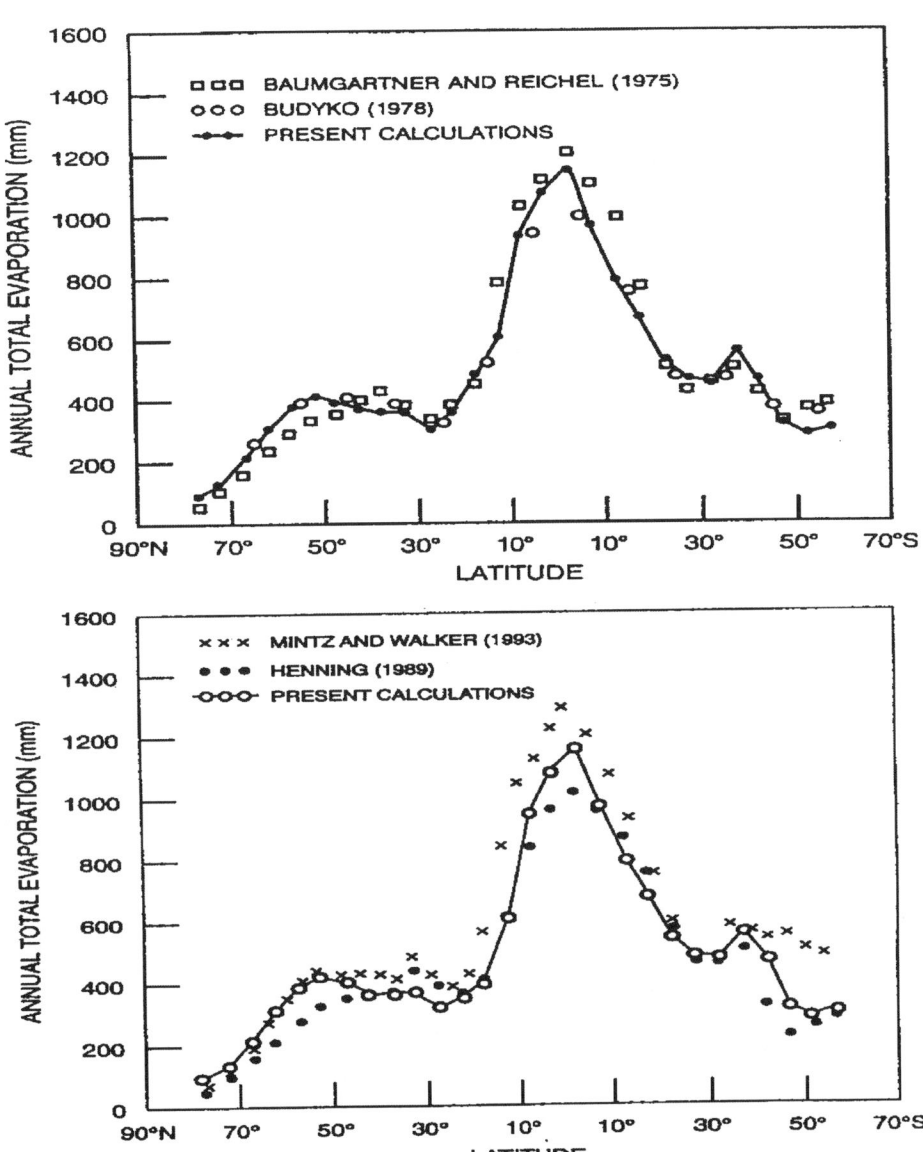

Figure 2. Comparison of zonal average values of annual total evaporation (1987- 1988 average) based on the present calculations with those calculated by, (a) Baumgartner and Reichel (1975) and Budyko (1978), and (b) Henning (1989) and Mintz and Walker (1993).

Our calculations agree with Baumgartner and Reichel in showing a minimum in evaporation for the latitude band 25°-30°N, although the magnitudes are different (we calculate 308 mm, while Baumgartner and Reichel give 342 mm). Overall, our values are generally within the range of previous estimates.

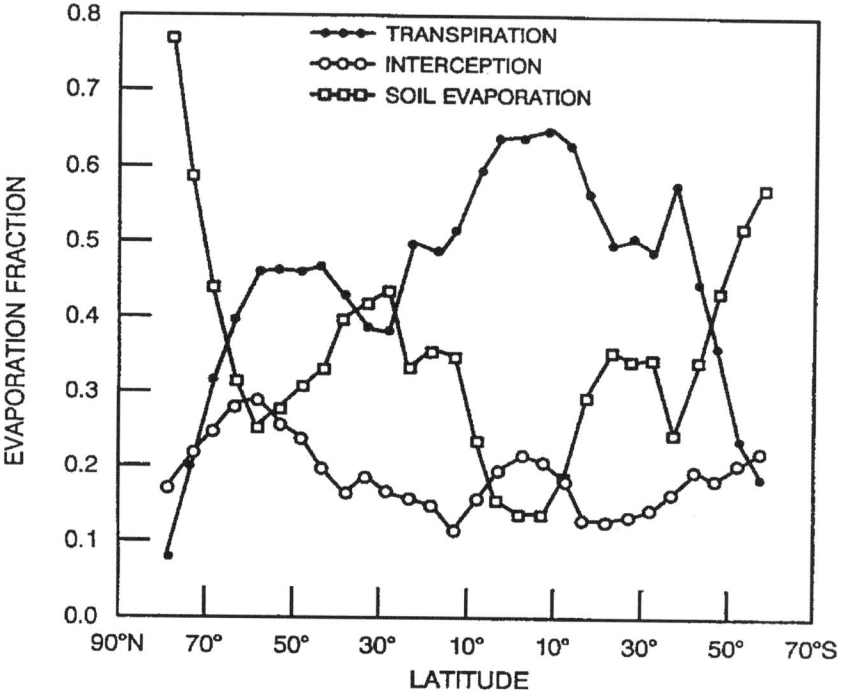

Figure 3. Zonal variations of the relative magnitudes of annual transpiration, interception, and soil evaporation.

The zonal variations of fractional transpiration, interception and soil evaporation are shown in Fig. 3. Transpiration is clearly the dominant fraction over most of the land surface, except at high latitudes where soil evaporation is the dominant fraction. Soil evaporation also appears as the dominant fraction in the 25°-35°N latitude band, where deserts occupy extensive areas (Sonaran, Chihuahuan, Sahara, Arabian, Iranian, and Thar deserts). The transpiration component decreases, while the soil evaporation component increases in the latitude band 20°-35°S, due to the deserts of southern Africa and Australia. Increase in transpiration in the latitude band 35°-40°S is due to forested areas, pasture and croplands in South America, Australia and New Zealand. South of 40°S, Patagonian desert appears as a significant determinant of the partitioning; transpiration decreases, while soil evaporation increases systematically. The pattern of zonal variation of soil evaporation fraction appears almost as a mirror image of transpiration fraction.

While transpiration is clearly the dominant component in most of the latitude bands (20 out of the 28 latitude bands in Fig. 3), interception exceeds soil evaporation only in four of the 5° latitude bands (55°-60°N and 5°-10°S). Soil evaporation and interception fractions are equal for the latitude band 10°-15°S. Thus, the rank ordering of the component dominating total evaporation, when all bands are taken together, would be transpiration, soil evaporation and interception. More specifically, the partitioning of annual total evaporation over the global land area is calculated to be 52% for transpiration, 28% for soil evaporation, and 20% for interception. We are not aware of any previous attempt to quantify the ranking of these components at regional or global scale based on spatially representative observations for specific years.

4. SUMMARY AND CONCLUSIONS

The present study has provided a framework for assessing land surface evaporation, energy balance and carbon accumulation using spatially representative data, several of which were derived from satellite observations. Future sensors will improve both the accuracy and the spatial resolution of these data, and are thus expected to improve the results.

5. ACKNOWLEDGEMENTS

Financial support for this work was provided by the NASA Earth Observing Systems (EOS) Project under interdisciplinary science investigations. Mr. N. DiGirolamo has provided computing assistance. This work would not have been possible without the support and assistance of Drs. J. Susskind, S. Gupta, and G. Asrar.

6. REFERENCES

Baumgartner, A., and Reichel, E. (1975) *The World Water Balance*, Elsevier, NY.

Budyko, M. I. (1978) The heat balance of the Earth, in J. Gribbin (Editor): *Climatic Change*, Cambridge University Press, N.Y., 85–113.

Choudhury, B. J. (1997) Global pattern of potential evaporation, *Remote Sensing of Environment*, **61**, 64–81.

Choudhury, B. J. and DiGirolamo, N. E. (1998) A biophysical process-based estimate of global land surface evaporation using satellite and ancillary data. I. Model description and comparison with observations, *Journal of Hydrology*, **205**, 164–185.

Henning, D. (1989) *Atlas of the Surface Heat Balance of the Continents*. Gebruder Borntraeger, Berlin.

Korzun, V. I. (Editor) (1978) *World Water Balance and Water Resources of the Earth*, UNESCO, Paris.

Mintz, Y. and Walker, G. K. (1993) Global fields of soil moisture and land surface evapotranspiration derived from observed precipitation and surface air temperature, *Journal of Applied Meteorology*, **32**, 1305–1334.

Oki, T., Musiake, K., Matsuyama, H., and Masuda, K. (1995) Global atmospheric water balance and runoff from large river basins, *Hydrological Processes*, **9**, 655–678.

Willmott, C. J., Rowe, C. M., and Mintz, Y. (1985) Climatology of the terrestrial seasonal water cycle, *Journal of Climatology*, **5**, 589–606.

Chapter 14

REMOTE SENSING OF LAND COVER AND LAND COVER CHANGE

B. K. Wyatt
Institute of Terrestrial Ecology, Monks Wood, United Kingdom.

1. INTRODUCTION

An extensive literature describing the use of remote sensing for land cover mapping has emerged since the launch of Landsat-1 in 1974. This includes numerous references to applications that purport to detect or measure change. However, many reported studies are either misleading, in that they describe limited environmental mapping exercises as "monitoring" or else are essentially case studies, limited to particular regions or time periods and designed principally to develop, test or demonstrate experimental techniques.

In Britain and elsewhere in Europe to the present time, analysis of changes in land use or in land cover that progress beyond such local studies have depended predominantly on data collected in the field (Barr, Bunce and Heal 1995; Bunce *et al.* 1996), or from air photography (Huntings Surveys and Consultants Ltd, 1986, Tudor, G.J. and Mackey, E.C., 1995). These are instances in which remote sensing has been used as an exploratory tool. For example, a limited assessment of the potential of remote sensing for change monitoring formed part of a national countryside survey which took place in Britain in 1988, entitled "Monitoring Landscape Change" (Deane, Griffiths and Greenwood 1987). The 'LACOAST' project, undertaken by the Space Applications Institute (SAI) of the Joint Research Centre (JRC) of the European Commission exemplifies probably the most extensive use of remote sensing for quantitative assessment of changes in land cover and land use at the landscape scale[1]. LACOAST utilized classified remotely sensed data (Landsat Multispectral Scanner, Landsat Thematic Mapper and SPOT-HRV) to construct a land cover time series from 1976 to 1995 for a 10 km buffer zone around the coastline of the European Union territories.

At continental and global scales, remote sensing has been used successfully to assess changes in land cover, due, for example to natural and anthropogenic disturbance (Hall et al. 1991; Belward et al. 1994; Bourgeau-Chavez, Harrell et al. 1997). However, at scales appropriate to the needs of most national, regional and local applications, the scientific literature is almost devoid of practical examples of the application of remote sensing to detect and monitor change. The reasons for this are not hard to discover. Change detection, measurement and monitoring pose theoretical and practical challenges that have yet to be fully met in the research context, far less in any operational setting. Given the importance of changes in land use as a driver of environmental quality, it is vital that these underlying research issues are addressed, and that the enormous potential of remote sensing is fully realized.

These research issues are numerous and highly inter-dependent; moreover, solutions which are optimal for one application may be wholly inappropriate for others. This paper addresses some of these issues, in the context of the future use of remote sensing as an operational tool for mapping land cover and for the detection and monitoring of change.

2. CHANGE DETECTION AND ANALYSIS

Most remote sensing systems rely on polar-orbiting satellites or on aircraft-mounted sensors. In consequence, data are typically captured as scenes that are instantaneous in comparison to the rates of change typically encountered in terrestrial landscapes. (Geo-stationary satellites have the capacity for genuinely continuous monitoring, but only at very coarse spatial resolutions). Methods for detecting or measuring change from remote sensing therefore invariably depend on comparisons between data sets acquired at intervals of time. This raises important issues regarding the accuracy of the data sets on which these comparisons are based. Various sources of potential error are identified below: they include spatial and temporal effects and the extent to which a given land cover class may be recognized unambiguously from its radiometric properties, perhaps under differing conditions of solar irradiation or atmospheric turbidity. The key point to bear in mind is that there is a limit to the capacity of any remote

---

[1] http://www.ais.sai.jrc.it/environment/lacoast/index.html

sensing system to detect change, and that this limit is related both to the accuracy with which land cover can be mapped at a point in time and also to the extent and rate of change on the ground. Four dimensions of change must be considered:
- Change in x- and y- (changes in extent);
- Change in t- (rate of change);
- Change in z- (degree of change, which might range, for example, from complete defoliation to qualitative changes, e.g., species composition).

3. LAND USE VS LAND COVER

The title of the paper addresses land cover specifically, although, in practice, most applications for land information require data on land use, rather than land cover. However, land use is rarely distinguishable from land cover by direct observation alone, either remotely or even in the field. What can be observed directly is land cover; it may then be possible to infer the underlying land usage from these observations. The implications of this generalization are explored in greater depth by Van Gils *et al.* (1991). An important consequence is that there is uncertainty in any maps or statistics derived from remote sensing which purport to describe either land use, or changes in land use. These uncertainties can only be resolved or eliminated by access to complementary data sets which provide more direct evidence of usage. Often, this will require reference to information on land tenure, field visits or even interview with land owners to establish purpose.

4. SPATIAL ASPECTS

Spatial factors may influence the capacity to detect change reliably in two ways. Firstly, it is important to consider the spatial resolution of the remotely sensed data in relation to the scale of the changes to be observed. Because of sensor design features, the minimum detectable area in a given image is determined by a number of factors in addition to the nominal spatial resolution of the sensor itself (Townshend 1981). It is important to consider not only the extent of the change that may be expected, but also the degree of fragmentation of the landscape within which change occurs. Highly fragmented landscapes may give rise to mixed image pixels, and the apparent degree of mixing will differ between images, even where there is no change on the ground.

Secondly, there will always be errors associated with the geo-registration of remotely sensed data. These errors are encountered, irrespective of whether the image is registered to a map base (which will itself contain error) or to a second image; they tend to be at a minimum in flat terrain. In consequence, it will never be possible to overlay two images precisely so that the pixelation matches exactly; a proportion of the apparent differences between images will therefore be due to mis-registration, rather than to real changes on the ground. These effects are more fully explored in Townshend, Justice *et al.* (1992), while Fuller, Barr and Wyatt (1998) demonstrate the dramatic consequences for classification accuracy of displacements of the order of a pixel or less.

5. TEMPORAL ASPECTS

It is also important to reconcile the time and frequency of acquisition of remotely sensed data with the rate of change in the features of interest. Changes in land use and land cover occur at a range of speeds. Users with an interest in the progress and condition of agricultural crops require data at key intervals in the growing season. Other changes in land cover, deforestation or flooding, for example, may take place almost instantaneously and near-synchronous data are required if the precise timing of the event is of interest. Many change processes take place only slowly, over years or even decades. In the case of optical imagery, the incidence of cloud is a major constraint on the frequency with which it is possible to acquire data (Legg 1991, Fuller *et al.* 1994). As a result, remotely sensed surveys over spatially extensive areas may derive from data acquired over extended periods, often of many months, and it is difficult to establish a precise baseline against which future change may be measured.

6. RADIOMETRIC CONSIDERATIONS

The successful detection of land cover from remote sensing is dependent on the land units of interest exhibiting distinct radiometric responses (spectral signatures) in the spectral regions covered by the sensor system. The same consideration applies both to optical systems and to SAR imagery, though the physical principles are very different. Many changes induce gradual, rather than sudden radiometric change, which renders their detection problematic, particularly at intermediate stages. The difficulty is especially acute in the case of qualitative changes, for example, in drainage, nutrition or species composition. The advent of hyper-spectral data sources such as CASI and MERIS (Babey and Anger, 1989, Rast, Bäzy et al. 1991) provides tools that may help to address the problem, though, at present, the use of these for change detection remains at best experimental. Seasonal changes in spectral signature, in principle, may be exploited to enhance class differentiation (Schriever and Congalton 1995), but this is obviously at the cost of detecting the seasonal changes themselves.

7. IMAGE SEGMENTATION METHODS

The most common approach to the segmentation and interpretation of multi-spectral remotely sensed data for land cover mapping utilizes a suite of probabilistic classification and clustering algorithms. Supervised classifications exploit the radiometric properties of known 'training' regions to identify areas elsewhere on the image with similar spectral properties. The hypothesis is that land cover of the training regions is identical to regions elsewhere in the scene with similar spectral characteristics. Kershaw and Fuller (1992) demonstrated the importance (and difficulty) of selecting training data that are truly representative of spectrally unique classes.

Unsupervised clustering methods have been successfully employed, especially for the segmentation of natural areas, where the choice of training regions is notoriously difficult (Belward, Taylor et al., 1990; Lark, 1995). However, the use of unsupervised methods for change detection presents unsolved problems, relating to the difficulty of generating comparable clusters from images with even slightly different radiometric characteristics. Spectral mixture modeling has been widely advocated (e.g., Adams, Sabol et al., 1995; Foody and Cox, 1994) as a means of dealing with continua, rather than distinct classes, but examples of operational applications are rare.

More advanced techniques, based to a greater or lesser extent on the use of artificial intelligence, include rule-based methods which recognize the context of pixels, rather than treating them in isolation (e.g., Alonso and Soria, 1991; Groom, Fuller and Jones, 1996; Binaghi, Madella et al., 1997) and methods based on neural networks and other fuzzy logic approaches (e.g., Gopal and Woodcock, 1996; Chen, Tzeng et al., 1995; Schaale and Furrer, 1995; Dreyer, 1993). Smith et al. (1997) have recently developed map- and image-based segmentation approaches, in which the classification unit is the land parcel, rather than the pixel; these appear to deliver substantially better performance than conventional per-pixel classifiers. Nevertheless, despite the richness of this recent literature and the claimed improvements in performance over conventional multi-spectral classification, examples of practical applications are so far almost non-existent.

8. CLASSIFICATIONS AND TERMINOLOGY

Given a set of objective procedures for classifying, or otherwise segmenting remotely sensed data, there is no reason why a classification scheme could not be applied to two or more radiometrically-corrected image sets, with entirely reproducible results. The difficulty arises when the remotely sensed data products come to be interpreted or compared with external data sets. Remotely sensed data do not record land cover directly; rather, land cover is inferred from observations of multi-spectral radiance or SAR back-scatter. Objective spectral classes are then interpreted in terms of land cover types that are often far from objective or reproducible. Wyatt *et al.* (1994) clearly demonstrate the impacts of classification differences on areal estimates of land cover in a variety of contemporary surveys, and hence the need for greater objectivity and consistency in defining the land cover classes that are to be mapped from remote sensing.

9. ALGORITHMS FOR QUANTIFYING CHANGE

Remarkably little attention has been given to the choice of algorithms for quantifying changes observed from remotely sensed data. Whatever the reason for this, the inference in much of the literature is that, once change has been detected, the subsequent steps needed for quantitative analysis are self-evident. The commonest approach is to measure differences between data sets (Green, 1984; Rees and Williams, 1997). One approach is to explore differences between the original multi-spectral data (which must first be radiometrically corrected); alternatively, change analysis may be based on differences between processed products, such as classified outputs.

There has been extensive use of Principal Components Analysis (e.g., Byrne, Crapper and Mayo, 1980; Fung and LeDrew, 1987) but it is not easy to ensure that the results from this method are reproducible. Sader *et al.* (1991) estimated change by use of ground reference data collected within a statistical sampling framework designed for the purpose. Lambin and Strahler, (1994a and b) successfully deployed a technique known as 'change vector analysis'.

10. CALIBRATION, VALIDATION AND INTEGRATION

It is possible to use correspondence or 'confusion' matrices to calibrate remotely sensed estimates of cover against validated reference data to derive estimates which are corrected for systematic bias in the remote sensing (e.g., Gonzalez-Alonso and Cuevas, 1993). Such techniques offer practical solutions to the estimation of change statistics. Other techniques, such as ratio-estimation and regression estimation, are well reported and understood (e.g., Cochran, 1977). They offer enormous and largely unexplored potential for refinement of monitoring methods.

11. PRESENTATION OF RESULTS

Rather little attention has so far been given to the important problem of how best to present and communicate information on land cover change. Typically, this information is conveyed as simple tabulations or maps, indicating 'before and after' extents of different land cover categories. If spatially explicit information is available, then it becomes possible to compute matrices of change, which, amongst other advantages, makes it possible to record successional pathways. Shi and Ehlers (1996) give examples of spatial representations of uncertainties in estimates of land cover and land cover change, and, generally, there is enormous scope for greater deployment of graphical devices, including 3-D projections and color, as a means of conveying the location, extent and rate of change. The existence of PC-based software, aimed at non-specialist users, (e.g., Haines Young et al. 1994), cries out for a more innovative approach to this task.

12. SUMMARY AND CONCLUSIONS

The measurement of land cover change from remote sensing is presently some way from operational status, despite a long history of the application of Earth Observation for land cover mapping. The fundamental challenge is to distinguish change from artifacts in the data with sufficient precision and consistency. Key limiting factors have been identified above. Increased spatial resolution from new sensors (e.g., IRS-1C) and those to be launched shortly will lead to increased accuracy in the mapping of land boundaries. Hyper-spectral systems offer the prospect of increased powers of discrimination and therefore of improvements in our ability to detect subtle changes in the quality of vegetation.

Nevertheless, we should recognize that land cover information has frequently been employed as a surrogate for physical variables that are not directly accessible by other means, rather than as a product in its own right. For example, meteorological models require estimates of aerodynamic roughness and land cover has been used as a proxy; hydrological models need maps of surface permeability in order to estimate through-flow; estimates of ecosystem productivity and bio-geochemical fluxes have similarly exploited land cover as a surrogate for more directly useful physical variables, such as rates of photosynthesis or evapotranspiration. Each of the above applications would be better served by representing the variable of interest as a continuous surface, rather than as an arbitrary map of an indeterminate classification of land cover. Given the very real prospect that it may soon be possible and preferable to extract the biophysical data of interest directly from Earth Observation, it may well be that the dominance of land cover mapping in terrestrial applications of remote sensing will soon be a phenomenon of the past, although it is likely that there will always be a requirement for information on land cover and land cover change to inform and direct the management and protection of landscapes.

13. REFERENCES

Adams, J. B., Sabol, D. E., *et al.* (1995) Classification of multi-spectral images based on fractions of end members: Application to land cover change in the Brazilian Amazon, *Remote Sensing of Environment*, **52**, 137–154.

Alonso, F. G. and Soria, S. L. (1991) Using contextual information to improve land use classification of satellite images in central Spain, *International Journal of Remote Sensing*, **12**, 2227–2235.

Babey, S.K. and Anger, C.D. (1989) A Compact Airborne Spectrographic Imager (CASI), *IGARSS 89/12th Canadian Symposium on Remote Sensing*, Vancouver. B. C.

Barr, C. J., Bunce, R. G. H., and Heal, O.W. (1995) Countryside Survey 1990: A measure of change, *Journal of the RASE,* 48–58.

Belward, A. S., Taylor, J. C., *et al.* (1990) An unsupervised approach to the classification of semi-natural vegetation from Landsat Thematic Mapper data. A pilot study on Islay, *International Journal of Remote Sensing*, **11**, 429–445.

Belward, A. S., Kennedy, P. J., *et al.* (1994) The limitations and potential of AVHRR GAC data for continental scale fire studies, *International Journal of Remote Sensing*, **15**, 2215–2234.

Binaghi, E., Madella, P., Montesano, G. and Rampini, A. (1997) Fuzzy contextual classification of multi-source remote sensing images, *IEEE Transactions on Geoscience and Remote Sensing,* **35**, 326–340.

Bourgeau-Chavez, L. L., Harrell, P. A., *et al.* (1997) The detection and mapping of Alaskan wildfires using a space-borne imaging radar system, *International Journal of Remote Sensing*, **18**, 355–373.

Bunce, R. G. H., Barr, C. J., *et al.* (1996) Land classification for strategic ecological survey, *Journal of Environmental Management*, **47**, 37–60.

Byrne, G. F., Crapper, P. F. and Mayo, K. K. (1980) Monitoring land-cover change by principal component analysis of multi-temporal Landsat data, *Remote Sensing of Environment*, **10**, 175–184.

Chen, K. S., Tzeng, Y. C., Chen, C. F. and Kao, W. L. (1995) Land cover classification of multi-spectral imagery using a dynamic learning neural network, *Photogrammetric Engineering and Remote Sensing*, **61**, 403–408.

Cochran,W.G. (1997) *Sampling techniques*, London, Wiley.

Deane, C. C., Griffiths, G. H., and Greenwood, N.D.S. (1987) Deriving landscape change statistics from air-photography and Landsat Thematic Mapper (TM) data in England and Wales, *Statistical Assessment of land use: The impact of remote sensing and other recent developments on methodology,* EUROSTAT, 347–376.

Dreyer, P. (1993) Classification of land cover using optimized neural nets on SPOT data, *Photogrammetric Engineering and Remote Sensing*, **59**, 617–621.

Foody, G. M. and Cox, D. P. (1994) Sub-pixel land cover composition estimation using a linear mixture model and fuzzy membership functions, *International Journal of Remote Sensing*, **15**, 619–631.

Fuller, R. M., Barr, C. J., and Wyatt, B.K. (1998) Countryside survey from ground and space - different perceptions, complementary results, *Journal of Environmental Management*, **54**, 101–126.

Fuller, R. M., Groom, G. B., *et al.* (1994) The availability of Landsat TM images of Great Britain, *International Journal of Remote Sensing*, **15**, 1357–1362.

Fung, T. and LeDrew, E. (1987) Application of principal components analysis to change detection, *Photogrammetric Engineering and Remote Sensing*, **53**, 1649–1658.

Gopal, S. and Woodcock, C. (1996) Remote sensing of forest change using artificial neural networks, *IEEE Transactions on Geoscience and Remote Sensing*, **34**, 398-404.

Gonzalez-Alonso, F. and Cuevas, J.M. (1993) Remote sensing and agricultural statistics: Crop area estimation through regression estimators and confusion matrices, *International Journal of Remote Sensing,* **14**, 1215–1219.

Green, K. M. (1984) Monitoring tropical forest ecosystems using remote sensing technology, *International Journal of Primatology*, **5**, 344.

Groom, G. B., Fuller, R., M, and Jones, A. R. (1996) Contextual correction: Techniques for improving land cover mapping from remotely sensed images, *International Journal of Remote Sensing*, **17**, 69–89.

Haines-Young, R.H., Bunce, R.G.H. and Parr, T.W. (1994) Countryside Information System: An Information System for Environmental Policy Development and Appraisal, *Geographical Systems*, **1**, 329–345.

Hall, F. G., Botkin, D. B., *et al.* (1991) Large-scale patterns of forest succession as determined by remote sensing, *Ecology*, **72**, 628–640.

Huntings Surveys and Consultants Ltd. (1986) *Monitoring Landscape Change*. Volumes 1 - 10. Department of the Environment and Countryside Commission.

Kershaw, C. D. and Fuller, R. M. (1992) Statistical problems in the discrimination of land cover from satellite images: A case study in lowland Britain, *International Journal of Remote Sensing*, **13**, 3085–3104.

Lambin, E. F. and Strahler, A. H. (1994a) Change-vector analysis in multi-temporal space: A tool to detect and categorize land-cover change processes using high temporal-resolution satellite data, *Remote Sensing of Environment,* **48**, 231–244.

Lambin, E. F. and Strahler, A. H. (1994b) Indicators of land cover change for change-vector analysis in multi-temporal space at coarse spatial scales, *International Journal of Remote Sensing*, **15**, 2099–2119.

Lark, R. M. (1995) A reappraisal of unsupervised classification. I. Correspondence between spectral and conceptual classes, *International Journal of Remote Sensing*, **16**, 1425–1443.

Legg, C. A. (1991) A Review of Landsat MSS Image Acquisition Over the United Kingdom, 1976-1988, and the Implications for Operational Remote Sensing, *International Journal of Remote Sensing*, **12**, 93–106.

Rast, M., Bézy, J. L., *et al.* (1991) The performance of the ESA Medium Resolution Imaging Spectrometer (MERIS), *Proceedings of the 5th International Colloquium - Physical Measurements and Signatures in Remote Sensing*, 14-18 January, Courchevel, France, European Space Agency (SP-319).

Rees, W. G. and Williams, M. (1997) Monitoring changes in land cover induced by atmospheric pollution in the Kola peninsula, Russia, using Landsat-MSS data, *International Journal of Remote Sensing*, **18**, 1703–1723.

Sader, S. A., Powell, G. V. N. and Rappole, J. H. (1991) Migratory bird habitat monitoring through remote sensing, *International Journal of Remote Sensing*, **12**, 363–372.

Schaale, M. and Furrer, R. (1995) Land surface classification by neural networks, *International Journal of Remote Sensing*, **16**, 3003–3031.

Shi, W. Z. and Ehlers, M. (1996) Determining uncertainties and their propagation in dynamic change detection based on classified remotely-sensed images, *International Journal of Remote Sensing*, **17**, 2729–2741.

Schriever, J. R. and Congalton, R. G. (1995) Evaluating seasonal variability as an aid to cover-type mapping from Landsat Thematic Mapper data in the northeast, *Photogrammetric Engineering and Remote Sensing*, **61**, 321–327.

Smith, G.M., Fuller, R.M., Amable, G., Costa, C. and Devereux, B.J. (1997) CLEVER Mapping: An implementation of a per-parcel classification procedure within an integrated GIS environment, *Proceedings of the Remote Sensing Society Annual Conference, Observations and Interactions: RSS97*, Remote Sensing Society, University of Nottingham, 21-26.

Townshend, J. R. G. (1981) The spatial resolving power of earth resources satellites, *Progress in Physical Geography*, **5**, 32–55.

Townshend, J. R. G., Justice, C. O., *et al.* (1992) The impact of misregistration on change detection, *IEEE Transactions on Geoscience and Remote Sensing*, **30**, 1054–1060.

Tudor, G.J. and Mackey, E.C. (1995) Upland land cover change in post-war Scotland, in D.B.A. Thompson, A.J. Hester and M.B. Usher (Editors): *Heaths and Moorland: Cultural Landscapes,* HMSO, Edinburgh.

Van Gils, H., Huizing, H., Kannegieter, A. and van der Zee, D. (1991) The evolution of the ITC system of rural land use and land cover classification (LUCC). *ITC Journal*, **3**, 163–167. ITC, Enschede, The Netherlands.

Wyatt, B. K., Greatorex-Davies, J. N., *et al.* (1994) Comparison of land cover definitions. *Countryside 1990 series.* London, Department of the Environment.

Chapter 15

LAND-COVER CATEGORIES VERSUS BIOPHYSICAL ATTRIBUTES TO MONITOR LAND-COVER CHANGE BY REMOTE SENSING

E. Lambin
Département de Géographie, Université Catholique de Louvain, Louvain-la-Neuve, Belgium.

1. INTRODUCTION

The objective of this commentary is to discuss methodological issues related to the monitoring of land-cover changes by remote sensing. The main emphasis is on the complex nature of land-cover change processes and on the requirement to adopt a continuous representation of the land surface attributes rather than a discrete representation of land cover categories, in order to grasp this complexity. This implies the further development of model inversion approaches to derive biophysical variables of the surface in a spatially explicit and temporally continuous ways. Improving the accuracy of the remote sensing-based methods currently used to derive these variables is viewed as a research priority to allow the scientific community to better measure, and therefore understand, processes of land-cover changes.

One generally distinguishes between land-cover *conversion*—the complete replacement of one cover type by another, and land-cover *modification*—the more subtle changes that affect the character of the land cover without changing its overall classification (Turner et al., 1993). Land cover modifications may result in degraded ecosystems, and are generally more prevalent than land-cover conversions. In principle, the monitoring of land-cover conversions (e.g., agricultural expansion or deforestation) can be performed by a simple comparison of successive land cover maps (e.g., derived by classification of remote sensing data or by field surveying).

However, the comparison of land-cover classifications for different dates does not allow the detection of subtle changes within land-cover classes. Even if some of the attributes of one class have changed, the magnitude of these changes will not always be large enough to justify a shift from one land cover category to another, unless the vegetation classification identifies a very large number of narrowly defined categories. Therefore, monitoring land-cover changes can only be achieved through repetitive measurements of biophysical attributes which characterize the land cover.

2. DISCRETE VERSUS CONTINUOUS REPRESENTATIONS OF LAND COVER

The land surface can be represented as a set of spatial units, each associated with one or more attributes. These attributes are either a single land cover category (i.e., leading to a discrete representation of land cover) or a set of values for continuous biophysical variables (i.e., leading to a continuous representation of land cover) (DeFries et al., 1995). The correspondence between these two representations can be established through a table which associates to each land cover category the average range of values for the biophysical variables.

In the continuous representation of land cover, the biophysical variables vary continuously not only in space but also in time, at the seasonal and inter-annual scale. By contrast, in the discrete representation of land cover, each spatial unit is represented by a single categorical value which is stable over a season. Inter-annual changes in the values of the biophysical attributes of the surface are described, in the discrete representation of land cover, as land-cover conversions only if the changes exceed the range of values which is characteristic of the land-cover category (i.e., if the magnitude of the change is such that the values of all biophysical attributes falls within the range of another land cover class during the entire seasonal cycle). By contrast, a land-cover modification—which is not detectable with the discrete representation of land cover—implies that variations in the values of the biophysical attributes remain within the range of values which is characteristic of the land-cover category. Processes leading to changes in the seasonal dynamics or in the fine scale spatial variability of biophysical attributes would also be described as land-cover modifications. Changes that would only affect the values of some of the biophysical attributes, for just part of the seasonal cycle would probably also enter in that category.

This suggests that monitoring land-cover changes by remote sensing requires the measurement of a set of indicators of the biophysical attributes of the surface, the seasonality of these attributes and their fine scale spatial pattern. These information requirements correspond to the three major information sources provided by remote sensing.

3. SURFACE ATTRIBUTES REQUIRED FOR CHANGE DETECTION

3.1 Biophysical variables

Rather than detecting changes on the basis of land-cover categories, change detection is better performed on the basis of the continuous variables defining these categories, whether these are reflectance values measured by a satellite sensor or biophysical attributes derived by model inversion. Coppin and Bauer (1996) review techniques used for this comparison, such as image differencing, image ratioing, multi-spectral or multi-temporal change vector analysis, image regression or multi-temporal linear data transformation. Empirical studies demonstrated that there is not a single optimal change detection technique but that different techniques are best suited for different change patterns.

3.2 Seasonal variations

Land-cover changes take place at a variety of temporal scales, e.g., short events with detectable effects only for a few months, modifications in seasonal trajectories of ecosystem attributes, processes that affect the land cover through several seasonal cycles and long-term, permanent changes. Land-cover changes may affect, and therefore be indicated by, the phenology of the vegetation cover. The analysis of the temporal trajectories of vegetation indices based on high temporal frequency remote sensing data allows to monitor vegetation phenology and biome seasonality (Justice et al., 1985).

Processes such as a shortening of the growing season, a de-phasing of the phenology of different vegetation layers or modifications of the cover due to disturbances such as fires can only be detected if inter-annual changes in the seasonal trajectories of vegetation covers are analyzed. For any landscape with a strong seasonal signal, the detection of inter-annual changes needs to explicitly take into account the fine scale temporal variations. If data from only one or a few dates a year are used to measure inter-annual changes, the under-sampling of the temporal series hinders the change detection accuracy and might lead to the detection of spurious changes (Lambin, 1996).

3.3 Landscape heterogeneity

A major attribute of a landscape is its spatial pattern, i.e., the arrangement in space of its different elements. The concept of landscape spatial pattern covers, for example, the patch size distribution of residual forests, the location of agricultural plots in relation to natural vegetation, the shapes of fields or the number, types and configuration of landscape elements (i.e., their spatial heterogeneity). Landscape spatial pattern is seldom static due both to natural changes in vegetation and human intervention. The spatial dynamics of landscapes interact with ecological processes which have important spatial components (Turner, 1989), such as flows of energy and matter between landscape components, biological productivity, bio-diversity, or the spread of disturbances. Remote sensing offers the possibility to analyze changes in spatial structure at the scale of landscapes. Indicators of the degradation of the vegetation cover can be derived from such measures (Jupp, Walker and Pendridge, 1986; De Pietri, 1995).

4. MODEL INVERSION

Physically based approaches, relying on model inversion, have been developed to derive biophysical surface attributes (e.g., Goel and Strebel, 1983; Li and Strahler, 1986; Pinty and Verstraete, 1992; Gobron *et al.*, 1997). In these methods, a scene model is constructed to describe the form and nature of the energy and matter within the scene and their spatial and temporal order (Strahler et al., 1986). The scene model is coupled with an atmospheric and a sensor model. This coupled model is then inverted against the remote sensing data to infer, from these measurements, some of the properties or parameters of the scene which were unknown (Strahler et al., 1986). In the case of complex landscapes, these methods are not yet sufficiently robust to be applied routinely to the large-scale monitoring of land-cover changes, due to the difficulty for representing adequately in a model the natural variability in landscape structure and composition.

In the future, modeling approaches should allow the scientific community to derive surface variables such as hemispherical albedo, land surface temperature and emissivity, soil moisture, snow cover, leaf area index, net primary production, total biomass, evapotranspiration, incident short-wave radiation, outgoing long-wave radiation, photosynthetically active radiation (PAR), and fraction of absorbed photosynthetically active radiation (FAPAR). These variables describe the nature of a land-cover category, which is just an aggregate concept encapsulating specific values of surface attributes.

5. NEW SENSORS AND MULTISENSOR APPROACHES

The coming years will see an increase in the number and a qualitative improvement in the performance of Earth Observation systems in orbit. These new sensors will include both very high spatial resolution systems (i.e., one to a few meters) with a low spatial coverage and low temporal frequency, and medium to low spatial resolution systems with broader and more frequent coverage in multiple spectral bands. These advanced sensors will often include on-board calibration and permit better geo-referencing of the data. The potential for multi-sensor analysis (i.e., combining data from different sensors) will increase greatly. Three types of combinations will be interesting:
1. high and low spatial resolution data, to combine a detailed view of the land surface over a few representative locations with an exhaustive view of the surface at a coarser spatial resolution;
2. data at approximately the same spatial resolution (fine or low) but from different sensors to increase the temporal frequency of coverage of a given area; and
3. data at approximately the same spatial resolution but in different spectral ranges (visible, thermal, microwave) to describe more comprehensively surface processes.

6. REFERENCES

Coppin P.R. and Bauer M.E. (1996) Digital change detection in forest ecosystems with remote sensing imagery, *Remote Sensing Reviews,* **13**, 207–234.

DeFries, R.S., Field, C.B., Fung, I., Justice, C.O., Los, S., Matson, P.A., Matthews, E., Mooney, H.A., Potter, C.S., Prentice, K., Sellers, P.J., Townshend, J.R.G., Tucker, C.J., Ustin, S.L. and Vitousek, P.M. (1995) Mapping the land surface for global atmosphere-biosphere models: Toward continuous distributions of vegetation's functional properties, *Journal of Geophysical Research,* **100**, 20,867–20,882.

De Pietri, D.E. (1995) The spatial configuration of vegetation as an indicator of landscape degradation due to livestock enterprises in Argentina, *Journal of Applied Ecology,* **32**, 857–865.

Gobron, N., Pinty, B., Verstraete, M.M. and Govaerts, Y. (1997) A semi-discrete model for the scattering of light by vegetation, *Journal of Geophysical Research,* **102**, 9431–9446.

Goel, N.S. and Strebel, D.E. (1983) Inversion of vegetation canopy reflectance models for estimating agronomic variables. I. Problem definition and initial results using Suits model. *Remote Sensing of Environment,* **13**, 487–507.

Jupp, D.L.B., Walker, J. and Pendridge, L.K. (1986) Interpretation of vegetation structure in Landsat MSS imagery: A case study in disturbed semi-arid eucalypt woodland. Part 2. model-based analysis. *Journal of Environmental Management,* **23**, 35–57.

Justice, C.O., Townshend, J.R., Holben, B.N. and Tucker, C.J. (1985) Analysis of the phenology of global vegetation using meteorological satellite data, *International Journal of Remote Sensing,* **6**, 1271–1318.

Lambin, E.F. (1996) Change detection at multiple temporal scales: Seasonal and annual variations in landscape variables, *Photogrammetric Engineering and Remote Sensing,* **62**, 931–938.

Li, X. and Strahler, A.H. (1986) Geometric-optical bidirectional reflectance modeling of a conifer forest canopy. *IEEE Transactions on Geosciences and Remote Sensing,* **GE-24**, 906–919.

Pinty, B. and Verstraete, M.M. (1992) On the design and validation of surface bidirectional reflectance and albedo models, *Remote Sensing of Environment,* **41**, 155–167.

Strahler, A.H., Woodcock, C.E. and Smith, J.A. (1986) On the nature of models in remote sensing, *Remote Sensing of Environment,* **20**, 121–139.

Turner II, B.L., Moss, R.H. and Skole, D.L. (1993) *Relating land use and global land-cover change: A proposal for an IGBP-HDP core project.* IGBP Report No. **24**, HDP Report No. **5**, International Geosphere-Biosphere Programme, Stockholm.

Turner, M.G. (1989) Landscape ecology: The effect of pattern on process, *Annual Review of Ecological Systems*, **20**, 171–197.

Chapter 16

A NEW APPROACH TO CHARACTERIZE GLOBAL LAND SURFACES
Preliminary results from AVHRR data

N. Gobron, B. Pinty and M. M. Verstraete
Space Applications Institute, Ispra, Italy.

1. INTRODUCTION

The upcoming generation of satellite sensors will provide more and much higher quality data (specifically much better spectral and angular sampling of the radiative fields emerging from terrestrial surfaces) than have been available so far. These data are of primary interest for providing an improved characterization of land surfaces which, for global scale investigations, have traditionally been based on the temporal analysis of a vegetation index (such as the Normalized Difference Vegetation Index) computed from AVHRR data. Indeed, the radiance fields measured by satellites do depend on the radiative properties of the surface (among other factors), which include a number of key variables such as the Leaf Area Index (LAI). Estimating the values of these surface variables on the basis of remote sensing data reduces to the solution of an inverse problem. An algorithm designed to identify the most probable solutions amongst a set of potential solutions that are predefined in a Look-Up Table (LUT) has been implemented and explored. These solutions have been derived from direct simulations achieved with a one-dimensional physical model of radiation transfer in vegetation canopies. In this approach, every pre-defined solution is associated to the set of values for all the physical variables required by the radiation transfer model used to generate the LUT. Taken together, these sets of variables constitute a quantitative characterization of the land surfaces fully compatible with the radiative properties measured by the particular satellite used in this application.

Depending on the resolution of the LUT and on the accuracy of the observations, one or more solutions can be obtained by the inverse procedure. The output of this method is thus twofold: an ensemble of values for the variables used by the radiative transfer model that created the LUT and the range of variations for the values of these variables. This information allows the production of land cover maps, together with the corresponding level of uncertainties on the retrieved variable values.

The model used to create the LUT and the inversion algorithm itself will be outlined. Preliminary results of the application of this technique on the AVHRR/GVI/LASUR data set (Berthelot et al, 1997) at the global scale are shown in the form of the maximum values of LAI retrieved for the months of January and June 1989. The aim of this exercise is to demonstrate the potential of such a new methodology for addressing land cover issues.

2. OVERVIEW OF THE ALGORITHM

2.1 Modeling approach

Assessing the land cover type reduces to the qualitative description or the quantitative characterization of the properties of the land surface. A series of physical, chemical and biological properties are usually associated with a given land cover type. Determining the latter therefore results in the assignment of at least approximate values of these variables. The essence of land cover classification on the basis of remote sensing data consists in similarly associating certain radiative characteristics of the environment to specific land cover types, so that the observation of the former leads to the reliable identification of the latter. This is best achieved when the properties of the environment are explicitly linked to the measurable observations gathered in space, and the proposed approach thus hinges on the simulation of top-of-atmosphere radiances typically measured by satellite sensors on the basis of canopy variables. The set of such variables (or scenarios), together with the resulting simulated reflectances, is then archived in a LUT, as these computations are executed once and for all. Once this LUT is available, the actual measurements gathered with the satellite sensors are compared with the simulated reflectances of the LUT, and all table entries "close enough" to the string of spectral and directional measurements are considered potential solutions of the problem, i.e., possible descriptions of the actual environment, as will be seen shortly. For the purpose of this paper, the LUT was generated with the semi-discrete radiation transfer model of Gobron et al. (1997) to represent the interaction of solar light with plant canopies. This model implements a statistical description of the discrete nature of the canopy to relax the usual continuous (turbid) medium assumption, and includes an explicit representation of architectural effects for homogeneous canopies. The first two orders of scattering (by the soil and by plant leaves) are calculated in three-dimensional space using an adaptation of the original discrete model developed by Verstraete (1987) for the extinction of the direct incoming solar radiation in vegetation canopies. This statistical representation of the canopy architecture permits an explicit representation of the hot spot phenomenon in these first two orders of scattering, adapted from Verstraete et al. (1990).

The multiple scattering contribution is calculated with a Discrete Ordinates Method using an azimuthally averaged expression of the anisotropic scattering phase function proposed by Shultis and Myneni (1988). Extensive comparisons against a Ray-tracing model (Govaerts and Verstraete, 1995) simulating homogeneous canopies have shown the high performance of this semi-discrete model (see also Gobron et al. 1997).

The choice of a one-dimensional model reflects 1) the fact that no reliable ancillary information is available globally to run three-dimensional models, 2) the need to limit computational costs to permit the processing of global data and 3) the requirement to consider a large enough selection of radiative transfer variables values.

2.2   Design of the inversion algorithm

The inversion algorithm then consists in identifying the most probable solutions amongst the set of possible solutions predefined in the LUT. It operates in three basic steps. The first step consists in calculating, for each radiative biome b in the λ channels, and for each pixel p, the following cost function:

$$\delta^2(p,b,\lambda) = 4 \frac{[\rho_{data}(\theta_0^p,\theta_v^p,\varphi,\lambda) - \rho_{lut}(\theta_0^p,\theta_v^p,\varphi,b,\lambda)]^2}{[\rho_{data}(\theta_0^p,\theta_v^p,\varphi,\lambda) + \rho_{lut}(\theta_0^p,\theta_v^p,\varphi,b,\lambda)]^2}$$

where θ_0^p and θ_v^p are the illumination and observation zenith angles, respectively, and φ is the relative azimuth angle between the horizontally projected solar and observation directions. In this equation ρ_{data} corresponds to the measured bi-directional reflectance factors and ρ_{lut} corresponds to the simulated bi-directional reflectance factors associated in the LUT with biome b. This cost function has been chosen so as to measure the quadratic distance between the measured bi-directional spectral reflectance factors and the corresponding model simulations, in each available channel separately.

In a second step, the algorithm ranks the pre-selected biomes in increasing order of the cost function separately in each channel, with a rank of 0 for the best solution(s). This step thus ends with an ordered set of the pre-defined solutions in each spectral channel individually.

In the last step, a quality index is evaluated to identify which entries in the LUT lead to the smallest cost function in all spectral channels simultaneously, and thus represent the most probable solutions. This quality criterion is evaluated as follows:

$$Q(b,n_B) = 1 - \frac{1}{n_B k} \sum_\lambda a_{\lambda k} R(b,\lambda_k) \qquad R(b,\lambda_k) \leq R_{MAX}$$

where n_B is the number of biomes defined as possible solutions of the inverse problem, $a_{\lambda k}$ is the spectral weight associated with the corresponding spectral band (set to 1 for all spectral bands in our application), and R_{MAX} specifies the highest allowed rank beyond which the probability to find a given biome b is a priori considered very low. The radiative biome(s) for which $Q(b,n_B)$ is closest to 1 is (are) selected as the most probable solution(s). Since multiple combinations of spectral ranks may give the same value for the function $Q(b,n_B)$, there is no guarantee of finding a unique solution. The number of solutions retrieved with an equal value of the quality criterion is controlled by various factors, including the radiation transfer regime itself. Two or more geophysical situations may lead to indistinguishable radiance fields measured by the satellite. The noise inherent to the measuring instrument will of course further blur the slight differences between the radiative signatures of similar environments, and will lead to the identification of multiple solutions.

3. EXPERIMENTS AND RESULTS

The LUT contains the values of bi-directional reflectance factors computed with the semi-discrete model for a large number of vegetation types (made up of healthy green leaves only) and for an appropriate set of illumination and observation angles. Table 1 lists the different "radiative biomes", which are defined in terms of the model parameters. The 35 biome types considered here were selected to cover most typical radiance values (in the red and near-infrared bands) actually observed with the AVHRR instrument.

16. A NEW APPROACH TO CHARACTERIZE GLOBAL LAND SURFACES

Table 1. Definition of the radiative biome types

LAI	Diameter of a single leaf	Height of the canopy	Leaf Angle Distribution	Soil brightness
1	Small	Small	Erectophile	Dark
1	Small	Small	Planophile	Bright and Dark
1	Large	Small	Erectophile	Dark and Bright
1	Large	Small	Planophile	Dark
1	Large	Large	Erectophile	Dark
1	Large	Large	Planophile	Dark
2	Small	Small	Erectophile	Dark
2	Small	Small	Planophile	Bright and Dark
2	Large	Small	Erectophile	Dark and Bright
2	Large	Small	Planophile	Dark
2	Large	Large	Erectophile	Dark
2	Large	Large	Planophile	Dark
3	Small	Small	Erectophile	Bright
3	Small	Small	Planophile	Bright and Dark
3	Large	Small	Erectophile	Bright and Dark
3	Large	Small	Planophile	Bright and Dark
3	Large	Large	Erectophile	Bright and Dark
3	Large	Large	Planophile	Bright and Dark
5	Small	Small	Planophile	Bright
5	Large	Small	Erectophile	Bright
5	Large	Small	Planophile	Bright
5	Large	Large	Erectophile	Dark and Bright
5	Large	Large	Planophile	Bright
0				Dark
0				Very Bright

Note: The two bare soil cases correspond to anisotropic surfaces with a single scattering albedo of 0.1 for Dark and 0.7 for Very Bright soils in the visible band and 0.2 for dark and 0.8 for Very Bright soils in the near-infrared band. The other two model parameters are the hot-spot parameter (0.1) and the asymmetry factor (−0.2). The small (large) diameter of a single leaf corresponds to a value of 0.01 (0.05) m. The small (large) height of canopy corresponds to 0.5 (2.0) m.

The search algorithm described above was applied to AVHRR/GVI/ LASUR data set, corrected for atmospheric effects for the two months of January and June 1989. Five (four) images in January (June) were used respectively, assuming that the land surfaces have not changed during the acquisition period in order to provide a sufficient angular sampling of the radiation fields. This accumulation of data in time was required to limit the number of solutions that were obtained for each pixel during preliminary tests. The relatively high level of noise in the AVHRR data set mitigates the constraining nature of multiple acquisitions on the model inversion, however.

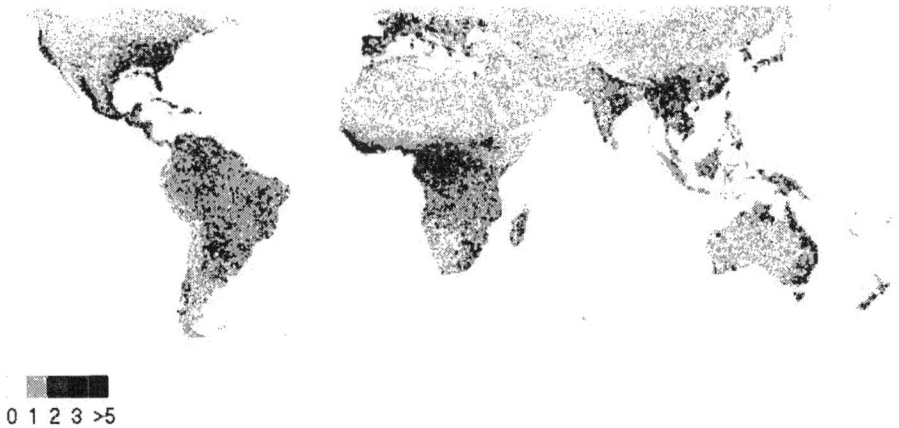

Figure 1. Maximal LAI values using 5 days of GVI/LASUR data in January 1989

As an example of the various products generated by this approach, Figures 1 and 2 exhibit the maps of the maximum values in LAI that were retrieved for the months of January and June, respectively.

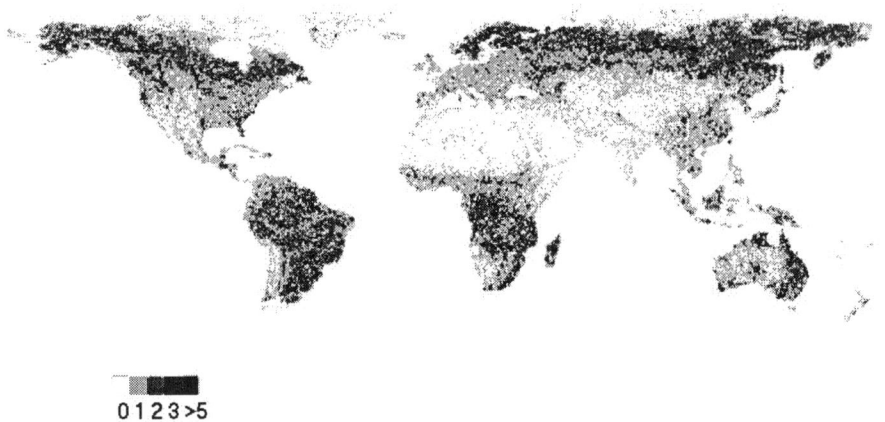

Figure 2. Maximal LAI values using 4 days of GVI/LASUR data in June 1989

Additional maps can be produced to represent the range of variations (or any other statistics) for any one of the model variables corresponding to the entry points in the LUT. Since each individual variable corresponds to a specific type of information, they can be analyzed jointly to build a land cover map on the basis of quantitative estimates of measurable variable values (Gobron, 1997). In essence, the level of uncertainty in the retrieved maps is also expressed by the number of solutions found for a given value of the R_{MAX}

16. A NEW APPROACH TO CHARACTERIZE GLOBAL LAND SURFACES

parameter intervening in the cost function. Since this algorithm permits to characterize land surface properties on the basis of the radiative properties controlling the radiation transfer regime, it is also possible to estimate, for all selected solutions, other geophysical variable of interest such the Fraction of Absorbed Photosynthetically Active Radiation and the spectral albedo.

4. CONCLUSIONS

This short paper outlines a new methodology to access information related to land cover from remote sensing data and shows the feasibility of this approach by exhibiting preliminary results. It aims at demonstrating the potential of simulation and inversion techniques based on the application of radiation transfer models used as interpretation tools. This approach has been applied to a pre-processed AVHRR global data set accumulating multi-angular data over a short period of time as if they had been acquired simultaneously. The proposed methodology is easy to implement, does not depend on any particular sensor or radiation transfer model and appears to be quite robust for operational applications. This type of algorithm will likely become more popular when analyzing high quality data such as will be provided by the next generation of sensors and in particular by the MISR instrument on the NASA EOS platform.

5. REFERENCES

Berthelot, B., L. Adam, L. Kergoat, F. Cabot, P. Maisongrande, and G. Dedieu (1997) A global data set of surface reflectances and vegetation indices derived from AVHRR/GVI time series for 1989-1990: The land surface reflectances (LASUR) data. *Proceedings of the 7th ISPRS International Symposium on Physical Measurements and Signatures in Remote Sensing*, Courchevel, France, 7–11 April 1997, CNES.

Gobron, N., B. Pinty, M. M. Verstraete, and Y. Govaerts (1997) A semi-discrete model for the scattering of light by vegetation, *Journal of Geophysical Research*, **102**, 9431–9446.

Gobron, N. (1997) *Caractérisation des surfaces terrestres par télédétection spatiale à partir de méthodes physiques avancées*, EUR S.P.I. 97.101.

Govaerts, Y. and M. M. Verstraete (1995) Evaluation of the capability of BRDF models to retrieve structural information on the observed target as described by a three-dimensional ray tracing code, *The European Symposium on Satellite Remote Sensing*, EUROPTO-SPIE Conference, Rome, 26–30 September 1994, 9–20, SPIE Volume **2314**.

Shultis, J. K. and R. B. Myneni (1988) Radiative transfer in vegetation canopies with anisotropic scattering, *Journal of Quantitative Spectroscopy and Radiation Transfer*, **39**, 115–129.

Verstraete, M. M. (1987) Radiation transfer in plant canopies: Transmission of direct solar radiation and the role of leaf orientation, *Journal of Geophysical Research*, **92**, 10,985–10,995.

Verstraete, M. M., B. Pinty, and R. E. Dickinson (1990) A physical model of the bidirectional reflectance of vegetation canopies. 1. Theory, *Journal of Geophysical Research*, **95**, 11,765–11,775.

Chapter 17

REMOTE SENSING REQUIREMENTS TO SUPPORT FOREST INVENTORIES

E. Tomppo
Finnish Forest Research Institute, Helsinki, Finland.

1. BACKGROUND OF FOREST INVENTORIES

The history of forest inventories goes back to the end of the Middle Ages, when the heavy use of forest resources created a shortage of wood. This shortage forced people to bring some form of planning to the forests nearby towns and mines. The first information collected for the purposes of this early planning was an assessment of the forest area. The term 'inventory' means the preparation of a detailed list of articles according to their properties and has a commercial origin.

Assessing the properties of an area of forest is an enormous task, given, for instance, that the total number of trees with a height of at least 1.3 m in Finland is about 65 billion. It is simply not possible to measure each tree, nor is it necessary. Measuring a part of tree population and deriving properties of tree populations from the sub-population, i.e., sampling is more rational.

Sampling, though, immediately raises a number of questions. For example:
- What is the smallest part of the population that can be measured and still provide reliable information?
- How can the mean properties of the population in the sample be assessed?
- How can one guaranteed that, on the average, the right parameter value is obtained and that no bias exists?
- How can the reliability of the parameter be assessed, i.e., what can be said of the real parameter value if the estimate is derived from a sample?

Inventory methods have been developed in order to be able to answer these and similar questions, and to present a clear picture of the state of forests and the amount of forest resources. Statistically designed forest inventories were introduced simultaneously in the three Nordic countries, Norway, Sweden and Finland at the beginning of the 1920's. In these countries, and especially in Sweden and Finland, the utilization of forests has been of vital importance to the national economy. The first operative, country-level satellite image-based inventory was carried out in Finland in the beginning of the 1990s.

2. TRADITIONAL TYPES OF FOREST INVENTORIES

Inventories can be divided into operative (small scale), strategic (large area) and surveying-type inventories.

Operative inventories produce information for localized (usually stand-level) planning of cuttings and silvicultural regimes. An operative forest management plan is based on this information and is a tool of a forest owner. The time interval between such inventories often varies from 2 to 15 years.

Inventories for strategic planning produce information for:
- large area forest management planning
- optional cutting possibilities with future scenarios of forests
- level of cuttings and silvicultural regimes
- sustainability of wood production
- planning of forest industry investments
- large area forest and nature conservation policy
- forestry legislation
- financial decisions
- assessing the status of nature conservation.

Surveying types of inventories produce an overview-type picture of the area of interest, for instance for planning a more thorough forest inventory.

Local forestry organizations are in many cases responsible for operative inventories, while national agencies, e.g., ministries, forest commissions or research institutes deal with national inventories. The United Nation's Food and Agriculture Organization (FAO) has compiled information on the world's forest resources from national data sources since 1947. The next assessment, FRA 2000, is currently ongoing.

3. RECENT INFORMATION NEEDS FOR FOREST INVENTORY

3.1 Timber production needs

Paper consumption has increased from about 130 million tons at the beginning of the 1970s to 276 million tons in 1995. It is expected to increase to 420–440 million tons by 2010 (Ranneby 1997). On the other hand, over one half the timber harvested each year is still used for cooking and heating, causing for instance deforestation in the dry tropics.

The principal purpose of forest inventories has been to provide accurate information for forest management planning and the planning of forest industry investments. Accordingly, optional cutting possibilities with future forest development scenarios have been computed on the basis of inventory results. These scenarios have formed the basis for forest policy and forest utilization. This is especially true in industrialized countries.

Timber processing and marketing practices have changed dramatically in these countries during the last decade. Earlier, there was a long time interval between harvesting and the delivery of the timber, panel or paper products to the customer. Harvesting decisions were mainly forest production driven, derived from long-term yield regulations and management plans, modified by operational logging schedules. Large timber and end product stocks had to be maintained and prices fluctuated rapidly as the market changed.

Today, mechanized mobile harvesting machinery, high speed processing facilities, and short timber drying or pulping schedules have reduced to less than one week the time between the decision to harvest and delivery of a specific timber end-product. Harvest planning has now the potential to become "market driven". On the other hand, forestry and harvesting increasingly have to meet strict environmental demands and to achieve "sustainability" in its widest sense. Today, forest certification or eco-labeling systems are often seen as a solution to enforce improved practices in forestry and forest industry. To achieve these requirements, it is necessary to have up-to-date geo-referenced forestry data systems with precise knowledge about log specifications and the forest ecosystem. If inventories based on field measurement techniques alone were employed, they would require a very high proportion of the trees to be measured, and this would simply be unrealistically expensive.

3.2 Ecological needs

Forest inventories have traditionally provided information related to the biological diversity of forests, such as the structure of growing stock, areas of site fertility classes and sometimes the distribution and abundance of plant species. An increasing concern about the loss of diversity, caused by, e.g., deforestation, human induced environmental and climate changes as well as the extinction of species, has promoted interest in the whole forest ecosystem and its biological diversity. It has been estimated that the present extinction rate is 100–10,000 times higher than would occur naturally. There is wide agreement that some components and indicators of forest biodiversity can only be measured efficiently in the context of large area inventories. Examples of such components or indicators are the composition and structure of landscape, the existence of 'ecological corridors' between different habitat types, the fragmentation of forests or land types, the areas and spatial distributions of important habitat types which support rich flora and fauna, as well as the amount of rotting wood in the Boreal zone.

3.3 Carbon fixation

Forests have also been seen as having a role in reducing the effects of global warming by sequestering some of the increasing amount of carbon dioxide in the atmosphere for periods of years (Kauppi et al. 1995). It has been estimated that the current forest area is only one half of the area that was forested 8000 years ago (under hypothetical similar climate conditions). The total carbon release in the beginning of 1990s was 6.2 gigatons. The estimated carbon storage in the world's trees is 360 gigatons. The recent annual carbon flux to trees is not well known because tree growth estimates for the whole Earth are unreliable. However, the figures 1 ± 1 gigatons per year are often given. A proper use of wood resources and an increase of forest area have a good potential to decrease the amount of atmospheric CO_2.

3.4 Forest health monitoring needs

Industrialization has resulted in transboundary air pollution effects. For several decades, acidic deposition has threatened forest trees and soil, as well as other environmental components, especially in Central and Eastern Europe. Awareness of wide area forest damages caused by air pollution arose at the beginning of the 1970s. As a result, more recent inventories and wide forest health monitoring programs have gathered information on new forest damage characteristics, as well as damages caused by processes other than air pollution.

3.5 Consequences

To satisfy the increasingly specific and diverse requests for scientifically substantiated information, efficient methods are needed to measure forest resources, their status, and the components of the whole forest ecosystem. This includes forest area, volumes, annual growth and their annual changes. It is not enough to know the volume of growing stock, although that single variable is already difficult to estimate for the whole Earth. The areal extent of forests is immense and the natural questions are: 'How can one measure forests in a cost-effective way?' and 'How can that be repeated with sufficient frequency?' The requirements can be summarized as follows:
- new methods are needed in national and global level forest inventories to answer to increasingly urgent demands
- in some areas or cases, the cost-efficiency of inventories should be improved
- the potential of satellite remote sensing must be further demonstrated and exploited.

4. REQUIREMENTS OF FOREST INVENTORIES FOR REMOTE SENSING

The utilization of airborne remote sensing has a long tradition in forest inventories. A common methodological approach has been the visual interpretation of images. The EFICS study, concerning forest inventories of 15 EU countries, Norway and Switzerland and some countries in Eastern Europe showed that space-borne remote sensing is used operationally only in Finland. Few other countries used satellite images for supporting forest inventories, e.g., for stratification the area for different sampling units. One reason for the minor utilization may the requirements of inventories. Usually estimates for about 100–400 different parameters are needed. There are also some other requirements which cannot easily be fulfilled with space-borne remote sensing data. On the other hand, investigators in forestry and remote sensing (especially those dealing with space technology) do not often meet.

Examples of forest inventory issues are:
- time frequency

- for national and global inventories, full image coverage occurs once every 1–5 years, in practice all areas should be covered annually
  - for operative inventories, once per year
  - clouds often prevent optical imaging with sufficient frequency
- spatial resolution
  - for national and continental inventories, 10–30 m
  - for stand-level inventories, 5–10 (20) m
- availability of ancillary data
  - DTM is almost always necessary
  - digital map data improve the accuracy, sometimes necessary
  - phenological and soil data improve the accuracy
- spectral range and resolution
  - the range 450–680 µm is important
  - the range 680–750 µm is very important
  - the range 780–810 µm is very important
  - the range beyond 850 µm is very important, i.e., Landsat TM bands, 3, 4, 5 (all bands should be used)
- radiometric resolution
  - in today's satellites, it is usually poor compared to the requirements
  - signal-to-noise ratio should be improved, especially at the low end of the spectrum
  - important especially for small scale estimates
- requirements for methods
  - define the inventory units
  - all important variables should be estimated for all units
  - pixel level errors may sometimes be high, in spite of low inventory units level errors

One of the crucial problems is image availability. Recent microwave data cannot compensate the lack of optical area remote sensing data (Tomppo et al. 1996).

5. ADVANTAGES OF REMOTE SENSING BASED INVENTORIES

There are some obvious advantages of remote sensing data for forest inventories:
- information can be produced with small additional costs for much larger areas than is possible with field measurements only
- information is more accurately geo-referenced than with field measurement-based inventories

*17. REMOTE SENSING REQUIREMENTS TO SUPPORT FOREST   157
INVENTORIES*

- it is possible to get some information at low cost that would be very expensive with field measurements, e.g., ecological information, landscape diversity, ecological corridors
- output data is a model of forests, it can be used for simulation studies, e.g., for simulation of sampling errors of different field sampling designs and for simulation alternative future scenarios of forests.

6. THE FINNISH MULTI-SOURCE FOREST INVENTORY

The Finnish Multi-Source National Forest Inventory (MS-NFI) is described shortly below because it is one of the few operative inventories utilizing space-borne remote sensing data. Finland's forest resources have been investigated by means of eight National Forest Inventories since the year 1921 (Ilvessalo 1927). The information has been utilized in large area forest management planning, such as determining the level of cuttings and other treatments needed, and has formed the information basis for official forest policy making and for the strategic planning of the forest industries.

During the eighth inventory (1986–1994), a multi-source inventory system was developed. It utilizes satellite images and digital map data in addition to ground measurements. For a real forest inventory, it is not sufficient to utilize remote sensing data in such a way that only some classes are derived, e.g., on the basis of tree species dominance. The image analysis and parameter estimation method has been designed in the Finnish MS-NFI so that it is possible to estimate all inventory parameters for areal units of about 40 hectares or larger (Tomppo, 1991, 1993 and 1998). The ninth inventory, which started in 1996, also contains the measurement of some additional characteristics describing forest biodiversity.

Examples of parameters measured in MS-NFI in the field are given below to provide an idea of the kind of estimates that should also be produced in a remote sensing based inventory. Stand level variables measured in Finnish MS-NFI are:
- general data of field plot cluster (inventory type, record type, crew leader, coordinates of cluster, date, inventory area),
- plot identification data (plot number, coordinates, administrative information, multiple use, plot size, etc.),
- site data (e.g., land class and its changes, direction and distance to the closest stand boundary, main site fertility type, mixture of site fertility types, specification of mire type, type of soil, quality and thickness of

organic layer, drainage accomplished or proposed, forest income taxation class, etc.),
- crown layer information (e.g., species of layer, development class, stand establishment, dominant tree species, species mixture, number of stem, quality, mean diameter or height, age, syndrome, originating time, cause and seriousness of damages),
- stand level information (e.g., damages, lichen survey, stand quality, accomplished measures and time, proposed measures and time, basal area).

Tree level data are measured with two intensities, at tally tree level and sample tree level. The tally trees variables are measured for each tree tallied:
- coordinates (only on permanent plots),
- tree species,
- diameter,
- timber assortment class and its precision,
- crown layer.

The additional variables measured at sample tree level (each 7th tally tree) are:
- origin of tree
- upper diameter (at the height of 6 m, on every 9th cluster)
- bark thickness (only on every 9th cluster)
- height of dead branches
- height of green branches
- height and length of broken part
- height increment
- diameter increment, at the height of 1.3 m
- age at 1.3 m and age - age at 1.3 m.

Damage information includes
- syndrome and time of origin
- cause and length of rotten part
- seriousness
- defoliation
- lengths and timber quality classes of each part of stem, timber assortments class, reasons for possible lowering.

The total number of parameters measured in the field is about 150. Several other parameters are derived in the subsequent computation phase. For multi-source parameter estimation, a non-parametric k nearest neighbor estimation method has been developed (Tomppo 1991, 1993 and 1997). The main advantages of the method are:
- much more detailed information about forests can be obtained with very low additional costs compared to the inventory methods which employ sampling and field measurements only

- the method is more statistically oriented than the old classification-based approach to use of satellite images
- in principle, all variables can be estimated for each computation unit, which is not possible with ordinary classification methods
- the method preserves the natural dependency structure between forest parameters
- the method can be applied with minor modifications to very different types of forests
- the method can directly be applied using different remote sensing material.

Airborne imaging spectrometer research is under development and can be integrated into the system (Mäkisara and Tomppo 1996). The multi-source inventory method developed by the Finnish Forest Research Institute has also already been tested in some other countries.

7. CONCLUSIONS

Inventories have produced large area forest resource information in some countries since the beginning of the 1920s. The first world wide forest resource assessment was compiled from national statistics by FAO in 1947. The requirements for forest inventories have increased during the last decades. New information sources and new methods make it possible to increase cost-efficiency of inventories and change them from a field measurement-based systems into a multi-source monitoring of the whole forest ecosystem, thereby providing information about the structure of forests, their health and their biodiversity status for small and large areas. One of the problems in further utilization of space-borne remote sensing data is that the advantages of these data are not often perceived. Traditional classification based approaches, which do not provide sufficient information for forestry purposes, have been proposed and tested by remote sensing community. The lack of communication between remote sensing and forest inventory communities is one of the obstacles in expanding the use of remote sensing data. Remote sensing has, however, obvious potential in national and global level forest inventories.

8. REFERENCES

European Forest Institute (1997) *Study on the European Forest Information and Communication System (EFICS)*, Final Report, EFI Internal reports **1/97**. Unpublished.

European Commission (1997) *Study on European Forestry Information and Communication System. Reports on Forestry Inventory and Survey Systems*, Volumes 1 and 2, Office for Official Publications of the European Communities, Luxembourg, pp. 1328.

Ilvessalo, Y. (1927) The Forests of Suomi (Finland). Results of the general survey of the forests of the country carried out during the years 1921–1924, *Communicationes Ex Instituto Quaestionum Forestalium Finlandiae*, Editae **11**.

Mäkisara, K. and Tomppo, E. (1996) Airborne Imaging Spectrometer in National Forest Inventory, in *Proceedings of the IGARSS'96*, 1996 International Geoscience and Remote Sensing Symposium, Remote Sensing for A Sustainable Future, Vol II, pp 1010–1013, IEEE Catalogue Number 96CH35875, Library of Congress Number, 95–80706.

Ranneby, B. (1998) New Methodologies to Assess Wood Production and Environmental Status of the World's Forests–a Vision, in *Managing the Resources of the World's Forests*, The Marcus Wallenberg Foundation Symposia Proceedings, **11**, 98-96.

Tomppo, E. (1991) Satellite Image-Based National Forest Inventory of Finland, *International Archives of Photogrammetry and Remote Sensing*, **28**, Part 7–1, 419–424.

Tomppo, E. (1998) Recent Status and Further Development of the Finnish Multi-Source Forest Inventory, in *Managing the Resources of the World's Forests*, The Marcus Wallenberg Foundation Symposia Proceedings, **11**, 53–69.

Tomppo, E., Goulding, C., Katila, M. (1999) Adapting Finnish Multi-Source Forest Inventory Techniques to the New Zealand Preharvest Inventory, *Scandinavian Journal of Forestry Research*, **14**, 182–192.

Kauppi, P., Tomppo, E., and Ferm, A. (1995) C and N storage in living trees within Finland since 1950s, *Plant and Soil*, 168–169: Kluwer Academic Publisher, Printed in Netherlands, pp. 633–638.

Tomppo, E., Mikkelä, P., Veijanen, A., Mäakisara, K., Henttonen, H., Katila, M., Pulliainen, J., Hallikainen, M. and Hyyppä, J. (1996) Application of ERS-1 SAR data in large area forest inventory, in Guyenne, T.-D. (Editor): *Proceedings of the Second ERS Applications Workshop*, London, UK, 6–8 December 1995, Noordwijk, 103–108.

Chapter 18

SOME RESEARCH AND APPLICATIONS IN THE CSIRO (AUSTRALIA) EARTH OBSERVATION CENTRE ON SCENE BRIGHTNESS DUE TO BRDF

David L. Jupp
CSIRO Earth Observation Centre, Australia.

1. INTRODUCTION

A range of Earth Observation research activities in CSIRO (Australia) are promoted and coordinated through a unit called the CSIRO Earth Observation Centre (EOC, see http://www.eoc.csiro.au). The EOC has been formed to coordinate CSIRO Earth Observation activities in generic science. It is developing and supporting activities aimed at establishing coordinated validation missions, and calibration and validation sites in Australia to develop long term high quality data and well characterized land surface parameters. The sites provide potential validation for a range of products of satellite global measurements programs.

CSIRO Earth Observation scientists engage in a number of related research and applications oriented projects involving land surface reflectance anisotropy. In particular, these involve the issue of scene brightness, or BRDF (Bidirectional Reflectance Distribution Function or variation in scene radiance with Sun and look position) effects. A fundamental theme of the work is to test/establish the existence of a stable underlying "typology" of BRDF directly attributable to land surface spectral and structural parameters that will allow sensible selections of the "shape functions" used in correction and inversion algorithms. This requires the existence and use of well-established field sites and a range of measurements other than radiometer data. There now exist established sites and there are important canopy and land cover missions planned in the future which may provide more of this type of data.

BRDF related activity in CSIRO EOC Research involves:
- Data normalization (AVHRR, scanners, videos, air photographs, calibrations sites, panels)
- Atmospheric correction (AVHRR, scanners)
- Determining land cover structure (photography, scanners), and
- BRDF characterization.

Land cover structure is known to be a major factor in controlling the BRDF shape, i.e., the magnitude of its reflectance variation with Sun and look angle. The opportunity to use BRDF to establish and/or monitor land cover structure is a major research and development task. However, there are a number of areas where scene brightness R&D is already operational and this paper is primarily concerned with them.

2. DATA NORMALISATION & ATMOSPHERIC CORRECTION

Data normalization and atmospheric correction activities are aimed at achieving "seamless" mosaicking of video and scanner data as well as normalization and standardization of satellite data—especially AVHRR and future environmental satellite data. In the case of satellite applications, calibration sites must have their BRDF characterized, and it is clear that field panels used for reflectance measurements must also be carefully characterized (EOC Discussion Paper, 1996). One pragmatic objective is to produce base series of data that users now happy to use image processing (such as filtering and classification) and are familiar with Spot and Landsat can use with impunity. This has not been true of airborne data nor AVHRR data in the past.

In current work of this kind around the world, there is an acceptance of the key role of the 'Kernel Function' approach for which many applications are near operational. In the kernel function approach, the land surface brightness variation is modeled statistically by a class of simple functions. Advanced examples of this is the AMBRALS model (which is really a set of kernels of choice) being applied for the MODIS BRDF product (Strahler *et al.*, 1995; Wanner *et al.*, 1995) and the RPV model being applied to the MISR BRDF product (Rahman *et al.*, 1993).

One outcome of the rising understanding of and capacity to model BRDF in the last 10 to 15 years has been the wider acceptance and recognition of its crucial role in atmospheric correction. The kernels approach has provided a tool for BRDF to be introduced into standard atmospheric models, which leaves us with the question "What kernels are needed to deal with the land surfaces we are interested in?"

3. VIDEO DATA

Video data are used for low cost environmental monitoring. Such monitoring requires consistent and standardized information. However, a great deal of cost has been added to video data to try and achieve this consistency in the face of the BRDF and atmospheric effects, which compound calibration and instrument problems. The extensive hotspot and BRDF effects are a *major* problem for central perspective sensors like videos and cameras (even digital ones). Reducing this added cost is an important *commercial* objective.

Australian groups who have used empirical methods with video data are the Perth Minesite Rehabilitation group (Ong *et al.*, 1995) and the Alice Springs rangelands group (Pickup et al., 1995). The Alice Springs work involves a variant on the kernel function approach applied to images averaged along a run. The model is then used to balance the brightness variations in the individual frames. The Perth Minesite Rehabilitation group have been working to overcome spatial and angular variation by an innovative use of a base image with little or no brightness variation—in their case a Landsat image.

At CSIRO Mathematical and Information Science (CMIS), work is in progress to statistically analyze video images against known kernels using robust statistical methods. The objective for the video data producer is (low cost) seamless mosaics approximating reflectance. However, there are a number of research questions. For example, "Are there a few universally representative functions? Do they change with land cover"? Since the archive of video data may supply their own answer to these questions, it has been decided to explore them with the range of existing kernels.

4. CAL/VAL SITES

Calibration sites [and reflectance panels!] need BRDF models if they are to be successfully used with satellite (or even airborne) data. An Australian high reflectance calibration and aerosols site at Tinga Tingana in northern South Australia has been modeled by Denis O'Brien and Ross Mitchell at

CSIRO Division of Atmospheric Research (DAR) using a kernel model due to Staylor and Suttles (1986), as discussed in Cosnefroy *et al.* (1996).

At Tinga Tingana, the high reflectance and temporal consistency of the target meant BRDF dominated the AVHRR variation as the Sun and view angles changed over a one-year period. Validation sites also need good BRDF models. Fred Prata (also from DAR) is characterizing sites at Uardry near Hay in NSW and Amburla near Alice Springs, using innovative ground and tower based measurements. These efforts are part of a network of validation sites characterization and a new site in the north of Australia is planned to be established.

At many validation sites modeled so far around the world, simple kernel functions seem sufficient. However, for more complex land surfaces, atmospheric and BRDF effects will need to be separated and it is not clear whether and how consistent results will be obtained. Questions that arise in this activity are:
- Can all corrections be done with simple functions?
- If not, are there a few simple "forms" for specific land cover types (i.e., is there a "Typology" of land cover BRDF and associated kernels)?
- How does one separate atmospheric and BRDF effects?

These questions are crucial since it is one thing to characterize the relatively simple land surface of a calibration sites, a bit harder for a validation site and possibly very difficult for a general land surface. They are also very pertinent at a time when people are keen to establish a consistent and standardized set of environmental data series. BRDF effects can dominate such series of AVHRR, Landsat and other data that are coming on-line (even airborne data).

5. NDVI COMPOSITING AND CONSISTENT AVHRR DATA TIME SERIES

In particular, there has been a very useful discussion recently concerning AVHRR NDVI data. These data are a primary long-term data series that many people wish to use for environmental monitoring or environmental reporting. However, producers and users in Australia (Richard Smith, WASTAC, DOLA, WA) are understandably worried by the greening of deserts in winter. Li *et al.* (1996) have shown that the scene brightness can account for up to 30% of the variation *of NDVI* for some land covers. They have since established the effectiveness of simple kernel models for reducing this effect—but the problem is land cover dependent.

Following on from this work, Qi *et al.* (1996) and Qi and Kerr (1997) have discussed how scene brightness corrections using kernels should

18. SOME RESEARCH AND APPLICATIONS IN THE CSIRO (AUSTRALIA) EARTH OBSERVATION CENTRE ON SCENE BRIGHTNESS DUE TO BRDF

interact with Maximum Value NDVI compositing and atmospheric correction. Basically, they concluded the best choice of method is to correct for "BRDF" first, then composit and finally atmospheric correct. The reason for this is that the BRDF effect increases the NDVI, which leads to problems with the compositing. Essentially, the compositing (selecting pixel with maximum NDVI over a period) was introduced to select against cloud and edge pixels since these tend to have reduced NDVI. The resulting selected pixel from the compositing period supplies the Maximum Value NDVI. However, since atmospheric correction alone *increases* the BRDF effect, it makes the compositing result worse and results in pixels from the edges or with greater Sun/look variations become the selected pixels.

It seems that any attempt to provide consistent and standardized AVHRR data–especially for environmental monitoring–*must* account for BRDF and should *not* be atmospherically corrected without account of BRDF.

There is little doubt of the value or need for scene brightness correction in areas as diverse as AVHRR NDVI and video data mosaicking. In each case, however, we still have a fundamental issue of whether there is a consistent and simple typology of BRDF? In particular:

- Is there a consistent typology of BRDF, which relates to land cover structure?
- Is it representable by simple BRDF (e.g., Kernel) functions that can provide consistency and standards?
- Can remote sensing consistently monitor changes in the coefficients of the function and/or changes in functional form?
- Do the changes recorded key in to significant structural changes in the surface cover?

The last point raises the land cover structure question.

6. LAND COVER AND STRUCTURE

Land cover structure concerns vertical and horizontal spatial variation of components of the vegetation—or the 'gappiness' of canopies. It is a key element in the ecology of a landscape and a key element in fluxes of water, heat and carbon. Both in models of the ecology and the remote sensing, biomass is not enough! Land cover derived from spectral data is dominated by cover. However, it is the spatial distribution, gappiness and variance of canopies that is needed for monitoring major system changes.

To study scene brightness and structure in forests, CSIRO has also spent effort on obtaining airborne and field data in the past. For example, low and medium wide-angle aerial photography was collected at a well-measured site at Goonoo near Dubbo in NSW. The experiments used 120-degree aerial photography with an anti-vignetting filter, which allowed assessment of models rather than instrument characteristics. The Li-Strahler model with two spectral components fitted well. At high resolution, variance dominated the BRDF 'mean' signal. This scene variance is important and often dominates a single frame requiring frame stacking to average.

To overcome the high spatial variance in the photography, Daedalus Data were also flown for Goonoo. By flying into or at 90° to the Sun, it was possible to sense in or across the principal plane. By taking along scan averages of up to 1000 lines, a common Sun-sensor geometry was achieved for each scan line and each average. The SWIR and Thermal provided good reference BRDF channels for modeling, as they were not affected by atmosphere.

Since then, photography and scanner data of this type, as well as video and AVHRR data are all being collated and documented to help test the feasibility of the proposed land cover BRDF typology, and to assess the Australian structural typology using simple and detailed models. The capacity and potential for these studies to help with validation and interpretation of the global products that will come on-line in the near future is a prime opportunity and one of particular interest to the CSIRO EOC.

7. SEARCH FOR A STRUCTURAL TYPOLOGY

The search for consistency among photography, video data, scanner data and even AVHRR data is being pursued in a project being undertaken by CSIRO Information and Mathematical Sciences. The pilot for this work used data from the Daedalus Goonoo flights with the following results.

The data for Goonoo State Forest were collected by a Daedalus or DATM scanner (DATM stands for Daedalus Advanced Thematic Mapper scanner) for CSIRO DWR as a test for the Li-Strahler model. Unfortunately, there was a problem with the blue, green and red bands so the models were originally tested using the SWIR (Short Wave InfraRed, DATM bands 9 (see Figure 1), band 10 and Thermal band (DATM band 11).

18. SOME RESEARCH AND APPLICATIONS IN THE CSIRO (AUSTRALIA) EARTH OBSERVATION CENTRE ON SCENE BRIGHTNESS DUE TO BRDF

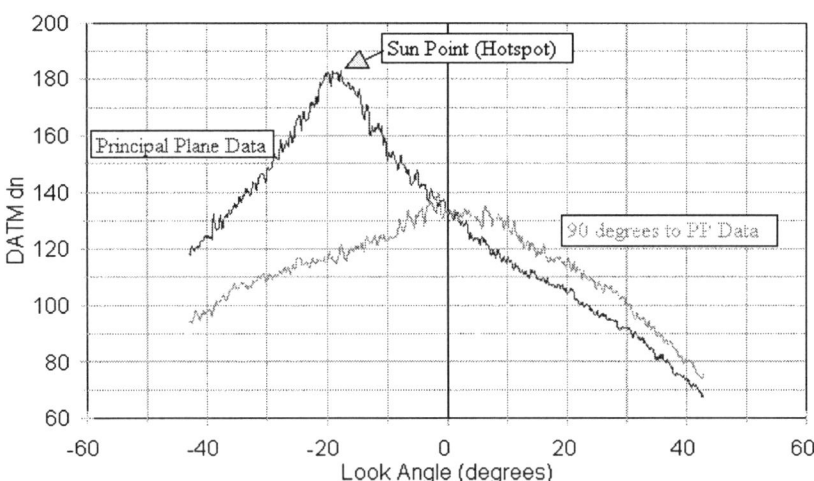

Figure 1. Bi-directional reflectance factors observed in the band 9 (SWIR) of the DATM at the Gonoo State Forest

However, this was no major problem as the model being tested assumed no atmospheric effects and no multiple scattering. The objective was to see if the geo-optical model was correct in the unadorned cases. The atmospheric effects in the SWIR and thermal are very small and do not contribute significantly to the directional signature. Also, in the SWIR and thermal the simple model with no multiple scattering was quite good. This is particularly so as both vegetation and shadows are respectively very dark and cool in these bands and have high contrast (at least at Goonoo in summer) with the sunlit background.

All of the models listed in Jupp and Strahler (1996) from Strahler *et al.* (1995) have been fitted for the current pilot study. In addition, the model due to Staylor and Suttles (1986) as discussed in Cosnefroy *et al.* (1996) has been fitted. This has had some favor with POLDER and atmospheric people. It was the kernel used by CSIRO scientists (Denis O'Brien and Ross Mitchell) to correct the Tinga Tingana data and should obviously therefore be included as a candidate in the testing.

The kernels are the "AMBRALS" set consisting of singly or combined models developed as described in Roujean *et al.* (1992) and Wanner *et al.* (1995) called:

- Ross Thin
- Ross Thick
- Roujean
- Li Sparse
- Li Dense
- Walthall

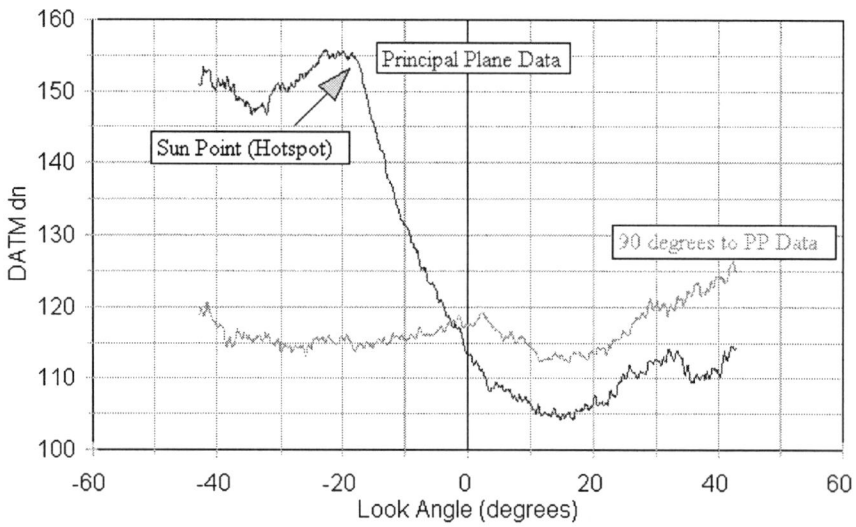

Figure 2. Bi-directional reflectance factors observed in the band 7 (NIR) of the DATM at the Gonoo State Forest

The NIR band(s) (see Figure 2) can also be used for kernel fitting and have been so. They were not originally used as the model being tested at Goonoo did not compute multiple scattering in the canopy. The SWIR and thermal provided ideal tests for the simple models and the multiple scattering present in the NIR is such that the Ross and Li-dense models should be much more useful than the simple Li-sparse which fits the SWIR Goonoo data very well. The main point is that these data are for the same landscape for which fieldwork had established that it was relatively low cover of trees with established structural properties.

18. SOME RESEARCH AND APPLICATIONS IN THE CSIRO (AUSTRALIA) EARTH OBSERVATION CENTRE ON SCENE BRIGHTNESS DUE TO BRDF

Table 1. Short-wave infrared data for Goonoo.

Model	k_0	k_1	k_2	SE	R^2	NSR
LS	177.2658	47.35126	0	4.936181	0.96135	0.53023
Rt+LS	182.3403	−24.989	49.96207	4.021908	0.97436	0.606878
RT+LS	185.8543	−114.387	56.59068	4.092269	0.973455	0.617782
SS	−352.131	−94.546	2225.06	10.25693	0.833239	1.18344
R	173.1894	138.8606	0	10.43852	0.827161	1.208809
Rt+R	168.2913	54.29317	136.8691	7.854958	0.902198	1.231747
RT+R	163.9187	181.4008	103.9104	8.051075	0.897253	1.265975
Rt+LD	221.0022	−96.1096	93.78764	8.39515	0.888284	1.326726
LD	188.122	68.27081	0	12.33672	0.758586	1.491802
W	135.3267	−83.1465	56.03012	11.84039	0.777776	1.999712
RT	129.2309	412.113	0	15.22707	0.632213	2.016965
RT+LD	181.0995	63.55604	59.79771	12.25507	0.761937	2.091149
Rt	114.9428	62.73437	0	23.81444	0.100411	7.915239

In Table 1, 'LS' stands for 'Li Sparse', 'Rt' for 'Ross Thin', 'RT' for 'Ross Thick', 'SS' for 'Staylor and Suttles', 'R' for 'Roujean', 'LD' for 'Li Dense', 'W' for 'Walthall'.

The statistics should be obvious except for the "*NSR*". The *NSR* or Noise to Signal Ratio is defined as:

$$NSR = \frac{100}{\sqrt{F}}$$

$$F = \frac{R^2}{1-R^2} \frac{M-p}{p}$$

where F is the usual F-Ratio with M as the number of data values, p as the number of parameters and this is used to apply the *NSR* criterion. The *NSR* criterion is an heuristic in which the model with minimum *NSR* is the best and/or most parsimonious model.

It seems that for the SWIR data, Li Sparse wins hands down possibly with the Ross Thin added (Figure 3 A). This is good as it is the one which fits this case from the physical point of view! There are some surprises (such as the changes in sign if the Ross Thin or Ross thick kernels are added to the Li Sparse) but it is early days in this testing yet. The next set of tests used the NIR DATM and in the future some models with and without atmospheres as well as video data will be tested in this way.

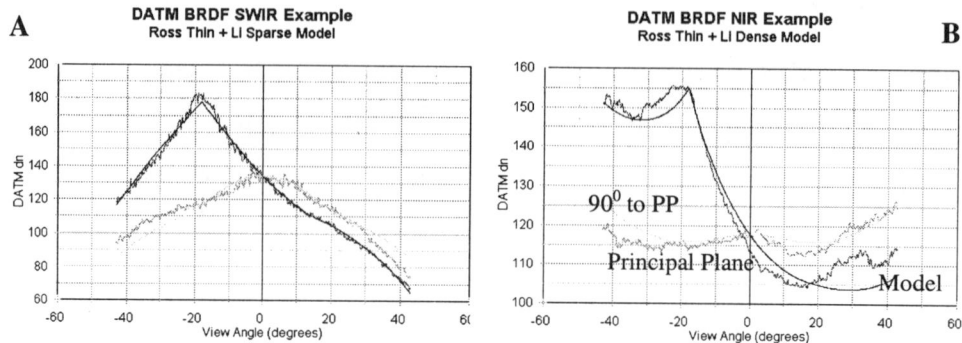

Figure 3. Evaluation of model performance against observations: A. SWIR Ross Thin + Li Sparse Model; B. NIR Ross Thin + Li Dense Model; dark gray line: principal plane data; light gray line: 90° to principal plane data; continuous smooth line: model

Table 2. Near infrared data for Goonoo

Model	k_0	k_1	k_2	SE	R^2	NSR
Rt+LD	147.079	47.78977	29.26081	3.502321	0.94246	0.924381
LD	163.4284	41.94886	0	5.698611	0.84756	1.12149
Rt+LS	131.0527	75.13833	12.64905	5.030711	0.881283	1.373089
SS	−198.333	−2.83037	1028.496	6.778957	0.784433	1.386745
RT+LD	159.7873	32.95353	37.55559	5.650526	0.850227	1.570175
Rt	113.9895	97.34751	0	7.784933	0.715508	1.667475
RT	127.2115	251.8626	0	8.01025	0.698802	1.736124
Rt+R	124.4304	95.69547	26.78684	6.42441	0.806392	1.833102
RT+R	116.577	322.5939	−31.8567	6.963711	0.772523	2.030073
W	117.6883	19.75226	38.71823	7.818818	0.713227	2.372214
RT+LS	127.8383	246.0352	0.626357	8.011406	0.698926	2.455387
LS	146.3111	20.49936	0	9.972013	0.533206	2.47427
R	133.0636	30.29701	0	13.71882	0.116527	7.281399

This time, there has been a big change. The Li Dense and Li-x plus Ross Thin (Figure 3 B) are the best choices. Note that these fits are for the same land cover. So it appears that we have an issue here. What do the different models mean? If it is the same land surface, why is there not one model?

We must face the fact that the two bands we have chosen resulted in such different kernels. To get a Li Dense in one case and a Li Sparse in the other when we know the forest is relatively sparse and structurally identical is somewhat confusing.

Table 3. Li Sparse + Walthall Fits for Goonoo

Data	Model	k_0	k_1	k_2	k_3	SE	R^2	NSR
NIR	LS+W	149.066	69.573	−4.8834	33.5135	2.51117	0.97044	0.79995
SWIR	LS+W	182.469	−8.2947	−9.4779	50.3514	4.05396	0.97397	0.74934

18. SOME RESEARCH AND APPLICATIONS IN THE CSIRO (AUSTRALIA) EARTH OBSERVATION CENTRE ON SCENE BRIGHTNESS DUE TO BRDF

How can we make sense of such a result when the objective is not just to fit data but to interpret the results in terms of the land cover? To try and overcome this, we also fitted the models with Li Sparse (which with its shape parameters at 2.0 matches the structure we know from associated field data, models and both DATM and photographic data) and the Walthall model. The reason for this was that the Walthall model could be seen as the tool to model the (symmetric) multiple scattering leaving the Li Sparse to fit the hotspot. The results are summarized as follows (noting that there is now an extra parameter making it a four parameter model):

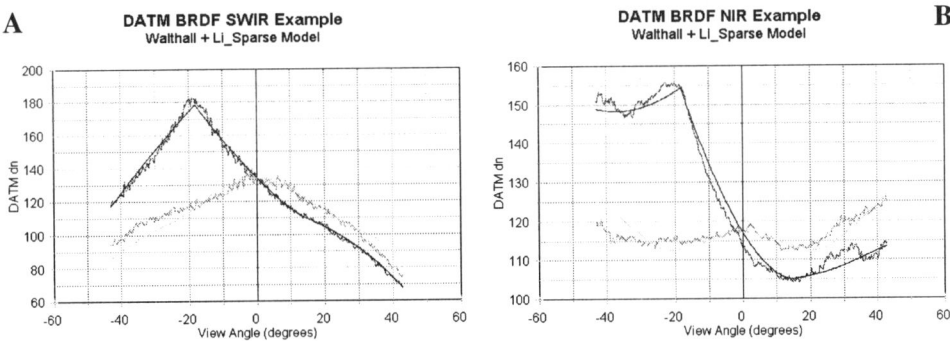

Figure 4. Evaluation of model performance against observations: A. SWIR Walthall + Li Sparse Model; B. NIR Walthall + Li Sparse Model; legend as in Fig. 3

The model provides a good fit in the NIR and the model is also competitive with the previous Li dense fit. The differences between the cases are simply noise-fitting, I suspect. There are some effects in these data such as the fact that the Principal Plane and 90° to PP data were not flown exactly together, nor perfectly in those directions. The result, however, is very satisfying in that the structure component is identical and the differences are due to the multiple scattering.

The resulting plots are shown as Figures 4 A and 4 B. If one separates the fitted components due to the Walthall and Li Sparse kernels (results not shown), it is clear that the Walthall model takes out a symmetric bowl shaped variation much like what is expected from the multiple scattering in the canopy. In the SWIR case, it is effectively flat and not present.

This shows that it *is* possible to use the 'same' structural model in more than one band—at least at Goonoo for this one case. Much more needs to be done to get a generally consistent result.

From this simple pilot study, it has been decided to develop a much more extensive study involving a wide range of data (including data sets from video systems, airborne scanners, Polder, ATSR and AVHRR) and analytic tools, which is now underway at CSIRO. For example, the testing methodology now also involves using Ridge and eigenvalue analysis to examine correlations between the kernels and the effects this introduces. This study will be reported and its findings may help in the development of a "global" BRDF typology in the future.

8. ACKNOWLEDGEMENTS

Norm Campbell's group in Perth, especially Harri Kiiveri, Fiona Evans and Suzanne Furby have developed a systematic testing process from which the results quoted here form a small part. Fred Prata, Denis O'Brien and Ross Mitchell have supplied data, knowledge and activity for this EOC Thread activity as have Vanessa Chewings, Peter Hick and Cindy Ong from their extensive holdings of video data of many land covers and many Sun and view angles.

9. REFERENCES

Cosnefroy, H., Leroy, M. and Briottet, X. (1996) Selection and characterization of Saharan and Arabian Desert sites for the calibration of optical satellite sensors, *Remote Sensing of Environment*, **58**, 101–114.

Jupp, D.L.B. and Strahler, A.H. (1996) *Image Brightness and BRDF Workshop Issues.* EOC Discussion Paper, http://www.eoc.csiro.au/.

Jupp, D.L.B. (1996) *Issues in Reflectance measurement.* EOC Discussion Paper, http://www.eoc.csiro.au/.

Li, Z., Cihlar, J., Zheng, X., Moreau, L. And Ly, H. (1996) The bidirectional effects of AVHRR measurements over Boreal regions, *IEEE Transactions on Geoscience and Remote Sensing*, **34**, 1308–1322.

Ong, C., Hick, P., Craig, M., Warren, P. and Newman, C. (1995) A correlative technique for correction of shading effects in digital multi-spectral video imagery, *Proceedings ISSSR*, Melbourne, November 1995.

Pickup, G., Chewings, V.H. and Pearce, G. (1995) Procedures for correcting high resolution airborne video imagery, *International Journal of Remote Sensing*, **16**, 1647–1662.

Qi, J. and Kerr, Y.H. (1997) On current compositing algorithms, *Remote Sensing Reviews*, **15**, 235–256.

Qi, J., Kerr, Y.H., Moran, M.S., Sorooshian, S. (1996) Bidirectional and atmospheric consideration in compositing multi-temporal AVHRR data over Hapex Sahel experimental sites, USDA-ARS Water Conservation Laboratory, Phoenix, Arizona; CESBIO, Toulouse, France, Dept. of Hydrology and Water Resources, The University of Arizona, Tucson, AZ, http://jadito.tucson.ars.ag.gov/~qi/projects/vgt_vgt/intrpt.html#APDXC.

18. SOME RESEARCH AND APPLICATIONS IN THE CSIRO (AUSTRALIA) EARTH OBSERVATION CENTRE ON SCENE BRIGHTNESS DUE TO BRDF

Rahman, H., Pinty, B. and Verstraete, M.M. (1993) Coupled surface-atmosphere reflectance (CSAR) model 2. Semi-empirical surface model useable with NOAA Advanced Very High Resolution Radiometer data, *Journal of Geophysical Research*, **98**, 20,791–20,801.

Roujean, J-L., Leroy, M. and Deschamps, P.Y. (1992) A bidirectional reflectance model of the Earth's surface for the correction of remote sensing data, *Journal of Geophysical Research*, **97**, 20,455–20,468.

Staylor, W.F. and Suttles, J.T. (1986) Reflection and emission models for deserts derived from Nimbus-7 ERB scanner measurements, *Journal of Climate and Applied Meteorology*, **25**, 196–202.

Strahler, A.H., Barnsley, M.J., d'Entremont, R., Hu, B., Lewis, P., Li, X., Muller, J-P., Barker Schaaf, Wanner, W. and Zhang, B. (1995) *MODIS BRDF/Albedo Product*, Algorithm Theoretical Basis Document, Version 3.2, NASA EOS, May 1995.

Wanner, W., Li, X. and Strahler, A.H. (1995) On the derivation of kernels for kernel-driven models of bidirectional reflectance, *Journal of Geophysical Research*, **100**, 21,077–21,090.

Chapter 19

REMOTE SENSING OF ALBEDO USING THE BRDF IN RELATION TO LAND SURFACE PROPERTIES

W. Lucht (1), C. Schaaf (1), A. H. Strahler (1) and R. d'Entremont (2)
(1) Center for Remote Sensing, Boston University, Boston, USA and (2) Atmospheric and Environmental Research, Inc., Cambridge, USA.

1. INTRODUCTION

Albedo characterizes the radiometric interface of the Earth's land surface with the atmosphere. It is defined as the fraction of downwelling solar shortwave radiation reflected back into the atmosphere from the surface. The numerical value of land surface albedo depends mainly on two things: the optical properties of the surface and its three-dimensional spatial structure. The question to be investigated in this paper is the following: Given that albedo may be retrieved from space using multiangular remote sensing data through the inversion and integration of a bidirectional reflectance distribution function (BRDF) model, can the magnitude of albedo and its dependence on solar zenith angle be related to land surface optical and structural properties?

2. THE ROLE OF ALBEDO IN THE GLOBAL SYSTEM

Land surface albedo plays an important role in the global biogeochemical, meteorological and climatological system for the following three reasons.

- Land surface albedo quantifies the energy reflected back into the atmosphere by the surface. It defines the radiometric lower boundary for any short-wave radiative transfer problem in the atmosphere, linking surface vegetation, soils and topography radiometrically to the meteorological system, and influencing the top-of-atmosphere energy balance, cloud formation and precipitation, feedback loops operating through atmospheric convective patterns, and general atmospheric circulation.
- Where the surface is vegetated, the value of albedo quantifies the short-wave solar energy input into the biosphere, its main source of energy. It influences the land surface energy balance through influencing the latent heat flux due to evapotranspiration from vegetation.
- Albedo is largely determined by the optical and structural properties of vegetated and non-vegetated surfaces. In vegetated areas, it may therefore potentially be used as an indicator of vegetation conditions, and changes in albedo may be related to changes in the vegetation, both structurally and in terms of green matter, quantified for example through the Leaf Area Index (LAI). The three-dimensional structural properties of vegetated and non-vegetated surfaces are of particular interest for an estimation of surface roughness length, which drives energy and momentum exchanges between the surface and the dynamic meteorological processes in the atmosphere.

In spite of this importance of land surface albedo in the global system, currently no global albedo data set approaching kilometer-scale resolution is available. Global circulation models use static data sets mostly driven by land cover type and derived from a variety of sources of often unknown quality and based on various modeling assumptions, or compute albedo from vegetation physiological parameters using very simple radiative models.

However, space sensors are now becoming available that allow the routine production of global albedo products in the very near future. These are the European POLDER, MSG and MERIS, and the American MODIS and MISR instruments. The algorithmic basis for deriving albedo from these sensors exploits new methods making use of the observed bidirectional reflectance distribution functions of the Earth's surface. With a few compromises on calibration, geolocation and accuracy of the atmospheric correction, these methods are also applicable to the AVHRR, which is especially important in view of the long data record available from that instrument.

3. DERIVING ALBEDO FROM MULTIANGULAR REFLECTANCE OBSERVATIONS

Intrinsic land albedo may be derived in a form decoupled from atmospheric conditions from integrating the BRDF over the hemispheres of viewing and illumination angle, yielding directional-hemispherical (black-sky) or bihemispherical (white-sky) albedo, respectively. The BRDF can be derived from atmospherically corrected multiangular remote sensing observations of reflectance through the use of a BRDF model that parameterizes the vegetation effects influencing the BRDF and albedo. Operationally, semi-empirical BRDF models describing the BRDF with three independent parameters are currently the most practical option at the kilometer scale. Several feasible and validated models are available (Roujean et al., 1992; Rahman et al., 1993; Wanner et al., 1995). Even though semi-empirical BRDF models describe the physics of light scattering in vegetation only very approximately or even mostly empirically, there is some evidence linking model parameter values with general land surface type (Engelsen, 1996; Hu et al., 1997) or vegetation properties (Hyman and Wanner, 1997; Roujean et al., 1997; Disney and Lewis, 1998).

BRDF and albedo of vegetated surfaces are a function of the optical properties and three-dimensional spatial structure of vegetation. In order to recover these from an inversion, one approach to the modeling is to separate leaf-level and crown-level effects. They reflect different although often coupled aspects of the underlying scattering process, one being related mainly to the LAI and one mainly to vegetation structure. This approach is taken by the Ambrals BRDF model, scheduled to produce the MODIS BRDF/albedo product globally with 1 kilometer spatial resolution in 7 spectral bands (spanning the visible and near-infrared), once every 16 days, after the launch of MODIS in December 1999 (Wanner et al., 1997).

The rationale of the modeling, developed by Roujean et al. (1992) specifically for remote sensing applications, is as follows. The reflectance of a scene is described as consisting of a linear superposition of the following three components.

- An overall isotropic scene reflectance is given as a Lambertian constant.
- Leaf-level volume scattering, dependent both on leaf optical properties (reflectance and transmission) and their spatial density or leaf area index (intra-crown gaps), is calculated from approximations to radiative transfer theory by Ross (1981). It is given by the so-called RossThick kernel function k_{vol} (Roujean et al., 1992).
- Crown-level surface scattering due to the three-dimensional structure of individual plants, either in the form of protrusions or gaps in otherwise

dense stands, leading to shadow casting and mutual shadowing, is given by approximations to the geometric-optical theory by Li and Strahler (1992). It is given by the so-called LiSparse kernel k_{geo} (Wanner et al., 1995).

The parameters of the model are the respective linear weights f of these kernels k in modeling the bidirectional reflectance of a scene:

$$R(\theta_i, \theta_v, \phi) = f_{iso} + f_{vol} k_{vol}(\theta_i, \theta_v, \phi) + f_{geo} k_{geo}(\theta_i, \theta_v, \phi)$$

where θ_i and θ_v are the viewing and illumination zenith angles, and ϕ is the relative azimuth between the viewing and illumination directions. The inversion of this model reduces to a matrix inversion. Since the model is linear, the well-developed apparatus of linear theory may be applied to the evaluation of model properties, parameter sensitivity etc. Note that the kernels themselves represent nonlinear functions of the angles.

Albedo in the form of integrals of the BRDF is given by a similar expression:

$$a(\theta_i) = f_{iso} + f_{vol} K_{vol}(\theta_i) + f_{geo} K_{geo}(\theta_i)$$

where K are the integrals of the kernels over the viewing hemisphere. Thus, the albedo model has the same parameters as the reflectance model, as the same scattering process determines both quantities. The functions $K(\theta_i)$ can be parameterized by third-order polynomials in θ_i, allowing computations of $a(\theta_i)$ from the model parameters f without direct reference to the BRDF model used.

4. ALBEDO FROM AVHRR AND GOES OBSERVATIONS

Albedo retrievals were performed building upon work by d'Entremont (1997). Daily NOAA-14 AVHRR red band data and hourly GOES-8 imager visible band data with nominal spatial resolutions of one kilometer were combined for a period in September 1995 (AVHRR, September 2-18; GOES, August 25 to September 6), over New England in the United States (40.6 to 44.2° North, 69.7 to 74.6° West). The data were carefully calibrated and mapped to 1-km pixels using orbital models, resulting in a 400 by 402 grid (160,800 pixels) with mapping residuals of around 0.4 km. The data were meticulously and conservatively cloud-cleared, using 8 spectral tests on all 5 bands of the AVHRR and temporal differencing on the hourly geostationary GOES-8 data. Cloud shadows and pixels immediately adjacent to clouds were also removed. Atmospheric corrections were performed using visibility data from ground stations in the area to characterize aerosol optical depth and the 6S atmospheric radiative transfer code (Vermote et al., 1997). The AVHRR data display an almost constant solar angle while the viewing zenith angle varies from one orbit to another with observations located at constant relative azimuth. The GOES data are characterized by a constant viewing zenith angle and varying solar zenith and relative azimuth angles.

The resulting multiangular reflectance observations data set was inverted using the reciprocal version of the RossThick-LiSparse BRDF model (Wanner et al., 1995; Lucht, 1998) on all pixels with at least 7 observations. Figure 1 shows, in panel A, the black-sky albedo derived from the BRDF model parameters for a solar zenith angle of 0 degrees. Since the observations were made at solar zenith angles between about 40 and 60 degrees, this image represents an extrapolation of the albedo observations to an angular range not originally sampled, through the inversion of a BRDF model. Black-sky albedo at all solar angles is relevant to land surface modeling with respect to diffuse-sky situations.

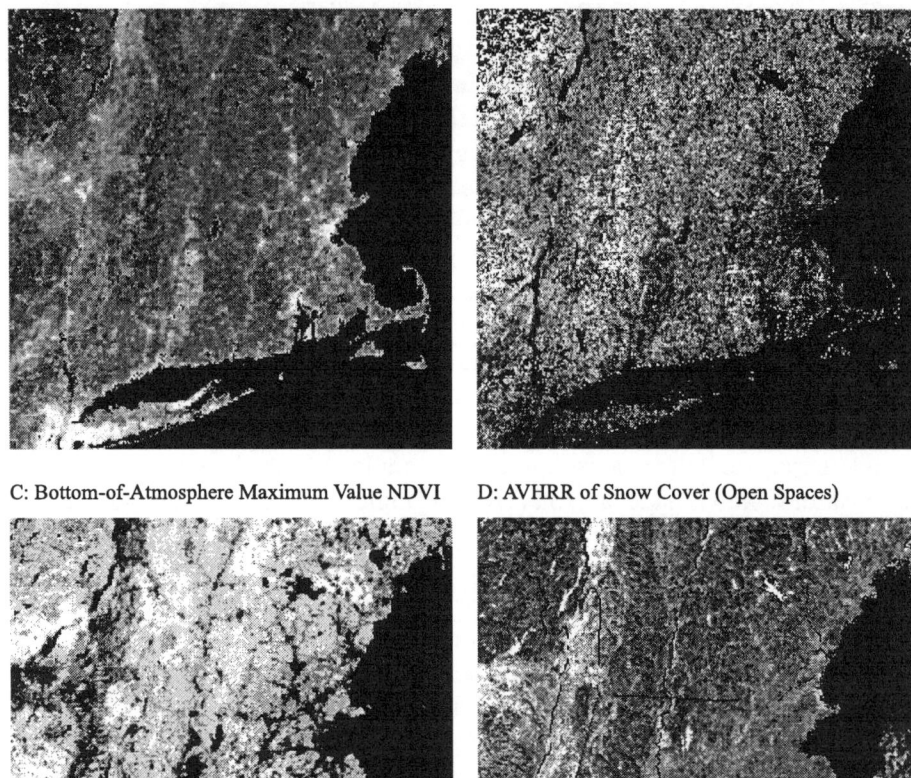

Figure 1. Ross-Li BRDF model inversions of combined AVHRR and GOES observations of New England from September 1995. Red band, nominal spatial resolution 1 km. A: Black-sky albedo for a solar zenith angle of 0°; linear gray scale from 0.0 to 0.8. Since the observations were made at solar zenith angles between 40 and 60°, this image demonstrates extrapolation in solar angle using the BRDF model. Black-sky albedo at all solar angles is relevant for albedo computations for diffuse or cloudy skies. Black areas are mostly ocean or inland water, and in a few cases computed albedos out of range. B: The ratio of black-sky albedo at 45 and at 0° solar zenith angle; linear gray scale from 0.8 to 1.8. C: Maximum-value composited AVHRR NDVI after atmospheric correction; linear gray scale from 0.8 to 0.95. D: AVHRR RGB (red/NIR/red) composite of 15 January 1996 18:15 GMT, gray scale, taken after a snowstorm covered the region. The image shows where barren and coniferous trees cover the region, and where the land is open.

At these coarse scales, the resultant albedos are nontrivial to validate with sufficient accuracy, but the spatial patterns display a clear relationship to features of the underlying landscape, indicating that the retrievals and subsequent considerable extrapolation in solar zenith angle do not lead to noisy or otherwise ill-determined results. For example, changes in the number of observations in different areas, due to differences in cloud cover, are not evident in the albedo. The bright areas seen are those of cities. For example, Boston may be identified on the central east coast, Providence at the end of an inlet west of Cape Cod, New York City and Long Island are on the lower left. A web of bright urban built-up areas along the major highways is visible. The relatively built-up Connecticut river valley in the lower center, the Hudson river valley running north-south along the left side of the image, and the Mohawk river valley with its agricultural lands running east-west on the western left edge of the image show bright surfaces as well. Dark areas also correspond to surface features: the dark region in the upper left corner is the Adirondacks area, which is heavily forested, and one can see the Berkshires with their forests running north-south between the Connecticut and the Hudson rivers. The woody Catskills are to the lower left. Generally, New England is densely vegetated, in many parts with dense forest. Small details, like the dark spur near the northern end of the Connecticut river valley, as well as bright spots representing small towns throughout the image correspond reliably to ground features as identified, for example, on Landsat imagery (Schaaf et al., 1998).

The physical properties of the landscape viewed are more reliably indicated by the images shown in panels C and D of Figure 1. The first shows the Normalized Difference Vegetation Index (NDVI), computed from the atmospherically corrected reflectances and maximum value composited. Some residual gaps remain in this image as the NDVI can be computed only for pixels which had at least one observation from the AVHRR. The image confirms the densely vegetated state of the region and the effects of human impact in urbanized areas, along highways, and in some agricultural areas associated with the river valleys. Panel D is a red-NIR composite of New England acquired by the AVHRR on January 15, 1996 after a snowstorm had blanketed the region. Trees in the region are mostly barren, although coniferous species grow in some areas or are part of a mixed forest. The snow has fallen from the trees to the ground, as is evident from the obscuration of the bright snow background in the vegetated areas. This image is a map of large-scale openness of the landscape, indicating where the background is freely visible and where it is hidden by trees. The image demonstrates that indeed the Connecticut river valley is more fractured in terms of canopy closure than the surrounding forests, that there are open areas along the Hudson river valley and in the Mohawk valley area. The city of Boston is not very much in evidence in this image, Cape Code is obscured by clouds.

Comparing panels C and D with the albedo shown in panel A confirms the conclusion that, from visual inspection, the albedo derived from per-pixel AVHRR and GOES inversions are reasonable in that the results obtained show a distinct relationship to landscape features. If the BRDF inversions were unreliable, the albedo image would be expected to be less clear, especially given the angular extrapolation shown. Similar images of albedo at other solar zenith angles show that albedo increases when the Sun approaches the horizon, and that contrast between different surfaces diminishes. Not unexpectedly, the white-sky albedo is very similar to the black-sky albedo at 45° solar zenith angle. An image of the ratio between nadir reflectance and albedo also shows that the transformation of reflectance into albedo is dependent on the surface features viewed, due to the associated differences in the BRDF. Using the modified Rahman-Pinty-Verstraete BRDF model arrives at somewhat higher values for albedo but otherwise similar results.

Of particular interest is panel B of Figure 1 which shows not the magnitude of albedo but its functional shape. The image shows the ratio of black-sky albedo at 45 and that at 0° solar zenith angle, demonstrating regional differences in the solar zenith angle-dependence of the albedo. These are also related to landscape properties, due to differences in the underlying BRDF. Dark areas indicate a weak dependence on solar zenith angle, bright areas an increase in the albedo with solar zenith. Generally, an increase is expected from volume scattering as from dense plant canopies while the shadowing effects of sparser canopies should lead to a smaller increase or even a slight decrease. This expectation is fully met by the results shown. The heavily wooded areas display a strong increase (values of the ratio of up to 1.8) while the urban and urbanized areas with their gaps in the vegetation cover and buildings show the effects of shadowing (ratios between 0.8 and 1.2). An interesting case is the Cape Cod region, which is the only area in the image that is vegetated but sparsely, due to sandy soils. It promptly shows decreased volume scattering.

Note that while the spatial distribution of the albedo ratio is supported by the NDVI and snow gap images, it is not a simply redundant retrieval of information. Details of the image show a wealth of spatial information that differs from that shown in panels C and D, reflecting the unique spatial distribution of surface reflectance properties of the visible band used. It is hoped that future work will allow evaluation of this surface-related information, making available an additional data dimension for the inference of land surface biophysical properties.

5. INTERPRETING ALBEDO IN TERMS OF SURFACE PROPERTIES: AN IDEA

The results shown above demonstrate that the shape of the BRDF, here expressed as the magnitude of the solar zenith angle-dependence of albedo, is related to the amount of vegetation present and to the gaps in the canopy as well as urban protrusions, causing either radiative transfer-type scattering by the canopy leaves or visibility of shadows in the scene due to its three-dimensional structure.

This leads to speculation as to whether the BRDF model parameters for volume and for geometric scattering may be directly interpretable in terms of these types of scattering. If this were the case, then one could attempt a subsequent step of renormalizing each type of scattering empirically to biophysical qualities of interest to land surface modeling. Particularly, the mathematical expressions for the kernels and the retrieved parameters reveal that they are related to LAI and surface structure (Wanner et al., 1995). The strength of volume scattering is theoretically related to the leaf area index as:

$$f_{vol} = C_1[1 - \exp(-C_2 LAI)]$$

Therefore,

$$LAI = -\ln(1 - f_{vol} / C_1) / C_2$$

where C_1 and C_2 are constants (overall brightness and extinction coefficient). The geometric-optical scattering coefficient is related theoretically to the three-dimensional structure of the surface as

$$f_{geo} = C_3 n d^2 / L^2$$

Therefore

$$SCI = d / (L / \sqrt{n}) = (f_{geo} / C_3)^{1/2}$$

where C_3 is a constant related to average crown reflectance, n/L^2 is crown, object or gap density (n crowns or gaps per area L^2) and d is a typical apparent crown, object or gap diameter as implied by the shadowing seen. The ratio, $d/(L/\sqrt{n})$, which we will call the shadow-casting index (*SCI*), may then be taken to be something like a roughness length, the ratio of typical object dimension (height) to the typical distance between objects or gaps.

For a given scene, expected maximal and minimal values of the *LAI* and of plant density could be used to empirically determine the factors C_1, C_2 and C_3 from the observed extremes of the parameters f. Using these constants, the BRDF model parameters would be nonlinearly rescaled to reflect more physical parameters.

In order to establish whether such a scheme would work in practice, the following requirements would have to be met:
- The spatial distribution of the model parameters would have to show a definite relation to the spatial distribution of surface properties.
- Coupling between model parameters due to partial non-linearity between kernels should be discounted as a source of the relationships seen, and
- Since LAI and surface structure are both wavelength-independent quantities, application of the method to the red and near-infrared band should produce similar results.

There is some evidence in the literature that a partial interpretation of kernel-based BRDF model parameters in this way may be possible (Barnsley et al., 1997; Hyman and Wanner, 1997; Roujean et al., 1997; Disney and Lewis, 1998), but the potential and limitation are still not fully understood. In all likelihood, we will have to wait for more advanced evaluation of POLDER data, and for the data stream from MODIS and MISR, before being able to give a definitive answer. For the New England data set shown here, volume scattering is found to be consistently low and geometric scattering high in the urban areas, the opposite holding for the dense forests, as expected and consistent with this proposal. However, GOES does not provide a near-infrared channel for an independent check of these results, and the question of kernel coupling under conditions of sparse angular sampling still has to be explored further. For this reason, we refrain from presenting these results here, but would like to suggest that this line of thought may warrant further investigation from an improved and perhaps more diverse data set providing at least two spectral bands.

6. CONCLUSIONS

It was the purpose of this paper to consider whether the biospherically and climatologically important quantity of land-surface albedo may be derived from multiangular remote sensing at 1 km nominal spatial resolution and how it may reflect land surface properties. We have given an example from our more extended analysis showing that the retrieved albedo magnitude and solar zenith angle dependence, both associated with the BRDF of the surface pixels viewed, follow land surface properties as identified by the NDVI of the region studied, a snow-based obscuration/gap-image, and the known location of cities and other human activity affecting the natural vegetation cover.

The findings indicate that the angular dependence of land surface reflectance may perhaps be utilized to investigate surface properties such as LAI through volume scattering or surface roughness through shadow casting effects in a semi-empirical way, whether directly through the ratio of albedo or reflectance at various angles, or indirectly through mappings of empirically rescaled model parameters. We therefore propose that these experimental results suggest that further inquiry into this area of research is warranted. The benefits could be substantial if techniques are successfully developed.

7. ACKNOWLEDGEMENTS

This work was supported by NASA's MODIS project under NAS5-31369.

8. REFERENCES

Barnsley, M. J., P. Lewis, M. Sutherland, and J.-P. Muller (1997) Estimating land surface albedo in the HAPEX-Sahel southern super-site: Inversion of two BRDF models against multiple angle ASAS images, *Journal of Hydrology*, **188–189**, 749–778.

Disney, M., and P. Lewis (1998) An investigation of how linear BRDF models deal with the complex scattering processes encountered in a real canopy, *Proceedings of the International Geoscience and Remote Sensing Symposium'98*, CD-ROM.

Engelsen, O., B. Pinty, M. M. Verstraete, and J. V. Martonchik (1996) *Parametric bidirectional reflectance factor models: Evaluation, improvements and applications*, Report, Joint Research Centre of the European Commission, EU **16426**, 114 pp.

d'Entremont, R.P. (1997) *Meteorological applications of surface bidirectional reflectance distribution functions retrieved from satellite data*, Dissertation, Boston University, Boston, MA 02215, 129 pp.

Hu, B., W. Lucht, X. Li, and A. H. Strahler (1997) Validation of kernel-driven models for global modeling of bidirectional reflectance, *Remote Sensing of Environment*, **62**, 201–214.

Hyman, A. and W. Wanner (1997) Relationships between semi-empirical BRDF model parameters and land cover type, *Proceedings of the Remote Sensing Society, UK Annual Study Conference'97*, 30–35.

P. Lewis and E. Vives Ruiz de Lope (1997) The application of kernel-driven BRDF models and AVHRR data to monitoring land surface dynamics in the Sahel, *Journal of Remote Sensing (China)*, **1**, Supplement, 155–161.

Li, X., and A. H. Strahler (1992) Geometric-optical bidirectional reflectance modeling of the discrete crown vegetation canopy: Effect of crown shape and mutual shadowing, *IEEE Transactions on Geoscience and Remote Sensing*, **30**, 276–292.

Lucht, W. (1998) Expected retrieval accuracies of bidirectional reflectance and albedo from EOS-MODIS and MISR angular sampling, *Journal of Geophysical Research*, **103**, 8763–8778.

Rahman, H., B. Pinty, and M. M. Verstraete (1993) Coupled surface-atmosphere reflectance (CSAR) model, 2, Semi-empirical surface model usable with NOAA Advanced Very High Resolution Radiometer data, *Journal of Geophysical Research*, **98**, 20,791–20,801.

Roujean, J. L., M. Leroy, and P. Y. Deschamps (1992) A bidirectional reflectance model of the Earth's surface for the correction of remote sensing data, *Journal of Geophysical Research*, **97**, 20,455–20,468.

Roujean, J. L., D. Tanré, F. M. Bréon, and J. L. Deuzé (1997) Retrieval of Land Surface Parameters from POLDER BRDF data during HAPEX-SAHEL, *Journal of Geophysical Research*, **102**, 11,201–11,218.

Ross, J. K. (1981) *The Radiation Regime and Architecture of Plant Stands*, 392 pp., Dr. W. Junk Publishers, The Hague.

Schaaf, C. B., W. Lucht, R. P. d'Entremont, and A. H. Strahler (1998) Relationship between land surface properties and BRDF/albedo parameters using satellite data, *Proceedings of the International Geoscience and Remote Sensing Symposium'98*, 1277-1279.

Vermote, E.F., D. Tanré, J.L. Deuzé, M. Herman, and J.-J. Morcrette (1997) Second simulation of the satellite signal in the solar spectrum, 6S: An overview, *IEEE Transactions on Geoscience and Remote Sensing*, **35**, 675–686.

Wanner, W., X. Li, and A. H. Strahler (1995) On the derivation of kernels for kernel-driven models of bidirectional reflectance, *Journal of Geophysical Research*, **100**, 21,077–21,090.

Wanner, W., A. H. Strahler, B. Hu, P. Lewis, J.-P. Muller, X. Li, C. L. Barker Schaaf, and M. J. Barnsley (1997) Global retrieval of bidirectional reflectance and albedo over land from EOS MODIS and MISR data: Theory and algorithm, *Journal of Geophysical Research*, **102**, 17,143–17,162.

Chapter 20

EXPERIMENTAL STUDY OF STATISTICAL CHARACTERISTICS OF PLANT CANOPY RADIATION REGIME

M. Sulev and J. Ross
Tartu Observatory, Toravere, Estonia.

1. INTRODUCTION

During the last decades, Swedish scientists have studied different aspects of willow forest growth. Some papers on radiation measurements, concerning methods of measurements (Perttu, 1970; Lindroth, Perttu, 1981) and light penetration in willow stand (Eckersten, 1984) have also been published.

In the joint Swedish-Estonian fast growing forest research project, an extensive study of the radiation regime of willow plantation was planned, including global radiation, photosynthetically active radiation (PAR) and net radiation measurements.

Radiation and phytometrical measurements were carried out during the growing periods of 1994–1997 in the willow Salix viminalis (clone 78021) plantation at Tartu Observatory ($\varphi = 58°$ 16', $\lambda = 26°$ 28'), Estonia. For more details see Ross, 1994 and Koppel et al., 1996.

Due to the great spatial variability of forest ecosystems, horizontal averaging of radiation measurements inside the plant canopy constitutes an important problem. Most radiative transfer approaches (e.g., Ross, 1981) consider radiation characteristics as a function of height z and suppose that averaging over xy-plane has been made with the necessary accuracy. However, measurements with one or two sensors at fixed points do not guarantee this. Different methods of horizontal averaging have been used (see reviews by Ross, 1981 and Campbell and Norman, 1989, Norman and Campell, 1989): linear sensors (Fassnacht et al., 1994), a set of sensors located at chosen sites inside the canopy (Pearcy et al., 1990; Sassenrath-Cole, 1995), or distributed along attached rod (e.g., ceptometer, Fassnacht et al., 1994), moving sensors on rod (Niilisk, 1969; Laisk, 1968, Tooming, reviewed in Ross, 1981), moving trams (Baldocchi et al., 1984; 1986, Chen, Black, 1992) or traversing system (Allen, Lemon, 1972).

In this work, linear averaging across the row direction using the moving sensor on a horizontal rod has been used. The following quantities were also measured on a 12-meter mast above the canopy: the irradiance, the global radiation Q, the photosynthetically active global radiation $PAR\ (Q)$, the corresponding reflected fluxes and net radiation, as well as the direct solar radiation S and $PAR\ (S)$.

2. MEASUREMENTS AND DATA PROCESSING

The measurement device consists of a 600 cm long horizontal bar, the height of which can be changed inside the forest (Fig. 1). A sensor carriage is moving at the speed of 3 cm/sec along the bar. The following instruments were mounted on the carriage: two Reemann pyranometers TR-3 to measure downward (Q) and upward fluxes (R) of global radiation, two LI-COR quantum sensors LI-190SA for the measurement of downward ($PAR\ Q$) and upward ($PAR\ R$) fluxes, a Reemann miniature net radiometer MB-1 for the measurement of net radiation (B), and a special instrument to measure sun flecks.

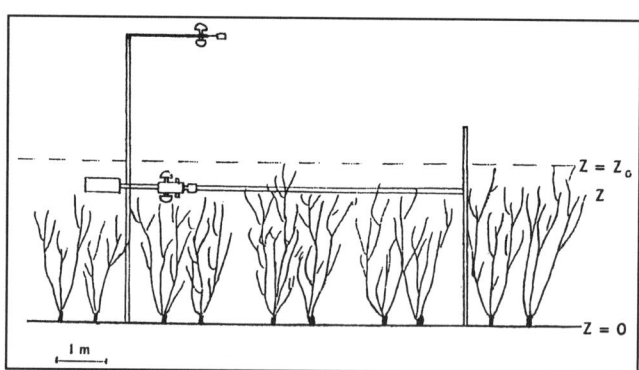

Figure 1. The radiation measurement system for detailed investigation of the radiation regime of the willow forest.

As indicated earlier, global radiation $Q(z_0)$, $PAR\ Q(z_0)$ and net radiation measurements are simultaneously recorded above the forest at the height of about 12 meters, with a Reemann pyranometer, a LI-COR quantum sensor and a Reemann net radiometer, respectively.

The data acquisition system consists of Delta-T Devices, a multi-channel data-logger and a Notebook PC U865x-25. The mean values, standard deviations and flux density distribution functions were calculated during a preliminary data processing phase. To calculate the flux densities in the PAR, the sensitivity of LI-COR quantum sensor provided with the instrument certificate was used. A conversion coefficient of $2.174\ 10^{-4}$ was used to convert the PAR flux densities expressed in $\mu mol\ s^{-1}\ m^{-2}$ to $kW\ m^{-2}$ (Sulev, Ross, 1996). Both the Reemann pyranometer and the net radiometer were calibrated against the Sun, using a normal pyrheliometer, i.e., their sensitivities were linked to the International Pyrheliometric Scale.

Simultaneously with these radiation measurements, air temperature and humidity at a height of 150 cm inside and outside the forest, wind speed above the canopy, and ground surface temperature beneath the forest were acquired.

3. RESULTS AND DISCUSSION

Measurements above the canopy at a height of about 12 meters enable to determine the daily course of incoming radiation fluxes, the fraction of the PAR in the global fluxes, and also the albedo of the plantation. An example of these data is presented in Fig. 2 and 3. The variability of these curves is caused by the presence of slight C_i clouds. The relative part of global PAR $PAR(Q)/Q$ exhibits a minimum of about 0.45 around noon, and increases to 0.5 at solar heights less than 10°. For direct solar radiation, the relation increases with solar height and reaches a maximum value of about 0.38 around noon. The relative part of PAR in the reflected radiation depends slightly on solar height and has a maximum of 0.10 for a solar height between 15 and 30°. The albedo for global radiation Ak has an obvious daily variation with a minimum value of about 0.24 at noon. The albedo in the PAR region is much less (about 0.04 at noon) and also increases when solar height decreases. The asymmetry of the curves is mainly due to azimuthal non-homogeneity of the willow plantation structure, and possibly also to changes in the plant structure and optical properties of leaves during the day.

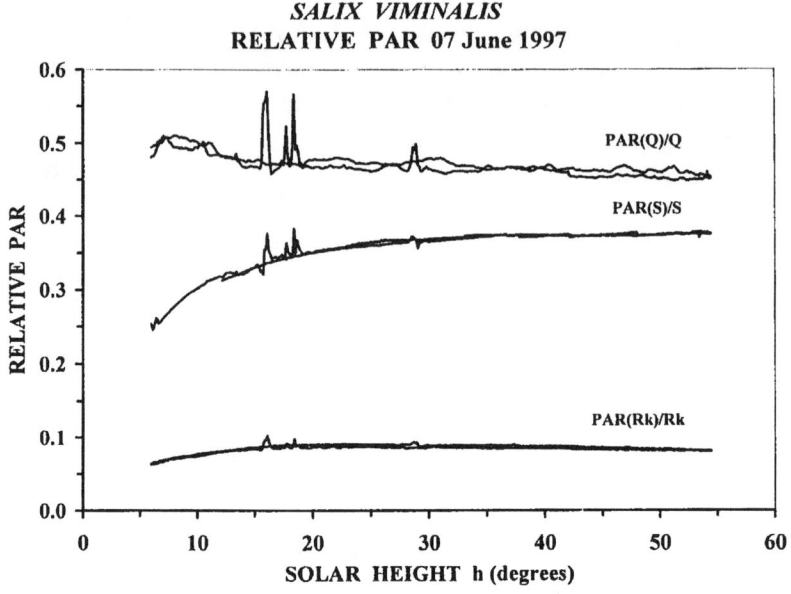

Figure 2. Relative radiation fluxes in the PAR region above the willow forest in Toravere, June 7, 1997. Clear sky with slight C_i clouds in the afternoon

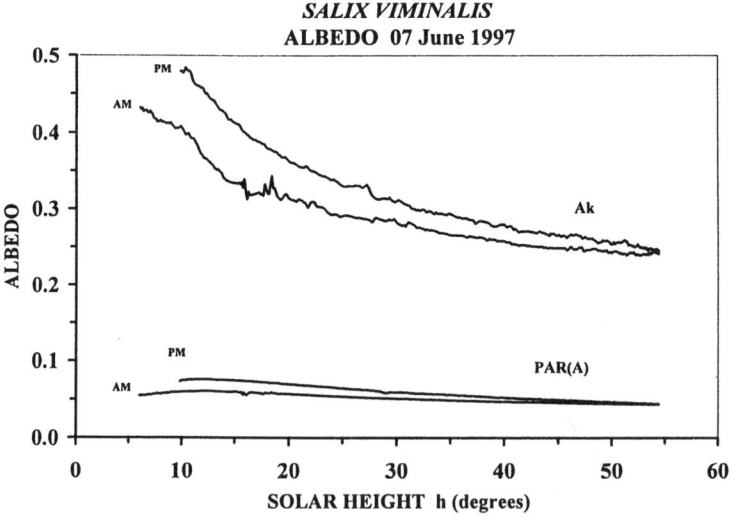

Figure 3. Albedo of the willow forest at Toravere, June 7, 1997. Clear sky with slight C_i clouds in the afternoon

20. EXPERIMENTAL STUDY OF STATISTICAL CHARACTERISTICS OF PLANT CANOPY RADIATION REGIME

The radiation field inside the canopy is extremely variable. The spatial and temporal variability is due to the architecture of the plantation as well as of individual plants, the passage of the Sun, and the movement of plants and leaves caused by the wind. An example of a record during one measurement scan is presented in Fig. 4. Different shapes of the Q and $PAR(Q)$ curves are caused by the different time constant of the sensors. Therefore the records generated by the pyranometer and the net radiometer do not give true information on the variability of the irradiance inside the canopy. The maximum values in Fig. 4 correspond to radiation values in sun flecks, while the minimum ones correspond to values in the shaded area. The penumbra created by narrow leaves can also represent an important area for radiation field studies inside the canopy. Fig. 7 presents the distribution functions of $PAR(Q)$ fluxes, calculated using the data shown in Fig. 4. From Fig. 7, it can be concluded that inside the canopy, the irradiance distribution function is bimodal: the primary maximum (high PAR (Q) values) corresponds to sun flecks while the secondary maximum results from lower values of irradiance in the shaded area. Between these lies a large penumbra area. This area depends on the spatial distribution of leaves, the leaves' linear dimensions and the distance between the sensor and leaves. The distribution function for global radiation Q is similar to that of $PAR(Q)$, but it is smoothed due to the greater time constant of the pyranometer.

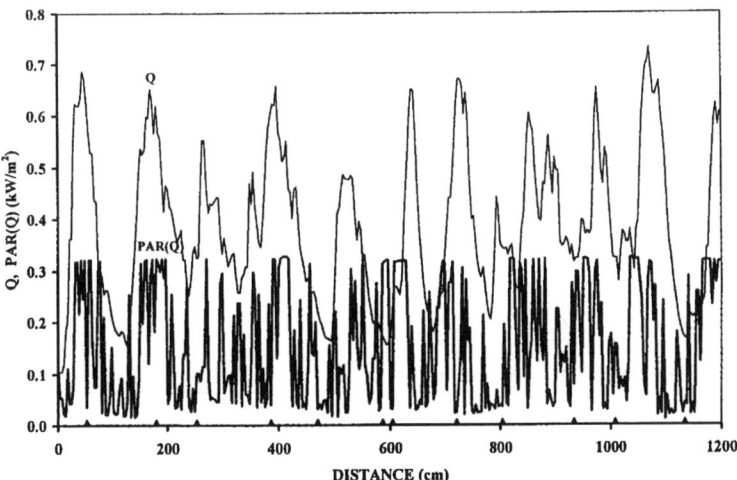

Figure 4. Record of the global and PAR irradiance at the height of 126 cm inside the willow energy forest. Toravere, Estonia, June 27, 1995. Height of the forest 290 cm, LAI = 2.1. The location of the rows is indicated by triangles

Vertical profiles of downward and upward radiation fluxes were determined (Fig. 5) using averaged results of the measurements at heights of 40, 126, 180, 240 and 320 cm. The rapid decreases of the downward fluxes of Q, PAR (Q), B and B_k occur in the layer between 120 and 250 cm, in which the largest part of the leaf area is concentrated. The profiles of net radiation B and short-wave net radiation B_k are quite similar, implying that the contribution of the long-wave radiation is small. In the lower forest $B_l > 0$ showing that the soil is cooler than foliage. The situation is reversed in the upper forest layers—the cloudless atmosphere is cooler than the foliage. In comparison with other fluxes, the reflected PAR is practically zero. The vertical distribution of the leaf area density in the canopy, expressed in $dm^2(leaf)/cm^3$, is shown on the left side of Fig. 5. The smoothed curves of short-wave radiation characteristics near noon for the entire forest canopy (Fig. 6) have been estimated using data from all measurements made on cloudless days during the 1994-1995 vegetation growing period. Errors in estimating such smoothed curves should be evaluated at ±5%. The analysis of Fig. 6 shows that the albedo of the system "willow forest-ground surface" slightly increased in both years, and, during the period of maximum growth, equaled 23%. The lower values of albedo in May 1995, compared to May 1994, are caused by the existence of the 'skeleton' part of the forest in 1995. At the beginning of the growing period in 1995, some 89% of incoming global radiation Q penetrated the forest skeleton (stems and branches without leaves), while 10% were absorbed, the albedo being 15%. In 1995, due to a drought period in July, the growth was disturbed and LAI started to decrease. The mean characteristics of the willow forest radiation regime at the end of the first and second growth periods are given in Table 1.

Table 1. Mean characteristics of the willow forest radiation regime for global radiation at the end of growth periods 1994-1997, Toravere, Estonia

	LAI	Albedo A	Penetration P	Absorption by foliage N	Absorption by ground N_0
1st growth year 1994	1.2	0.23	0.50	0.30	0.47
2nd growth year 1995	2.6	0.24	0.27	0.54	0.22
3d growth year 1996	4.2	0.25			
4th growth year 1997	5.4	0.26			

20. EXPERIMENTAL STUDY OF STATISTICAL CHARACTERISTICS OF PLANT CANOPY RADIATION REGIME

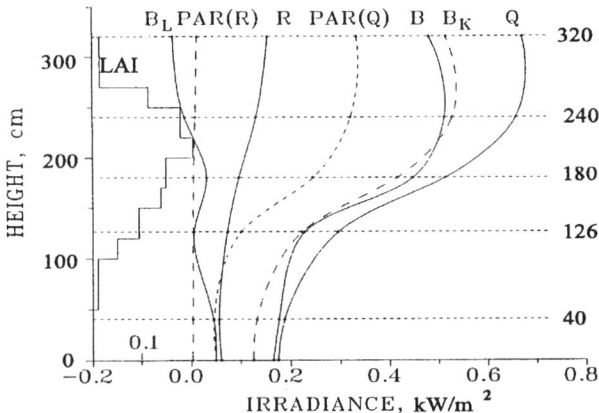

Figure 5. Vertical profiles of radiation fluxes inside the willow forest. July 28, 1995. Forest height -300 cm, LAI = 2.6. Measurements made around noon. Q and PAR Q - downward fluxes, R and PAR R - upward fluxes of global radiation and photosynthetically active radiation, B - net radiation, B_k and B_L -short-wave (integral solar) and long-wave (thermal) radiation, correspondingly.

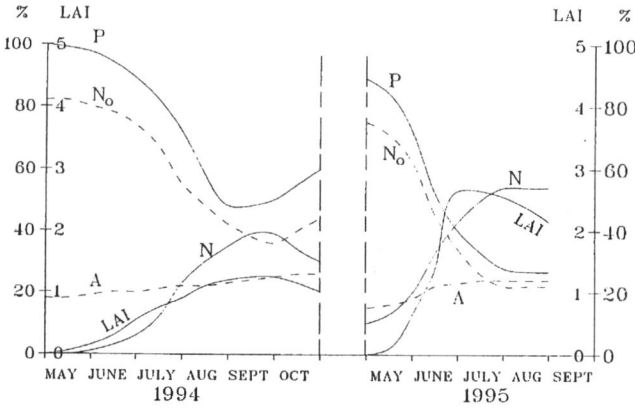

Figure 6. Smoothed curves of characteristics for the whole willow forest in case of global radiation during the vegetation periods of 1994 and 1995, Toravere, Estonia. Dynamics of the leaf area index LAI is also presented. During 1994 the height of the forest increased from 10 till 230 cm and during 1995 from 230 to 400 cm. A – albedo, N absorbed by canopy radiation, N_0 –absorbed by soil radiation, P penetrated radiation, LAI – leaf area index

4. A NEW STATISTICAL TREATMENT OF THE GLOBAL PAR INSIDE THE PLANT CANOPY

The horizontal variability of the global PAR irradiance Q inside the plant canopy is very large. Within sun flecks, the irradiance exceeds by more than 20 times the irradiance in the umbra (Fig. 4), and a statistical treatment of Q is necessary.

Fig. 7 presents the global PAR irradiance probability density function (IPDF) f_Q, which is calculated from the data presented in Fig. 4. In accordance with Fig. 7, the random variable global *PAR (Q)* can be considered a superposition of three random variables:

$$Q = Q^{umbra} + Q^{pen} + Q^{sf}$$

where Q^{sf}, the irradiance within sun flecks, is given by IPDF f_Q^{sf}, Q^{pen}, the irradiance in penumbra, is given by IPDF f_Q^{pen}, and Q^{umbra}, the irradiance in the shade, is given by IPDF f_Q^{umbra}.

We define:
- A sun fleck as an area inside the plant canopy in which the Sun's disc is not shaded by phytoelements. Within sun flecks, the irradiance of the direct solar radiation S is a nonrandom value, i.e., $S = S_0$ where S_0 is the irradiance of the direct solar radiation above the canopy,
- The penumbra as an area inside the plant canopy in which the Sun's disc is partly covered by phytoelements. In penumbra, S is a random value and $0 < S < S_0$, and
- The umbra as an area inside the plant canopy in which the Sun's disc is fully covered by phytoelements and $S = 0$.

The problem is to determine the three IPDFs: f_Q^{sf}, f_Q^{pen}, and f_Q^{umbra} from the measurements obtained with the PAR pyranometers, and to elaborate the procedure for estimating their characteristics. Schematically, the IPDFs in umbra, penumbra and sun flecks areas are expressed in Fig 8.

20. EXPERIMENTAL STUDY OF STATISTICAL CHARACTERISTICS OF PLANT CANOPY RADIATION REGIME

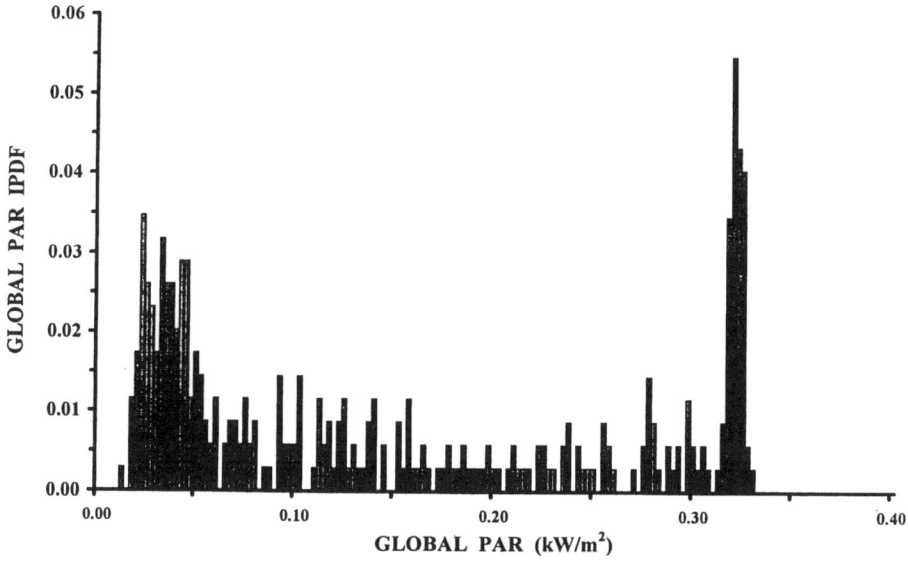

Figure 7. Global PAR irradiance probability density function (IPDF) fQ, calculated from the data, presented in Fig.4

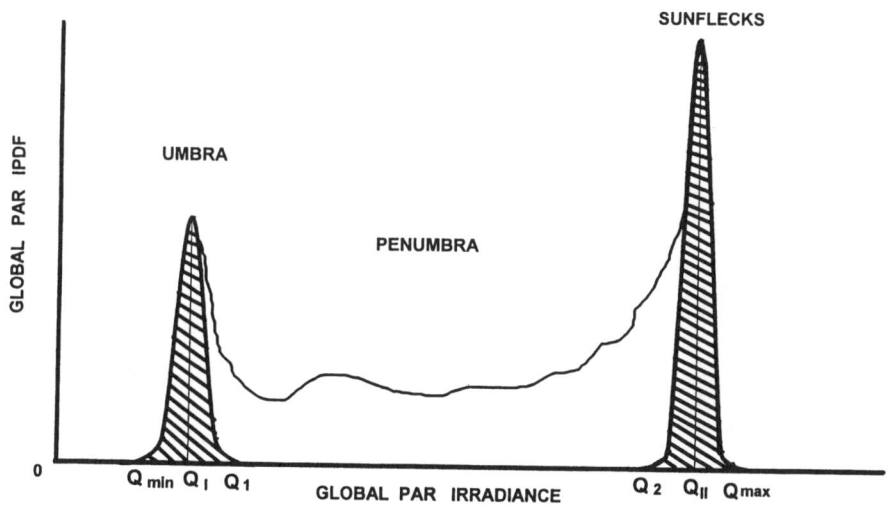

Figure 8. Schematic presentation of the irradiance probability density function in umbra, penumbra and sun flecks

The umbra IPDF, f_Q^{umbra}, is determined in the interval (Q_{min}, Q_I) and reaches its maximum value at the point Q_I; the penumbra IPDF, f_Q^{pen}, is determined in the interval ($Q_I - Q_{II}$), and the sun flecks IPDF, f_Q^{sf}, in the interval (Q_2, Q_{max}), with its maximum value at the point Q_{II}. The values Q_{min}, Q_I, Q_{II} and Q_{max} can be estimated from measurements with a PAR pyranometer, but the estimation of Q_1 and Q_2 needs special measurements or calculations.

Proportional areas of umbra k_U, penumbra k_P and sun flecks k_S are given by the following formulae

$$k_U = \int_{Qmin}^{Q1} f_Q^{umbra} dQ, \qquad k_S = \int_{Q2}^{Qmax} f_Q^{sf} dQ,$$

and

$$k_p = 1 - k_U - k_S$$

Within the umbra and the sun flecks, the shape of IPDF is relatively constant and can be fitted with a normal distribution. The situation is more complicated in penumbra, in which the shape of f_Q^{pen} varies rapidly. Among the various theoretical distributions available, only the beta-distribution is able, more or less satisfactorily, to describe the variability of the penumbra IPDF. The beta-distribution is given by formula

$$\beta(x, \nu, \omega) = \frac{x^{\nu-1}(1-x)^{\omega-1}}{B(\nu, \omega)},$$

where $0<x<1$, the parameters $\nu>0$ and $\omega>0$, and where $B(\nu, \omega)$ is the beta function given by

$$B(\nu, \omega) = \int_0^1 u^{\nu-1}(1-u)^{\omega-1} du.$$

To fit the IPDF in umbra, we can only use the values of Q in the interval between Q_{min} and Q_I, while the interval between Q_I and Q_2 is an overlapping area of umbra and penumbra. Similarly, to fit the IPDF in sun flecks, we can use the interval between Q_{II} and Q_{max}, the interval between Q_2 and Q_{II} being an overlapping area of penumbra and sun flecks.

20. EXPERIMENTAL STUDY OF STATISTICAL CHARACTERISTICS OF PLANT CANOPY RADIATION REGIME

Let us assume that Q_I corresponds to the maximum of the IPDF in umbra, and that Q_{II} corresponds to the maximum of the IPDF in sun flecks. The IPDF in umbra f^{umbra} is then characterized by a normal distribution, determined with the mean value Q_I and a dispersion σ_U. By analogy, the IPDF in sun flecks f^{sf} is characterized by a normal distribution determined with the mean value Q_{II} and a dispersion σ_s. In penumbra, the fitted beta distribution is constrained by two parameters, v and ω.

The characteristics of these IPDFs depend on solar elevation h and on the depth of canopy, expressed by the downward cumulative leaf area index L. These characteristics can be replaced by a joint parameter, the sunray's path length expressed as the leaf area $\tau = L/\sin h$.

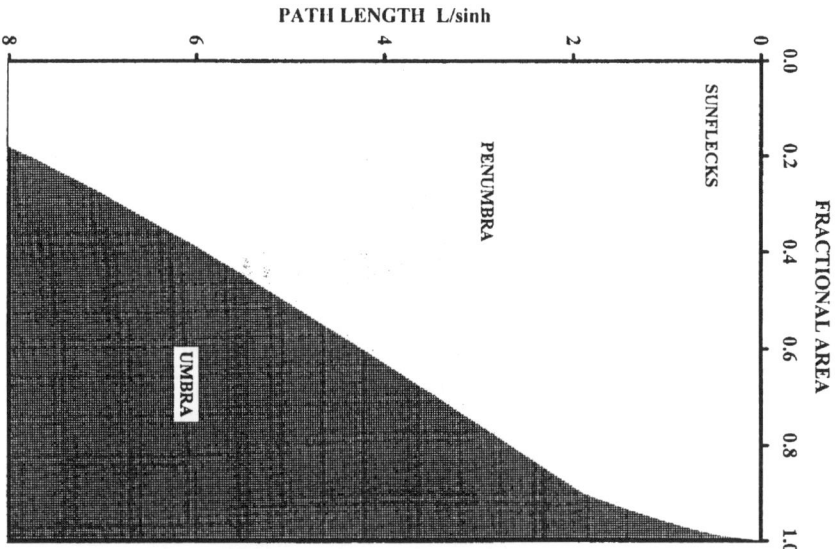

Figure 9. Vertical profiles of the sun flecks, penumbra and umbra fractional areas in willow energy forest

Fig. 9 expresses the distribution of the umbra, penumbra and sun flecks fractional areas at different depths within the willow forest. Sun flecks exist only in the upper part of the forest (where the path length $\tau<3$) and rapidly decrease with depth. The proportion of penumbra area increases correspondingly, and is dominating in the middle layers. Starting from a depth of $\tau=1$, the umbra area linearly increases with depth and is dominating in the lower layers.

The mean values of the global PAR relative irradiance in umbra Q_I/Q_0 (Fig. 10) decrease with the path length, although the dispersion is relatively constant $\sigma_U = 0.002$.

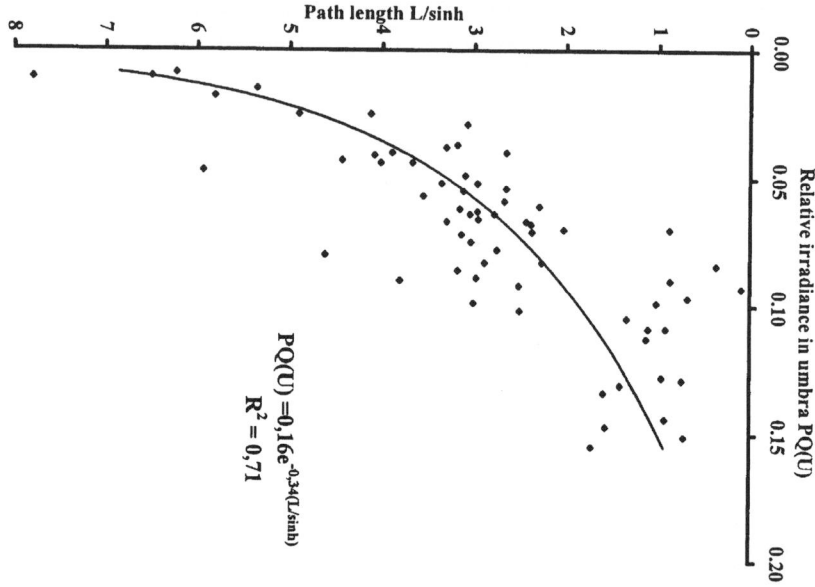

Figure 10. Vertical profiles of the global PAR relative irradiances in umbra

The mean values of the PAR relative irradiances in sun flecks Q_{II}/Q_0 slowly decrease from 1 at the upper forest level ($L=0$) to 0.92 at $\tau=3$, with $\sigma_s=0.002$. A decrease of the diffuse component of the global irradiance is responsible for this diminution.

In penumbra, the IPDF has a U-shape determined by the parameters ν and ω. The vertical profile of the parameter ω is exhibited in Fig. 11. The parameter ν is almost independent of τ and equals 0.6. In the upper layers of the forest, the right part of the U-shape (high irradiances) is dominating while the left part is regenerated. The situation is reversed in the lower layers of the canopy.

20. EXPERIMENTAL STUDY OF STATISTICAL CHARACTERISTICS OF PLANT CANOPY RADIATION REGIME

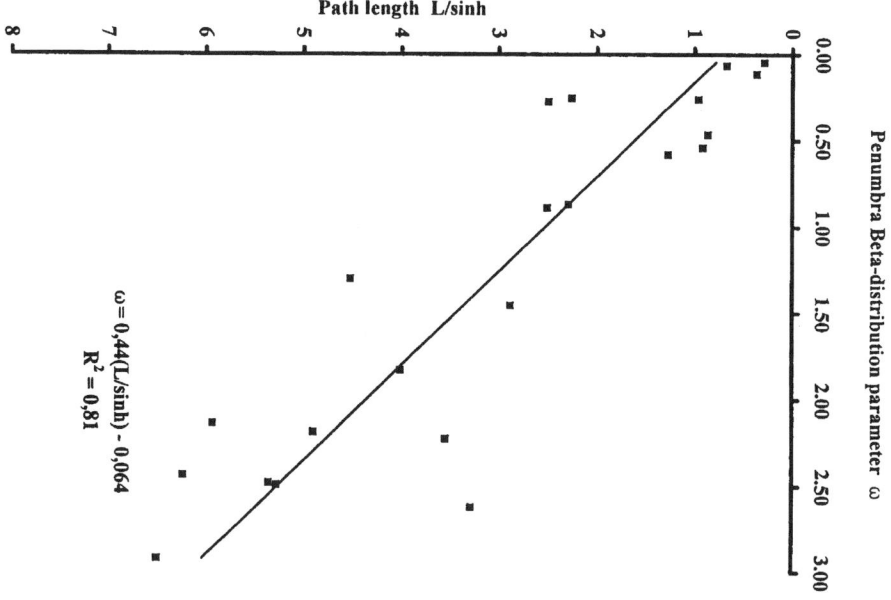

Figure 11. Vertical profiles of the penumbra beta distribution parameter ω

5. ACKNOWLEDGEMENTS

We thank Prof. Lars Christersson and Assistant Prof. Kurth Perttu for their constructive comments, Peeter Saarelaid, Enn-Märt Maasik and Maido Sulev for assistance in field measurements and data processing. This research was supported by ETSA Grant No 251, ISF Grants No LGG 000 and LKG 100 PECO-NIS-LTEEF project, Supplementary Agreement No ERB-CIPDCT94-0012.

6. REFERENCES

Allen, L.H. and Lemon, E.R. (1972) Net radiation frequency distribution in a corn crop, *Boundary Layer Meteorology*, **3**, 246–254.

Baldocchi, D.D., Matt, D.R., Hutchison, B.A., and McMillan, R.T. (1984) Solar radiation within an oak-hickory forest: An evaluation of the extinction coefficient for several radiation components during fully-leafed and leafless periods, *Agricultural and Forest Meteorology*, **32**, 307–322.

Baldocchi, D.D., Matt, D.R., Hutchison, B.A., and McMillan, R.T. (1986) Seasonal variation in the statistics of photosynthetically active radiation penetration in an oak-hickory forest, *Agricultural and Forest Meteorology*, **36**, 343–361.

Campbell, G.S. and Norman J.M. (1989) The description and measurement of plant canopy structure, in B. Marshall and P.G. Jarvis (Editors): *Plant Canopies: Their Growth, Form and Function*, Cambridge University Press, pp. 1–19.

Chen, J.M. and Black, T.A. (1992) Foliage area and architecture of plant canopies from sun fleck size distributions, *Agricultural and Forest Meteorology*, **60**, 249–266.

Eckersten, H. (1984) Light penetration and photosynthesis in a willow stand, in K. Perttu (Editor): *Ecology and management of forest biomass production systems*, Dept. Ecology and Environmental Research, Swedish University of Agricultural Science, Uppsala, Report **15**, 29–45.

Fassnacht, K., Gower, S.T., Norman, J.M., and McMurtrie, J.M. (1994) A comparison of optical and direct methods for estimating foliage surface area index in forests, *Agricultural and Forest Meteorology*, **71**, 183–207.

Koppel, A., Perttu, K., and Ross, J. (1996) Estonian energy forest plantations-general information, in K. Perttu and A. Koppel (Editors): *Short rotation willow coppice for renewable energy and improved environment*, Swedish University of Agricultural Science, Uppsala, Report **57**, 15–24.

Laisk, A. (1968) Statistical character of light extinction in plant communities, in *Solar Radiation Regime in Plant Canopy*, Institute of Physics and Astronomy, Academy of Sciences ESSR, Tartu, 81–111 (in Russian).

Lindroth, A. and Perttu, K. (1981) Simple calculation of extinction coefficient of forest stands, *Agricultural Meteorology*, **25**, 97–110.

Norman, J.M.and Campbell, G.S. (1989) Canopy structure, in R.W. Pearly, J. Ehleringer, H.A. Mooney and P.W. Rundel (Editors): *Plant Physiological Ecology: Field Methods and Instrumentation*, New York, Chapman and Hall, pp. 301–326.

Pearcy, R.W., Roden, J.S., and Gamon, J.A. (1990) Sun fleck dynamics in relation to canopy structure in a soybean (Glycine max (L.) Merr) canopy, *Agricultural and Forest Meteorology*, **52**, 359–372.

Perttu, K. (1970) Radiation measurements above and in forest, *Stud. Forestalia Suecica*, **72**, 1–49.

Ross, J. (1981) *The radiation regime and architecture of plant stands*, Dr W. Junk Publishers, The Hague-Boston-London, 391 pp.

Ross, V. (1994) Phytometrical measurements of the structure of the willow forest, in P. Aronson and K. Perttu (Editors): *Willow Vegetation Filters for Municipal Wastewaters and Sludges: A Biological Purification*, Swedish University of Agricultural Science, Uppsala, Report **50**, 199–204.

Sassenrath-Cole, G.F. (1995) Dependence of canopy light distribution on leaf and canopy structure for two cotton species, *Agricultural and Forest Meteorology*, **77**, 55–72.

Sulev, M. and Ross, J. (1996) Conversion factor between global solar radiation and photosynthetic active radiation, in K. Perttu and A. Koppel (Editors): *Short rotation willow coppice for renewable energy and improved environment*, Swedish University of Agricultural Science, Uppsala, Report **57**, 115–121.

Chapter 21

LIGHT SCATTERING MODELS AND REFLECTANCE MEASUREMENTS IN REMOTE SENSING OF SNOW

R. Kuittinen
Finish Geodetic Institute, Masala, Finland.

1. INTRODUCTION

The snow pack characteristics are never stable because metamorphism occurs all the time in different ways. This metamorphism can be destructive or constructive, and both affect the crystals and grains of snow. The size and shape of the grains and crystals as well as the structure and temperature of the snow pack have great effects on the emittance and reflectance of electromagnetic radiation from snow.

During the snow cover period three different phases can be distinguished if the snow depth or water equivalent is the variable. These phase are accumulation-, stable- and ablation phase. The weather circumstances and snow characteristics in different phases can be described as follows.
1. Accumulation phase.
   Snow surface temperature is zero or below zero centigrade, grain size is small, no diurnal variation of the snow surface temperature exist, albedo is at its maximum, snow covers the terrain including trees, snow-less patches do not exist and the thickness of the snow can vary from a thin layer to its maximum. Solid precipitation occurs. Several accumulation phases can exist during winter.
2. Stable phase.
   Snow surface temperature is below zero, grain size increases, density increases, diurnal variations of the snow surface temperature can exist,

albedo decreases, snow can cover partly or totally the terrain, snow thickness can vary from a thin layer to its maximum. During winter several stable phases can exist.
3. Ablation phase.
Snow surface temperature is zero or below zero and its diurnal variation can be large, grain size is large, density is high, liquid water exists in snow, albedo is smaller than in the accumulation phase and is constant or decreases, snow covers totally or partly the ground but not trees, snow thickness can vary and snow-less patches on ground can exist. If snow-less ground exist its temperature is zero or over zero and it is wet.
During winter a few ablation phases exist.

These three phases and typical snowpack characteristics are presented with some generalization in Table 1 as follows.

Table 1. The characteristics of snowpack in different phases of snow

Snow characteristics	Phases		
	Accumulation	Stable	Ablation
crystal/ grain size	usually large/ -	decreasing/ increasing	- / increasing or large
liquid water content	usually no	no	exist
density	low	increasing	increasing or maximal
temperature	lower than zero	lower than zero	zero
water equivalent	increasing	constant	decreasing
albedo	high	decreasing	low

During the stable and ablation phases remote sensing in visible and infrared regions of electromagnetic spectrum can be used because the presence of clouds dictates the possibilities to use optical remote sensing techniques.

2. PROBLEMS

Information about snow cover are collected for many purposes but in most cases the following information is needed:
- snow covered area (km^2),
- depth (m) or water equivalent of snow (mm),
- albedo of snow (ρ),
- snow melt (mm/d).

The interpretation of these characteristics is based on the reflectance of snow which is much larger than the reflectance of other terrain features. In some cases the reflectance of the soil surface and vegetation are used together with the snow reflectance in interpretation. Many problems arise because of the metamorphism of snow and the varying geometry of the Sun -

snow/terrain - sensor system. The reasons of these problems can be listed as follows:
- The reflectance of snow depends on the grain size and the presence of liquid water in snow. These parameters varies due to the metamorphism of snow which mainly is dependent on the age and temperature of snow.
- The reflectance of snow depends on the incidence angle of measurement and thus topography, Sun position, and sensor location affect on results of measurements.
- The reflectance of snow is different in different parts of the electromagnetic spectrum and thus the channels of the sensor are very relevant in snow cover monitoring.
- The spatial resolution of sensors causes that in many cases the sensor measures the reflectance of an area which only partly is snow covered. This so called transition zone should be interpreted correctly, too.
- Vegetation affects the reflectances of the terrain because it usually is not covered by snow.

Due to the cloud cover, the repeat cycles of satellites and different swaths of sensors it is in practice not possible to get images having a similar imaging geometry. Thus to obtain useful results we have first to be able to correct the effects of the incidence angle on reflectances, and secondly to correct the effect of snow metamorphism on reflectances. After this it is possible to model the reflectance of different terrain types covered partly or totally by snow.

3. SNOW COVER MODELLING

3.1 General

It is very important to know the angular dependence of snow reflectance when remote sensing is used for snow characteristics determination. What actually is needed is the bidirectional reflectance distribution function (BRDF) because many sensors used in the visible and infrared parts of the spectrum have wide swaths which means varying incidence angles for radiance measurements. Wiscombe and Warren (1980) as well as Wald (1994) have presented results concerning the angular dependence of snow reflectance. Figure 1 presents the results. We see that the reflectance increases with increasing incidence angle. The largest absolute variations happen in the region of 0.9–1.4 µm. This is a very important piece of information for satellite image interpretation because the effect of varying incidence angle are to be corrected to obtain accurate snow reflectances in all parts of images.

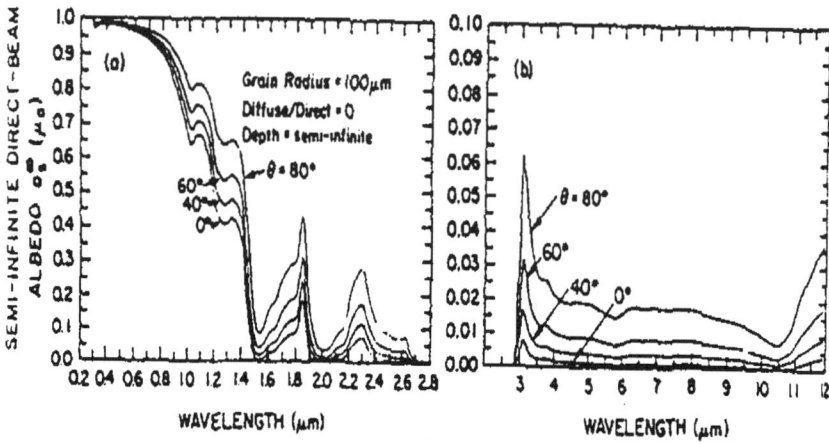

Figure 1. Spectral reflectance of snow for several values of incidence angle, Wiscombe and Warren (1980)

The basic relationship for scanner - surface - Sun interaction is presented in Figure 2.

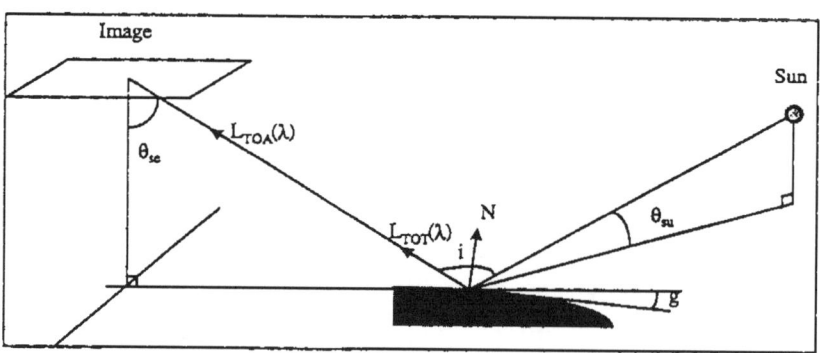

Figure 2. The basic geometry of sensor -terrain - Sun interaction with top of terrain (TOT) and top of atmosphere (TOA) radiances

21. LIGHT SCATTERING MODELS AND REFLECTANCE MEASUREMENTS IN REMOTE SENSING OF SNOW

The sensor registers the top of atmosphere radiance $L_{TOA}(\lambda,i,t)$ for each point of the image. This radiance consists of the components coming from atmosphere, target and the surroundings of the target. To obtain the top of terrain radiance, $L_{TOT}(\lambda,i,t)$, the effects of the atmosphere have to be corrected. Parameter t refers to snow metamorphism because the reflectance is different for different snow types (Table 1). This can be made by using available software codes and for input the locations of Sun and sensor as well as atmospheric parameters are needed. The main problem is the lack of basic information for corrections (aerosols, humidity, pressure) which usually is available only in places of permanent ground stations. Usually average values of atmospheric characteristics must be used.

To calculate the incidence angle i digital terrain model is needed in addition to the locations of Sun and sensor. Incidence angle is the key parameter in modeling the reflectance because BRDF is a function of this angle. If the incidence angle and the BRDF are known it is possible to calculate e.g., nadir radiances for all parts of the image. This allows the comparisons between different images.

The BRDF is a function of the wavelength and in practice it is not possible to measure the whole range of the spectrum. The BRDF can be approximated by measuring only the most important parts of the spectrum because the shape of the reflectance curve remains very similar even the incidence angle changes (Figure 1).

3.2 Modeling of snow

The bidirectional reflectance distribution function contains a lot of information and many measurements are needed for determining it. Thus it is important to analyze the possible incidence angles before starting measurements and modeling. For most of the snow covered areas in March-May in Finland the following ranges can be given:

Table 2.

Variable	Symbol	Range (deg.)
Sun elevation	θ_{su}	$5 < \theta_{su} < 40$
Look angle	θ_{se}	$-45 < \theta_{se} < +45$
Spectral range	λ	$900 < \lambda < 1400$, and $3000 < \lambda < 3500$ nm
Terrain slope	G	$0 < g < 40$

This selection reduces the amount of measurements needed for determining the BRDF of snow for practical applications.

In order to get most of the benefit from the BRDF the modeling of BRDF has to be made for three types of snow, namely:
- dry new snow
- dry old snow, and
- wet old snow.

These are the most common types of snow during winter and are closely connected with the three different phases of snow. In order to be able to make practical determination of snow reflectance for different incidence angles a reference reflectance curve, e.g., $i=55$ deg., has to be determined accurately for the whole selected range of the spectrum. The other parts of the BRDF can then be determined with less measurements because the shape of the reflectance curve seems not change much Wiscombe and Warren, (1980). The three cases have to be modeled to get a complete understanding of the behavior of the snow and to get basic information for the modeling of the reflectance of the snow covered terrain.

When these BRDF's are available the incidence angle image can be used to interpret the reflectance in different parts of the image. This opens the possibility to increase the accuracy of the interpretation of the snow following characteristics: dry and wet snow, albedo age. As mentioned earlier estimated nadir radiances could be useful method to compare different images in snow characteristics interpretation.

3.3 Modeling of the snow covered terrain

In practice the terrain is very seldom totally snow covered. The area inside the instantaneous field of view (IFOV) consists of trees, bushes, bedrock and soil which can be partly or totally snow covered. When the IFOV increases the possibility to get mixtures of these objects increases. The next step in modeling is very demanding because we should be able to model the reflectances of terrain which is partly or totally covered by snow. To make this modeling the priority could perhaps be given to the following cases marked with x in the table below.

Table 3.

Snow characteristics	Percentage of snow cover (%)			Density of forest (m^3/ha)		
	100	70	30	0	100	200
dry new	x			x	x	x
dry old	x	x		x	x	x
wet old	x	x	x	x	x	x

The size of the IFOV affects very much the modeling. It seems reasonable to make the modeling for the following sizes of the IFOV. The last one in the table below is the most important because these sensors provide the highest frequency in observing the snow. As a result of this modeling of snow-covered area and snow water equivalent can be interpreted.

Table 4.

Size of the IFOV (m^2)	Satellites/ Sensors for which modeling can be used	Remarks
25 × 25	Landsat/TM, SPOT/ HRV, JERS/OPS	not very useful sensors in snow cover monitoring
250 × 250	ENVISAT/MERIS, IRS-D/WIFS, RESURS/MSK	relatively useful sensors in snow cover monitoring
1000 × 1000	ERS/ATSR, ENVISAT/AATSR and MERIS, NOAA/AVHRR, METOP/VIRSR and ADEOS II/GLI	the most useful sensors for snow cover monitoring

As can be seen the work for modeling is huge and it only can be solved by starting from some in situ measurements of the BRDF and then continuing with mathematical approximations and by finishing with tests and control by using satellite images and the estimated incoming radiances.

4. DISCUSSION AND CONCLUSIONS

The most important optical sensors for snow cover monitoring will be in future MERIS and AATSR in ENVISAT, GLI in ADEOS II and VIRSR in METOP-1. The estimated radiometric accuracy figures of MERIS and AATSR shows that the sensors themselves will be sensitive enough for registering the reflectance changes caused by snow metamorphism. This opens the way for more accurate snow cover monitoring (presence, depth, albedo and melt) if only we can estimate the BRDF of different snow types and increase the accuracy of atmospheric correction of satellite images.

In snow measurements most efforts shall be concentrated in estimating the BRDF of snow by starting with in situ measurements. When this has been done or simultaneously with this, present satellite data can be analyzed. A very valuable sensor in this modeling is the POLDER on ADEOS I and coming ADEOS II, because some parts of the BRDF can be estimated over large areas (10×10 km^2) using this sensor.

When the BRDF of snow is known the modeling of the reflectance of snow covered terrain shall be studied. Ordinary satellite images will help in this modeling even they do not give very many representative incidence angle values for modeling. When this modeling gives useful results it is possible to monitor the snow cover much better than earlier with optical satellite sensors.

5. REFERENCES

ADEOS and ADEOS II (1999) http://www.nasda.go.jp/index_e.html, National Space Developing Agency of Japan.

Kramer, H.J. (1994) *Observation of the Earth and Its Environment, Survey of Missions and Sensors*, Springer-Verlag, Berlin.

ESA (1996) *MERIS: The Medium Resolution Imaging Spectrometer*, ESA, **SP-1184**.

Wiscombe, W.J. and Warren, S.G. (1980) A model for spectral albedo of snow. I: Pure snow. *Journal of the Atmospheric Sciences*, **37**, 2712–2733.

Chapter 22

RAY OPTICS APPROXIMATION FOR RANDOM CLUSTERS OF GAUSSIAN SPHERES

K. Muinonen
Astronomical Observatory, Uppsals University, Uppsala, Sweden.

1. INTRODUCTION

Scattering of light by natural particle clusters is one of the most important current scattering problems. It is the goal of the present work to provide first understanding about the differences in scattering by, on one hand, isolated single particles and, on the other hand, clusters made of such single particles.

The fundamentals of the lognormal probability distribution are concisely summarized by Aitchison and Brown (1963), including the multiplicative analogues to the Central Limit Theorem for the normal probability distribution. The lognormal distribution has been extensively used in studies of various kinds of small particles. Only recently has the lognormal distribution been applied in shape modeling (Muinonen et al. 1996, Muinonen 1996, Peltoniemi et al. 1989). In these approaches, scattering by random particles was studied in the ray optics approximation. Peltoniemi et al. adopted a Markovian approach based on propagation probabilities for stochastically rough particles. Muinonen et al. developed a spherical harmon method for the generation of Gaussian random (or stochastically rough) particles, and traced rays deterministically for sample particles, generating a new sample particle for each ray. It has been hypothesized that, e.g., the shapes of asteroids and cometary nuclei can be modeled by the Gaussian random sphere (Muinonen 1997, Lagerros 1997).

Peltoniemi and Lumme (1992) showed important first results for scattering by closely packed media. In close resemblance, but slightly rephrasing the research goals, the current approach is directed toward understanding the gradual changes in scattering characteristics when more and more member particles are accrued into clusters.

In Section 2, the ray optics approximation is briefly discussed for randomly oriented particles. Section 3 provides a brief summary of the clustering algorithm and the Gaussian random sphere, and Section 4 shows preliminary results from simulations. Conclusions close the paper in Section 5.

2. RAY OPTICS APPROXIMATION

The ray optics treatment is described in Muinonen et al. (1989) and (1996) and, in what follows, a short summary is given on the most essential parts of the treatment. The cluster size is described by the size parameter $x = ka_c$, where k is the wavenumber and a_c, is the radius of an equal-projected-area sphere.

The scattering phase matrix P relates the Stokes vectors of the incident ($I_i = (I_i, Q_i, U_i, V_i)^T$) and scattered light ($I_s = (I_s, Q_s, U_s, V_s)^T$); for random clusters of Gaussian particles

$$I_s = \frac{\sigma_{sca}}{4\pi r^2} P \cdot I_i$$

$$\int_{4\pi} \frac{d\Omega}{4\pi} P_{11} = 1$$

where σ_{sca} is the scattering cross section and P_{11} is the scattering phase function.

For particles larger than the wavelength, the scattering cross section and phase matrix can be divided into the forward diffraction and geometric optics parts (superscripts D and G),

$$\sigma_{sca} = \sigma_{sca}^D + \sigma_{sca}^G$$

$$P = \frac{1}{\sigma_{sca}} (\sigma_{sca}^D P^D + \sigma_{sca}^G P^G)$$

22. RAY OPTICS APPROXIMATION FOR RANDOM CLUSTERS OF GAUSSIAN SPHERES

$$\int_{4\pi} \frac{d\Omega}{4\pi} P_{11}^D = \int_{4\pi} \frac{d\Omega}{4\pi} P_{11}^G = 1$$

It is here strictly required that

$$\sigma_{sca}^D = \langle A \rangle,$$

$$\sigma_{ext} = \sigma_{abs} + \sigma_{sca} = 2\langle A \rangle$$

where σ_{ext} and σ_{abs} are the extinction and absorption cross sections, and $\langle A \rangle$ is the ensemble-averaged cross-sectional area. The absorption cross section is solely due to geometric optics: $\sigma_{abs} = \sigma^G_{abs}$. The geometric optics single-particle albedo ϖ and the asymmetry parameter g are

$$\varpi = \frac{\sigma_{sca}^G}{\langle A \rangle}$$

$$g = \int_{4\pi} \frac{d\Omega}{4\pi} \cos\theta P_{11}$$

where θ is the scattering angle. The asymmetry parameter can be divided into the forward diffraction and geometric optics parts g^D and g^G as in Eq. (2) for the scattering phase matrix.

It is presently assumed that the particles are very large compared to the wavelength so that the forward diffraction part can be approximated by a Dirac delta function,

$$P^D = 4\pi\delta(\Omega) \cdot \mathbf{1}$$

$$g^D = 1$$

where $\mathbf{1}$ is the 4×4 unit matrix.

3. CLUSTERS OF GAUSSIAN SPHERES

Particle clusters are generated by locating a seed particle at the origin, and utilizing a simple method of aggregating member particles from random directions with random offsets, with the assumption that all collisions lead to a capture. The aggregation is based on a pre-assigned hard-sphere-radius a_h for the member particles. The aggregation algorithm thus works correctly for perfectly spherical particles, but the irregular Gaussian spheres, depending on how a_h relates to their radii, may "overlap" or "levitate" in the cluster. In what follows, a short review is given on the shapes of the member particles, the Gaussian random sphere.

Gaussian random spheres can be generated using spherical harmonics series for the logradius $s = s(\vartheta, \varphi)$ (Muinonen 1996),

$$r(\vartheta,\varphi) = a\exp\left[s(\vartheta,\varphi) - \frac{1}{2}\beta^2\right]$$

$$s(\vartheta,\varphi) = \sum_{l=0}^{\infty}\sum_{m=-l}^{l} s_{lm} Y_{lm}(\vartheta,\varphi)$$

where a is the mean radius, β is the standard deviation of the logradius and related to the relative standard deviation σ by $\sigma^2 = \exp(\beta^2) - 1$, and Y_{lm}'s are the orthonormal spherical harmonics with Condon-Shortley phase (Arfken 1970). The logradius is real-valued so that

$$s_{l,-m} = (-1)^m s_{lm}^*, \quad l = 0,1,\ldots,\infty, \quad m = -l,\ldots,-1,0,1,\ldots,l$$

implying Im $(s_{l0}) \equiv 0$.

If the real and imaginary parts of the spherical harmonics coefficients s_{lm}, $m \geq 0$, are independent Gaussian random variables with zero means and variances

$$\text{Var}(\text{Re}(s_{lm})) = (1+\delta_{m0})\frac{2\pi}{2l+1}c_l\beta^2$$

$$\text{Var}(\text{Im}(s_{lm})) = (1-\delta_{m0})\frac{2\pi}{2l+1}c_l\beta^2$$

the logradius will be normally distributed with zero mean and covariance function $\beta^2 C_s$ (δ_{m0} is the Kronecker symbol),

22. RAY OPTICS APPROXIMATION FOR RANDOM CLUSTERS OF GAUSSIAN SPHERES

$$C_s(\gamma) = \sum_{l=0}^{\infty} c_l P_l(\cos\gamma)$$

where P_l's are the Legendre polynomials.

The two perpendicular slopes

$$s_\vartheta = \frac{r_\vartheta}{r} \qquad \frac{1}{\sin\vartheta} s_\varphi = \frac{r_\varphi}{r\sin\vartheta}$$

are noncorrelating Gaussian random variables with zero means and standard deviations

$$\rho = \frac{\beta}{l}$$

Here, l is the correlation length that can be further related to a correlation angle Γ:

$$l = \frac{1}{\sqrt{-C_s^{(2)}(0)}} = 2\sin\left(\frac{1}{2}\Gamma\right)$$

where $C_s^{(2)}$ is the second derivative of the correlation function with respect to γ.

For example, the modified Gaussian correlation function depends on a single parameter and can be phrased as

$$C_s(\gamma) = \exp\left(-\frac{2}{l^2}\sin^2\frac{1}{2}\gamma\right)$$

This function is a proper correlation function, since its Legendre coefficients are all non-negative (Muinonen 1996),

$$c_l = (2l+1)\exp\left(-\frac{1}{l_{\frac{1}{2}}^2}\right) i_l\left(\frac{1}{l_{\frac{1}{2}}^2}\right)$$

where l=0,...,∞ and i_l is a modified spherical Bessel function.

4. FIRST RESULTS AND DISCUSSION

Figure 1 presents three sample clusters consisting of 2, 10, and 50 Gaussian sample spheres generated assuming $\sigma = 0.1$ and $\Gamma = 30°$, and making use of the modified Gaussian correlation function in Eqs. (13) and (14) with cutoff degree $l_{max} = 11$ in Eq. (9).

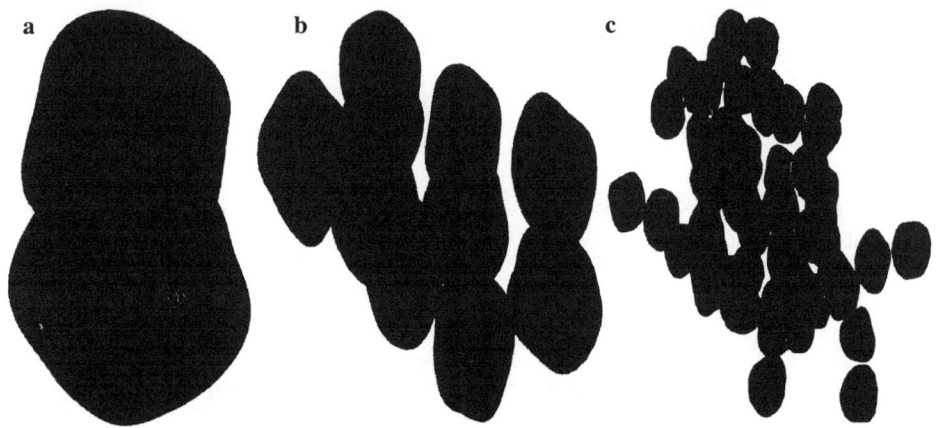

Figure 1. Sample clusters of Gaussian particles with relative standard deviation of radius $\sigma = 0.1$ and correlation angle $\Gamma = 30°$ assuming the modified Gaussian correlation function: a) 2-particle, b) 10-particle, and c) 50-particle clusters

When generating the clusters, the hard-sphere radius a_h was set at $a_h = a(1 + \sigma)$ so some of the member particles may not be in contact with the others. For the purposes of the present preliminary study of light scattering by clusters of Gaussian spheres, this simple algorithm of cluster generation was found satisfactory. It is to be expected that more precise cluster generation algorithms will not change the preliminary conclusions of the present study.

Figure 2 shows the geometric optics parts of the scattering phase matrices for 2-particle, 10-particle, and 50-particle clusters assuming a refractive index of 1.55 for member particles. Up to $p_{max} = 21$ ray chords (van de Hulst 1957) and 10^5 rays and sample clusters were included in ray tracing, and the relative cutoff flux was set at 10^{-5}.

22. RAY OPTICS APPROXIMATION FOR RANDOM CLUSTERS OF GAUSSIAN SPHERES

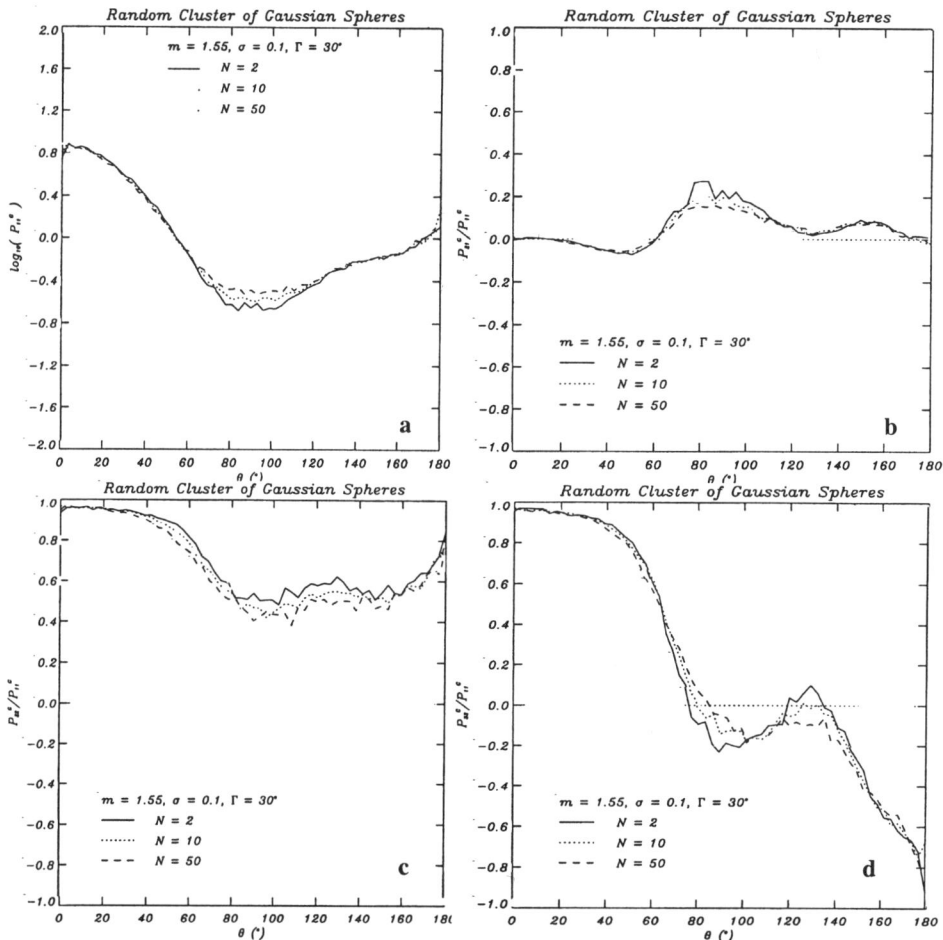

Figure 2 (first part). The ensemble-averaged scattering phase matrices for 2-particle, 10-particle, and 50-particle clusters of Gaussian particles with statistical parameters as in Fig. 1 and: a) P_{11}^G; b) $-P_{21}^G / P_{11}^G$; c) P_{22}^G / P_{11}^G; d) P_{33}^G / P_{11}^G.

The immediate preliminary result is that the scattering characteristics are qualitatively similar for all three cases. For increasing number of member particles, the scattering phase function becomes slightly flatter, with increasing scattering into intermediate scattering angles for increasing number of member particles. The degree of linear polarization slightly decreases for the same angular domain, and no evidence is seen for negative polarization close to the backward direction (cf. Muinonen 1994). As for the other scattering phase matrix elements, fairly small though definite changes can be distinguished between the three different random clusters.

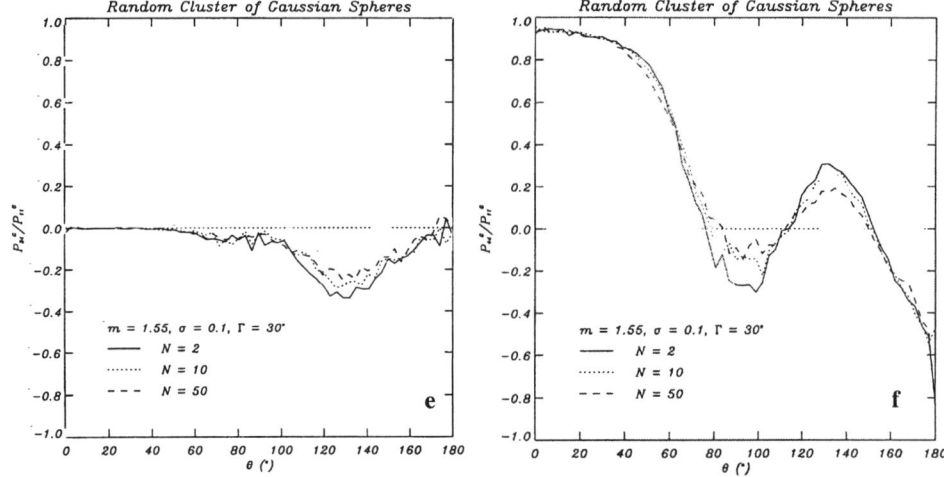

Figure 2 (second part). The ensemble-averaged scattering phase matrices for 2-particle, 10-particle, and 50-particle clusters of Gaussian particles with statistical parameters as in Fig. 1 and e) P_{34}^G / P_{11}^G; f) P_{44}^G / P_{11}^G

5. CONCLUSIONS

Future studies will include an improvement of the cluster generator in order to bring the "levitating" and "overlapping" member particles into grazing contact with the cluster. It is, however, to be expected that the cluster generation then slows down by a factor of several orders of magnitude.

The simple but efficient clustering algorithm can be further developed in order to generate clusters that are more closely packed. For example, every incoming and colliding particle could "shake off" particles, whose origins lie farther away from the seed particle origin. The algorithm becomes somewhat slower, but resulting clusters will have manageably differing close-packing properties.

It is straightforward to extend the current software to simulate light scattering by clusters consisting of particles with differing optical and shape properties. The present treatment allows a detailed comparison of ray optics and radiative transfer solutions for equivalent scattering media.

6. ACKNOWLEDGMENTS

The author thanks J. S. V. Lagerros for valuable suggestions during the course of the work.

7. REFERENCES

Aitchison, J. and Brown, J. A. C. (1963). *The Log-normal Distribution*, University Press, Cambridge, U.K.

Arfken, G. (1970) *Mathematical Methods for Physicists*, Second Edition, Academic Press, New York.

Lagerros, J. S. V. (1997) Thermal physics of asteroids. III. Irregular shapes and albedo variegations, Astronomy and Astrophysics, in press.

Muinonen, K. (1994) Coherent backscattering by solar system dust particles, in A. Milani, M. Di Martin, A. Cellino (Editors): *IAU Symposium No. 160, Asteroids, Comets, Meteors 1993*, 271–296 (Kluwer Academic Publishers, Dordrecht).

Muinonen, K. (1996) Light scattering by Gaussian random particles, *Earth, Moon, and Planets*, **72**, 339–342.

Muinonen, K. (1997) Introducing the Gaussian shape hypothesis for asteroids and comets, Submitted to *Astronomy and Astrophysics*.

Muinonen, K., Lumme, K., Peltoniemi, J. I., and Irvine, W. M. (1989) Light scattering by randomly oriented crystals, *Applied Optics*, **28**, 3051–3060.

Muinonen, K., Nousiainen, T., Fast, P., Lumme, K., and Peltoniemi, J. I. (1996) Light scattering by Gaussian random particles: Ray optics approximation, *Journal of Quantitative Spectroscopy and Radiative Transfer*, **55**, 577–601.

Muinonen, K., Lamberg, L., Fast, P., and Lumme, K. (1997) Ray optics regime for Gaussian random spheres, *Journal of Quantitative Spectroscopy and Radiative Transfer*, **57**, 197–205.

Peltoniemi, J. I., and Lumme, K. (1992) Light scattering by closely packed particulate media, *Journal of the Optical Society of America*, **A 9**, 1320–1326.

Peltoniemi, J. I., Lumme, K., Muinonen, K., and Irvine, W. M. (1989) Scattering of light by stochastically rough particles, *Applied Optics*, **28**, 4088–4095.

van de Hulst, H. C. (1957) *Light Scattering by Small Particles*, Wiley, New York.

Chapter 23

BACKSCATTERING OF LIGHT BY SNOW
Field measurements

J. Piironen (1), K. Muinonen (1, 2), S. Keranen (1), H. Karttunen (3) and J. Peltoniemi (4)

(1) Observatory, University of Helsinki, Finland, (2) Astronomical Observatory, Uppsala, Sweden, (3) Tuorla Observatory, University of Turku, Finland and (4) Finnish Geodetic Institute, Masala, Finland.

1. INTRODUCTION

We measured the scattering of light by snow in order to compare the small phase angle characteristics of snow and planetary objects. The phase angle range in our measurements is within the range of the opposition effect and opposition spike of Saturn's rings, asteroids, and planetary satellites without atmospheres.

The opposition effect is a gradual brightening of an object with decreasing phase angle toward the backscattering geometry, while the opposition spike is a sharp increase of brightness usually at sub-degree phase angles. The phase angle is the angle between the direction of illumination (light source) and the direction of observation (detector), as seen from the target. The opposition effect has been studied in the laboratory by, e.g., Hapke and van Horn (1963), Egan (1969), and Capaccioni et al. (1990), while the opposition spike has been measured by Oetking (1966), Simonelli and Veverka (1978), Pleskott (1981), Smythe et.al (1986), and Buratti et.al (1988).

One of the most interesting studies was made by Oetking (1966), who measured several types of materials at very small phase angles. These measurements showed that bright materials, e.g., MgO, $CaCO_3$, and sugar crystals, exhibit well-defined opposition spikes. All these measurements were usually made without measuring the size, shape, or optical properties of single particles. Thus, from the measurements alone, it is impossible to draw other than qualitative conclusions. Phase curves of snow had been measured earlier by Knowles Middleton and Mungall (1952). The smallest phase angle in their measurements was 2 degrees. The measurements have been studied in depth by Veverka (1970, 1973), and Verbiscer and Veverka (1990). The main result is obvious: snow scatters light according to Lambert's law at large phase angles. The results agreed well with those obtained by space probes for icy satellites of the solar system. The reflectance and transmittance of snow have been further studied, e.g., by Kimball and Hand (1930), Gerdel (1948), Dunkle and Bevans (1955), and Hyvärinen and Lammasniemi (1987). These studies were mostly carried out to estimate the melting of the snow cover. However, they also include basic information on different types and grain sizes of snow.

The measurements indicate the presence of an opposition spike, which behaves like that for the icy moons (e.g., Lockwood et al. 1980) and Saturn's rings (e.g., Franklin and Cook 1965). The opposition spike of snow is qualitatively similar to that of the high-albedo asteroids (44) Nysa (Harris et al. 1989) and (64) Angelina (Poutanen 1983).

In what follows, we describe the device used for the measurements of snow, and present the results of the three measuring campaigns carried out in 1979-81 (I), 1995 (II), and 1997 (III).

2. DEVICE FOR SNOW MEASUREMENTS

The first set of field measurements of snow and snowballs (campaign I) were measured in 1979-81 (Piironen 1994) by making use of a device designed at the Department of Astronomy, University of Oulu. Finland (Figure 1). The mechanical part of the system was built at the workshop of the Department of Physics, University of Oulu, and the photometer and the strip chart recorder were the instruments used for the photometry of planetary objects at the Aarne Karjalainen Observatory, Kiiminki.

23. BACKSCATTERING OF LIGHT BY SNOW

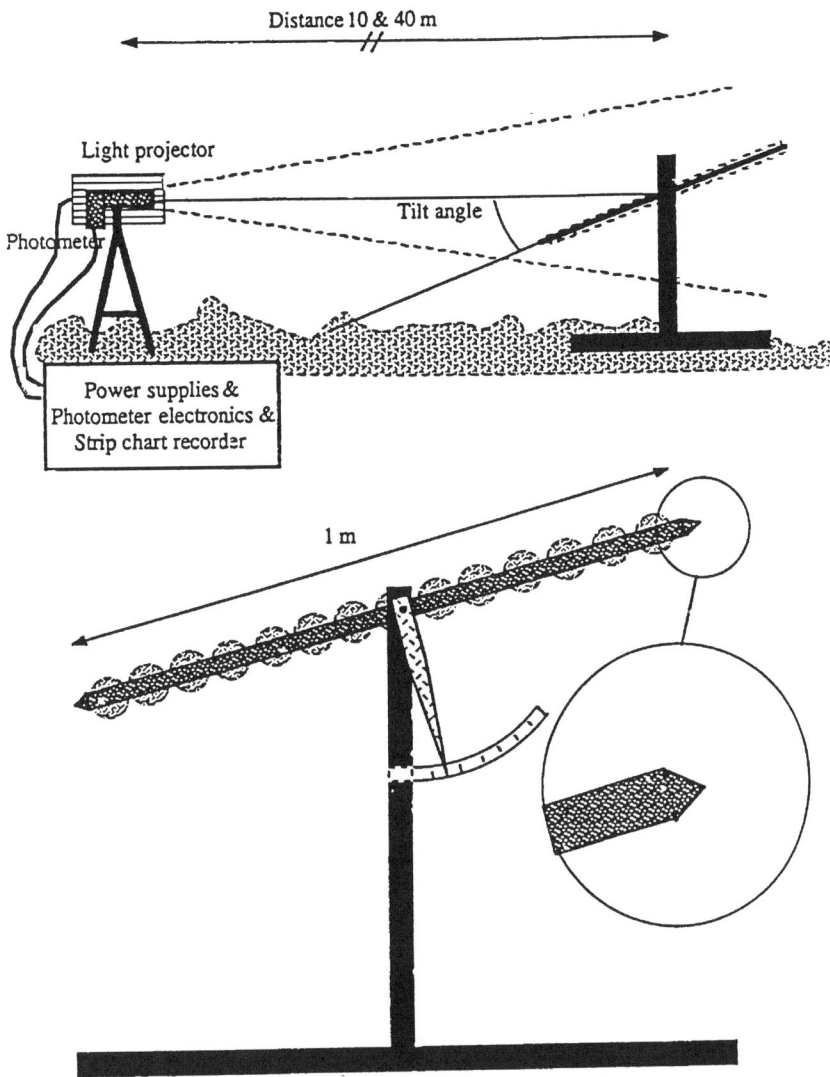

Figure 1. Schematic drawing of the device for snow measurements in 1979–81 at the Aarne Karjalainen Observatory, University of Oulu, Ylikiiminki, Finland

The measurements were made outdoors, which allowed a long baseline of 10 to 40 meters. The smallest possible phase angle was 0.3 degrees, which is well within the opposition spike of bright planetary objects. The main disadvantage was the maximum phase angle, which was only about 3 degrees when the baseline was 40 meters. However, the phase curves measured using the 10-m baseline extended the range of phase angles to 10 degrees. The measurements were carried out in the winters of 1979-80 and 1980-81 during nighttime and in as dry weather conditions as possible.

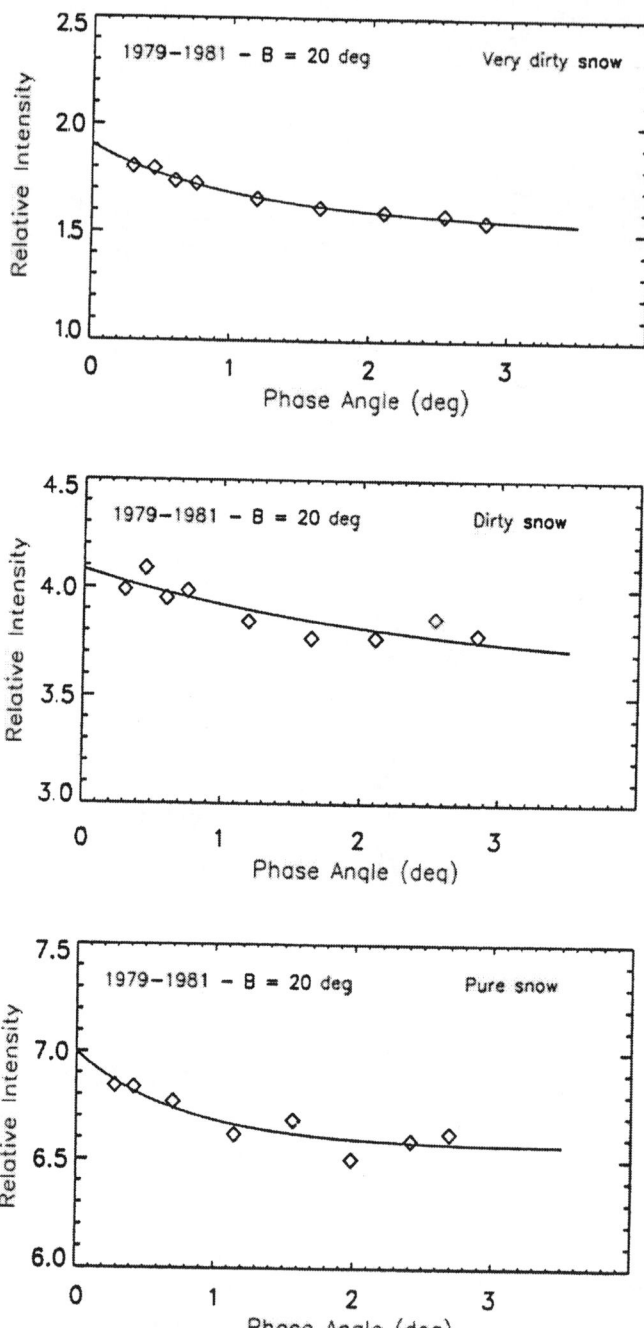

Figure 2. Phase functions measured in 1979–81 with a linear-exponential fit; tilt angle B=20°; (top) very dirty snowballs – boron carbide as impurity; (center) dirty snowballs – boron carbide as impurity; (bottom) pure snowballs – smaller ice crystals

23. BACKSCATTERING OF LIGHT BY SNOW

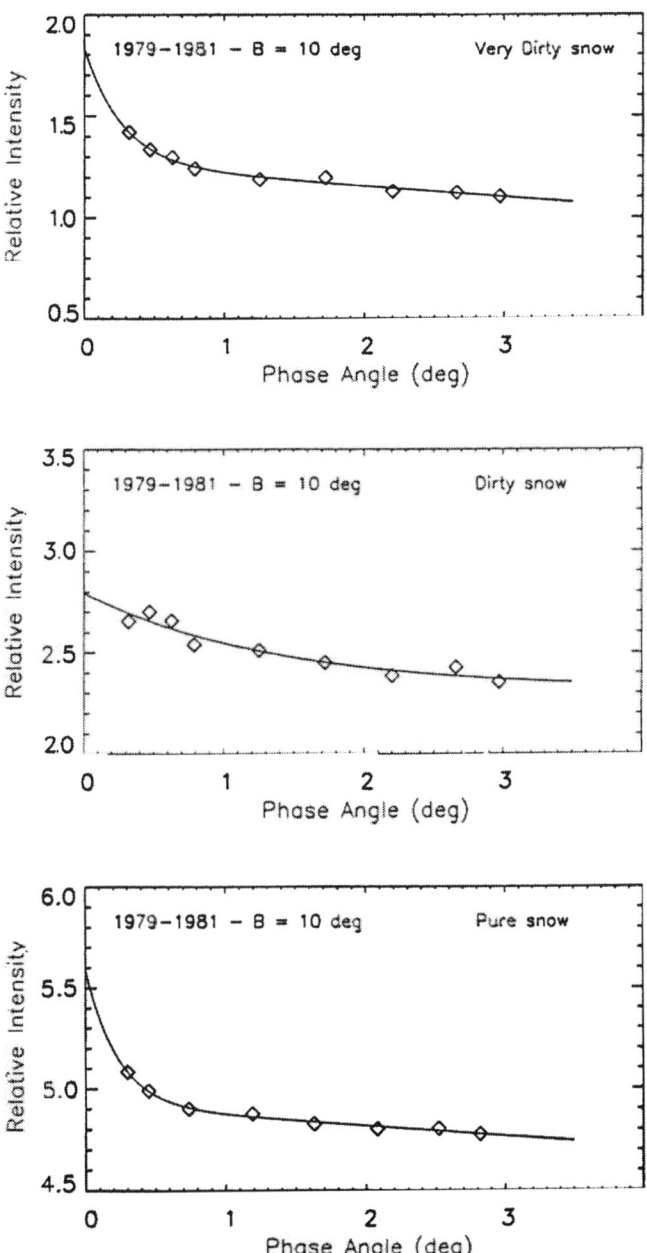

Figure 3. Phase functions measured in 1979–81 with a linear-exponential fit; tilt angle B = 10°; (top) very dirty snowballs – boron carbide as impurity; (center) dirty snowballs – boron carbide as impurity; (bottom) pure snowballs – smaller ice crystals

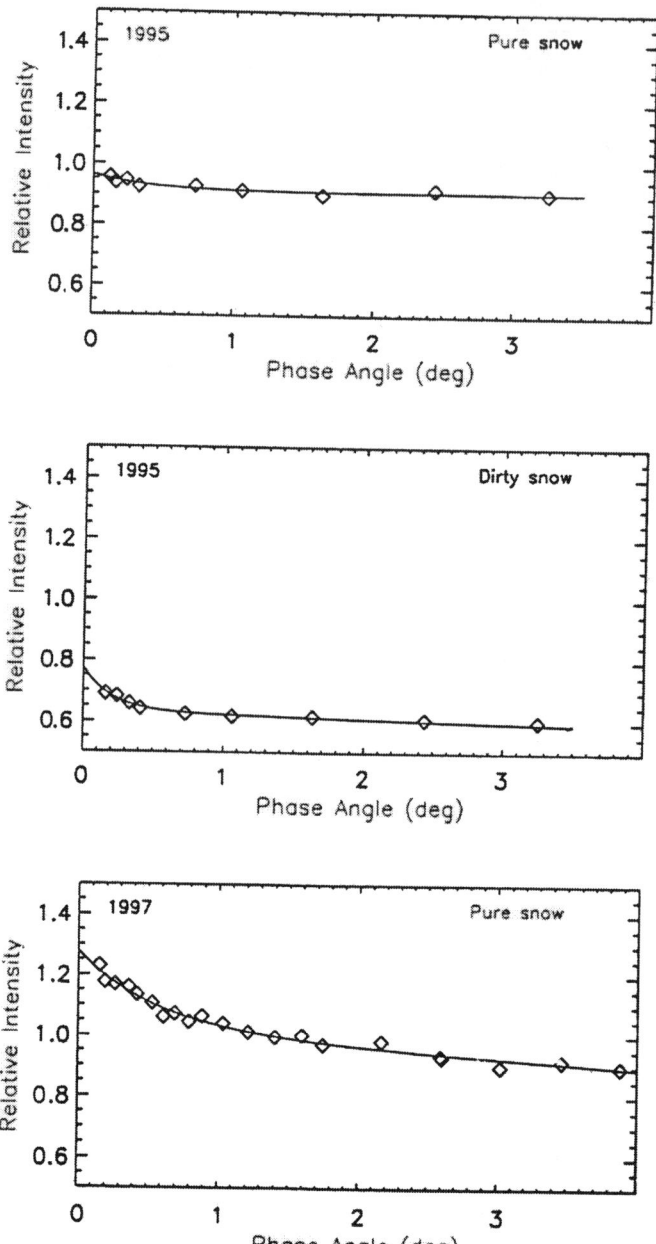

Figure 4. Phase functions measured in 1995 and 1997 with a linear-exponential fit; (top) pure snow, 1995, silicon carbide as impurity; (center) dirty snow, 1995, silicon carbide as impurity; (bottom) pure snow, 1997, large ice crystals

The diameter of the light source was about 15 cm, which corresponded to about 0.9° and 0.2° degrees at the 10-m and 40-m baselines respectively.

The beam diameter at the distance of 40 meters was limited to 1.5 meters, and the divergence of the beam was about 2°. The measurements were accomplished with an aperture that was slightly larger than the support for snow. Extra care was taken not to include the front edge of the sample in the aperture.

We measured either pure snow or snow with fine-grained boron carbide as impurity. The reflectivity of pure snow was between about 0.5 and 0.7 and, for the dirtiest snow, it dropped to 0.05-0.10. The crystals of snowgrains were from few tens of microns to few hundreds of microns in diameter (Figure 4). In this campaign we used mostly fresh new snow.

Campaigns II and III were carried out in 1995-1997 at the Metsähovi Observatory, Kirkkonummi, Finland. The light source was the same as in campaign I. The only difference was the use of a halogen lamp instead of a filament in order to guarantee a more homogeneous illumination of the surface. The baseline during the new measurements was 70 meters. Thus, we were able to measure the snow down to the minimum phase angle of 0.16 degrees. We measured older snow with grains larger than those in campaign I. As impurity we used silicon carbide powder with 25-50 micron particle size. The final reflectivities were higher than in campaign I.

3. MEASUREMENTS AND RESULTS

In order to express the backscattering characteristics in a simple way. we used a combination of linear and exponential functions (see Piironen 1994). Such a choice is reasonable and the fit is good. Furthermore, we do not want to bias toward any physical explanation at the moment. Now the phase curve is characterized by the width and the height of the surge, and the coefficient of the linear part:

$$I(\alpha) = I_s \exp(-0.5\alpha/l) + I_b + k\alpha$$

where α is the phase angle, I_s is the part of the intensity caused by the opposition spike, l is the width of the spike, and I_b and k are the background and the slope of the linear part of the intensity. The half-width of the spike is $2l\log_e 2 \approx 1.386l$. In Table 1, we present the parameters of the measured snowfields and in Figures 2 and 3 we present phase curves of the 1979-81 measurements. We give examples of the 1995 and 1997 data in Figure 4. In Table 2, we present, for comparison, parameters for selected planetary objects.

The phase curve measurements of snow do show in most cases an opposition spike, which is more pronounced for the dirtier snow (Figures 2 and 3. The spike can also be seen in the phase curve of the pure snow. Darkening the snow clearly pronounces the spike by lowering the intensity of long-range multiple scattering. So we conclude that either the surface hoar or microstructure of an individual snow grain could produce the spike. This agrees also with the laboratory measurements of sugar with small-particle sugar hoar on the surface (Oetking, 1966). In addition, there is a well-defined trend in the measurements: the opposition spike is larger for smaller tilt angles. Tilt angle is the angle between the sample mean plane and the light source.

Table 1. Results from the least-squares linear-exponential fits to the snow measurements. B is the tilt angle (angle between the sample mean plane and the observer), R is the reflectivity, I_s is the part of the intensity caused by the opposition spike, l is the width of the spike I_b and k are the background and the slope of the linear part of the intensity, and $(I_s + I_b)/I_b$ is the relative intensity of the surge.

Target	R	I_s	l	I_b	$(I_s + I_b)/I_b$	k
Pure snow - $B = 10°$	0.20	0.675	0.114	4.91	1.14	-0.0486
Pure snow - $B = 20°$	0.27	0.432	0.367	6.57	1.07	-0.0005
Pure snow - $B = 30°$	-	-	-	7,84	-	-0.1150
Dirty snow - $B = 5°$	0.075	0.475	0.0818	1,84	1.26	-0.110
Dirty snow - $B = 10°$	0.11	0.409	0.627	2.38	1.17	-0.014
Dirty snow - $B = 20°$	0.16	0.405	0.400	3.77	1.11	-0.000
Very dirty snow - $B = 5°$	0.045	0.515	0.177	1.03	1.50	-0.0376
Very dirty snow - $B = 10°$	0.05	0.569	0.136	1.26	1.45	-0.0534
Very dirty snow - $B = 20°$	0.07	0.304	0.459	1.60	1.19	-0.0189

Table 2. Results obtained for the selected bright planetary objects. Here p is geometric albedo, the other parameters are as in Table 1.

Target	p	I_s	l	I_b	$(I_s + I_b)/I_b$	k
Saturn's rings	-	0.228	0.223	0.804	1.29	-0.025
Callisto-trailing	0.20	0.200	1.47	0.877	1.23	-0.016
Callisto-leading	-	0.295	0.875	0.870	1.34	-0.022
64 Angelina	-	0.277	0.493	0.849	1.33	-0.012
44 Nysa	0.49	0.404	0.562	1.42	1.28	-0.023

4. CONCLUSIONS

The measurements may indicate the presence of coherent backscattering (e.g., Shkuratov and Muinonen 1991, Muinonen, 1994, and references therein). Conclusions should be, however, drawn cautiously. The snow particles in measurements were too large to support coherent backscattering. There is also a possibility that, in the case of dirty snow, shadowing may contribute to the opposition spike. However, shadowing produces normally a wider opposition effect. The opposition spike increases with increasing angle of incidence and when the material is darkened with the help of boron carbide. Because the size range of ice crystals inside the snowballs was few hundreds of micrometers, the opposition spike could be introduced by surface structure or microstructure of the fractures inside the crystals. Also we could not measure snow at extremely small phase angles (less than 0.16 degrees), which prevents us from making conclusions of the behavior of the opposition spike near zero phase angle. We did not measure particle size and shape distribution, single scattering properties, refractive indices of snow and boron carbide, or packing density of the target. We took photographs of the samples, which are accurate enough to allow a rough estimation of the properties above. The size estimation of the snow crystals is in good agreement with the study by Hyvärinen and Lammasniemi (1987).

We can conclude that the mechanism that produces the opposition spike is similar in all the cases above. This conclusion needs further theoretical considerations to be fully approved, because the comparison between the integrated brightness of planetary objects and the surface brightness of laboratory targets should be made cautiously.

An interesting conclusion can be drawn from the measurements: the opposition spike of snow could be used to determine the properties of different types of snows. With the help of small-phase-angle observations, one could determine the accumulation of dust on the snow surface and even estimate the particle size of pure snow. This could help in estimating the purity of the snow, and melting rate, and water content of snow. Such observations could help in estimating snow purity, melting rate, and water content, which are of great importance for, e.g., agriculture, and for estimating the purity of fresh water supplies.

5. ACKNOWLEDGEMENTS

J. Piironen is grateful to Ulla for patience during the instrument calibration in the 1979-81 campaign. We like to thank Olof Hernius and Markku Poutanen in helping to measure accurately the baseline during 1995 and 1997 campaigns.

6. REFERENCES

Capaccioni, F., Cerroni, P., Barucci, M.A. and Fulchignoni, M. (1990) Phase curves of meteorites and terrestrial rocks: Laboratory measurements and application to asteroids, *Icarus*, **83**, 325–348.

Dunkle. R.V. and Bevans, J.T. (1955) An approximate analysis of the solar reflectance and transmittance of a snow cover, *Journal of Meteorology*, **13**, 212–216.

Egan, W.G. (1969) Polarimetric and photometric simulation of the martian surface, *Icarus*, **10**, 223–227.

Franklin, F.A. and Cook, A.F. (1965) Optical properties of Saturn's rings II. Two-color phase curves of the two bright rings, *Astronomical Journal*, **70**, 704–720.

Gerdel, R.W. (1948) Penetration of radiation into the snow pack, *Transactions of the American Geophysical Union*, **29**, 366–374.

Hapke, B.W. and van Horn, H. (1963) Photometric studies of complex surfaces with application to the Moon, *Journal of Geophysical Research*, **68**, 4545–4570.

Harris, A.W., Young, J.W., Contreiras, L., Dockweiler, T., Belkora, L., Salo, H., Harris, W.D., Bowell, E., Poutanen, M. Pinzel, R. P., Tholen, D. J. and Wang, S. (1989) Phase relations of high albedo asteroids: The unusual opposition brightening of 44 Nysa and 64 Angelina, *Icarus*, **81**, 365–374.

Hyvärinen, T., and Lammasniemi. J. (1987) Infrared measurements of free-water content and grain size of snow, *Optical Engineering*, **26**, 342–348.

Kimball, H.H. and Hand, I.F. (1930) Reflectivity of different kind of surfaces, *Monthly Weather Review*, **58**, 280–282.

Knowles Middleton, W.E. and Mungall, A.G. (1952) The luminous directional reflectance of snow, *Journal of the Optical Society of America*, **42**, 572–579.

Lockwood, G. W., Thompson, D.T. and Lumme, K. (1980) A possible detection of solar variability from photometry of Io, Europa, Callisto and Rhea, 1976–1979, *Astronomical Journal*, **85**, 961–968.

Muinonen, K. (1994) Coherent backscattering by Solar System dust, Asteroids, Comets and Meteors 1993, in A. Milani, M. Di Martino, and A. Cellino (Editors): *Proceedings of 160th International Astronomical Union*, held in Belgirate, Italy, June 14–18, 1993, Kluwer Academic Publishers, Netherlands, 271–296.

Oetking, P. (1966) Photometric studies of diffusely reflecting surfaces with applications to the brightness of the Moon, *Journal of Geophysical Research*, **71**, 2505–2513.

Piironen. J. (1994) Ph.Lic. Thesis, University of Helsinki.

Pleskott, L. (1981) Ph.D. Thesis, UCLA, Los Angeles.

Poutanen, M. (1983) UBV photometry of asteroid 64 Angelina, in C.-I. Lagerkvist ja H. Rickman (Editors): *Asteroids, Comets, Meteors*, Uppsala Universitet, Reprocentralen HSC, Uppsala, 45–48.

Shkuratov, Yu. G., and Muinonen, K. (1991) Interpreting asteroid photometry and polarimetry using a model of shadowing and coherent backscattering, in A. W. Harris and E. Bowell (Editors): *Asteroids, Comets, Meteors 1991*, 549–552, Lunar and Planetary Institute.

Verbiscer. A.J. and Ververka. J. (1990) Scattering properties of natural snow and frost: Comparison with icy satellite photometry, *Icarus*, **88**, 418–428.

Veverka, J. (1970) Ph.D. Thesis, Harvard Universtity, Cambridge, Massachusetts.

Veverka, J. (1973) The photometric properties of natural snow and snow-covered planets, *Icarus*, **20**, 304–310.

Chapter 24

AUSTRALIAN SITES FOR THE VALIDATION OF SATELLITE RETRIEVALS OF THE RADIATIVE PROPERTIES OF LAND SURFACES

I. F. Grant (1), A. J. Prata (1), G. Rondeaux (2) and M. D. Steven (2)
(1) CSIRO, Division of Atmospheric Research, Australia and (2) Department of Geography, University of Nottingham, United Kingdom.

1. INTRODUCTION

The radiation budget at the Earth's surface is an important component of the climate system. Climate research needs global observations of the surface radiation budget (SRB) parameters in order, for instance, to validate the output of numerical climate models and to improve the parameterization of radiation in models. Satellite-based retrievals of SRB parameters over land must address the directional reflectance properties of the land surface, as well as spatial inhomogeneity, the effect of the atmosphere, and spectral and temporal sampling by the satellite sensor.

The long-term validation of SRB retrievals is difficult because ground-based measurements of the upwelling radiation fluxes are seldom representative of the 1-km^2 pixels of many current and planned satellite sensors. Therefore, Australia's CSIRO has established the Continental Integrated Ground-truth Site Network (CIGSN) for the validation of satellite retrievals of surface radiation fluxes and related parameters such as albedo and land surface temperature.

2. THE CONTINENTAL INTEGRATED GROUND-TRUTH SITE NETWORK

CIGSN is a network of Australian land sites that is being developed for the calibration and validation of satellite data and products, with a particular emphasis on the components of the SRB and the surface properties that control those components and influence their measurement from Earth Observation satellites.

The project's goals are
- development of techniques for retrieving land surface radiation parameters from satellite data
- calibration of satellite data and products
- validation and improvement of numerical climate models
- acquisition of a long time series of surface radiation measurements.

The emphasis is on newer sensors, such as ATSR, IMG and POLDER, and those planned for launch soon, including AATSR, GLI, ASTER and MODIS.

Two sites have been established in extremely flat and homogeneous natural terrain (Figure 1). Both are characterized by low atmospheric aerosol and water vapor amounts. The Uardry site (145.305° E, 34.392° S), shown in Figure 2, is located near the town of Hay in pasture that is typical of much of eastern Australia; here, data have been collected since mid-1992. Prata (1994) gives more detail on the Uardry site. The Amburla site (133.119° E, 23.385° S), near Alice Springs in central Australia, is located in arid grazing land and has operated since early 1995.

24. AUSTRALIAN SITES FOR THE VALIDATION OF SATELLITE RETRIEVALS OF THE RADIATIVE PROPERTIES OF LAND SURFACES

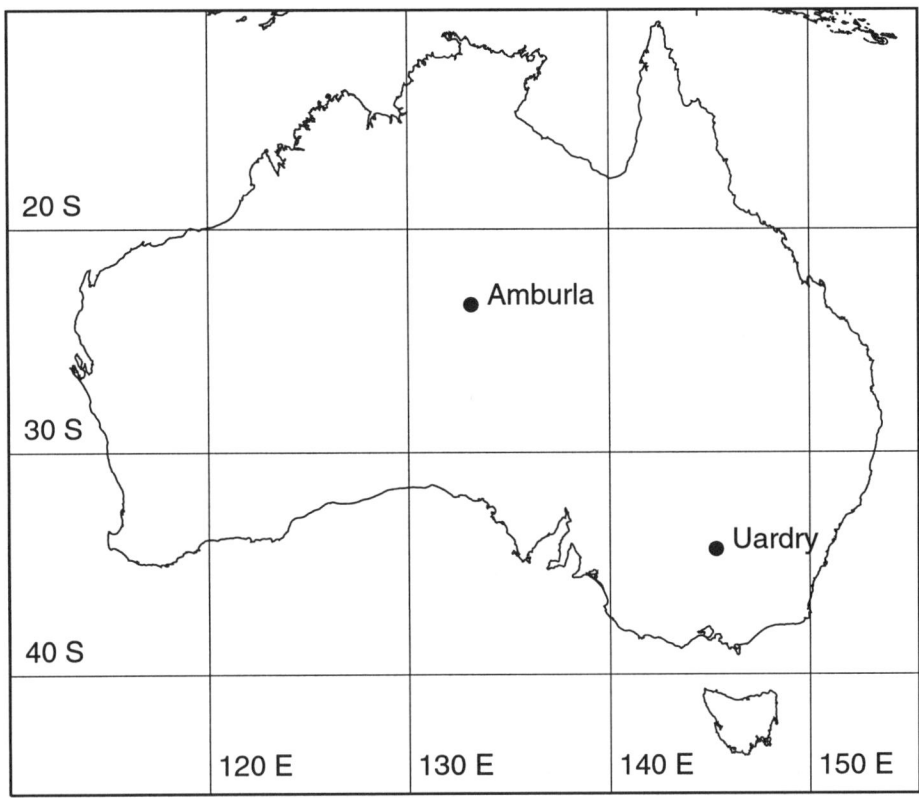

Figure 1. A map of Australia showing the location of the CIGSN sites

Figure 2. A measurement station at Uardry with a downward-looking pyranometer

The CIGSN sites continuously monitor
- solar (short-wave) radiation, downwelling and upwelling
- thermal (longwave) radiation, downwelling and upwelling
- land surface temperature
- air temperature, humidity, pressure, and wind speed at 2 and 15 m above the surface.

During campaigns the following variables are measured:
- directional spectral radiance
- tropospheric temperature and moisture profiles
- visible and thermal upwelling radiances from aircraft.

Aerosol optical depth and water vapor column amount are monitored continuously at Uardry and during campaigns at Amburla.

3. SITE UNIFORMITY

Both CIGSN sites were chosen for their high spatial uniformity over several square kilometers so that ground-based measurements would be representative of the area covered by a 1-km satellite pixel. Furthermore, the Uardry site was placed on the largest plain in Australia in the expectation

24. AUSTRALIAN SITES FOR THE VALIDATION OF SATELLITE RETRIEVALS OF THE RADIATIVE PROPERTIES OF LAND SURFACES

that it would be representative of the region at scales up to 100-200 km, which is the typical size of a grid box in a numerical global climate model. This section quantifies the spatial uniformity of the Uardry site using airborne and ATSR-2 images.

Figure 3 shows a 520-600 nm image (~20-m pixels) of the Uardry site taken by the Daedalus airborne multi-channel scanner on 26 April 1997. The overlaid histograms and statistics quantify the uniformity of each 1-km square. No calibration has been applied.

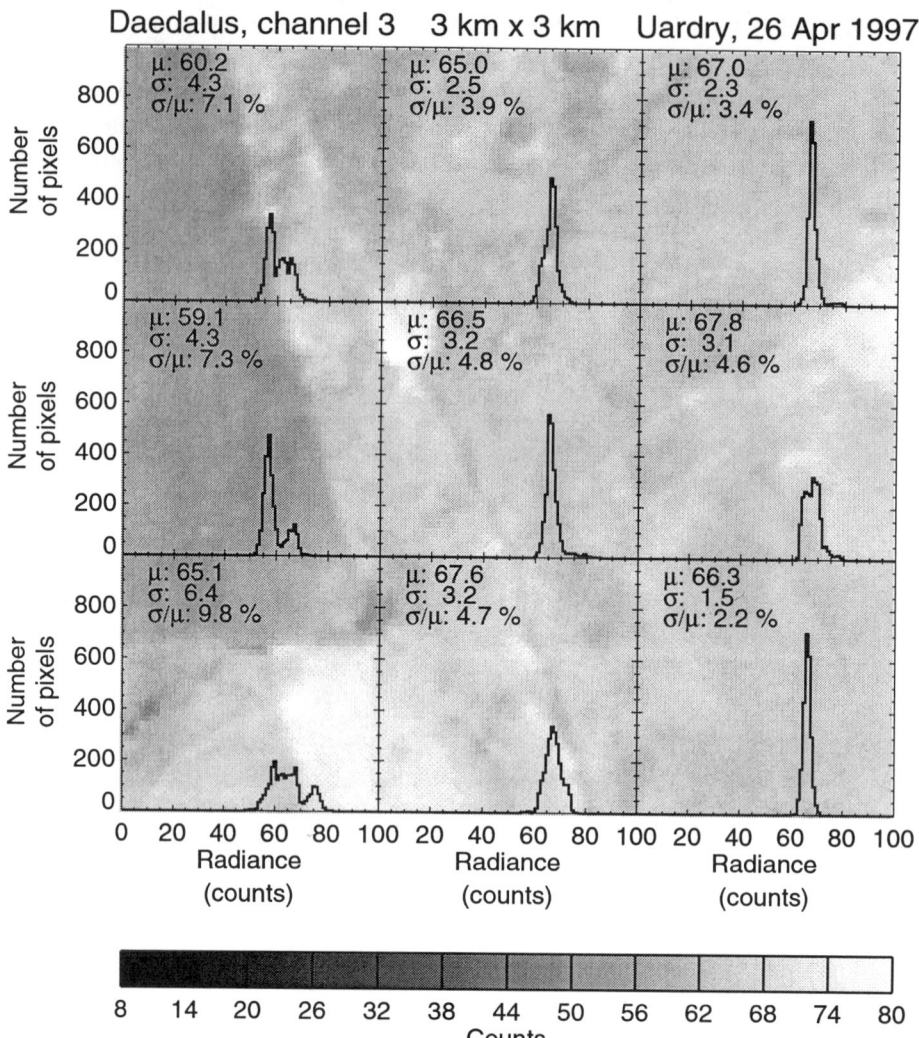

Figure 3. A Daedalus image of the Uardry site taken on 26 April 1997. Histograms and statistics of the pixel raw digital numbers (counts) are overlaid

The coefficient of variation (standard deviation / mean) is typically 5% and as low as 2% for some squares. This parameter measures how close a ground measurement of a randomly chosen 20-m pixel will be to the average over the 1-km² pixel of a sensor such as the ATSR-2.

Figure 4 follows the mean and standard deviation of a 3×3 grid of successively larger pixels formed by averaging aggregates of 20-m pixels. Also plotted for scales larger than 1 km are the same parameters derived from the 555 nm channel of an ATSR-2 image acquired two weeks later. The Daedalus means have been arbitrarily scaled to match ATSR-2 at 1 km.

24. AUSTRALIAN SITES FOR THE VALIDATION OF SATELLITE RETRIEVALS OF THE RADIATIVE PROPERTIES OF LAND SURFACES

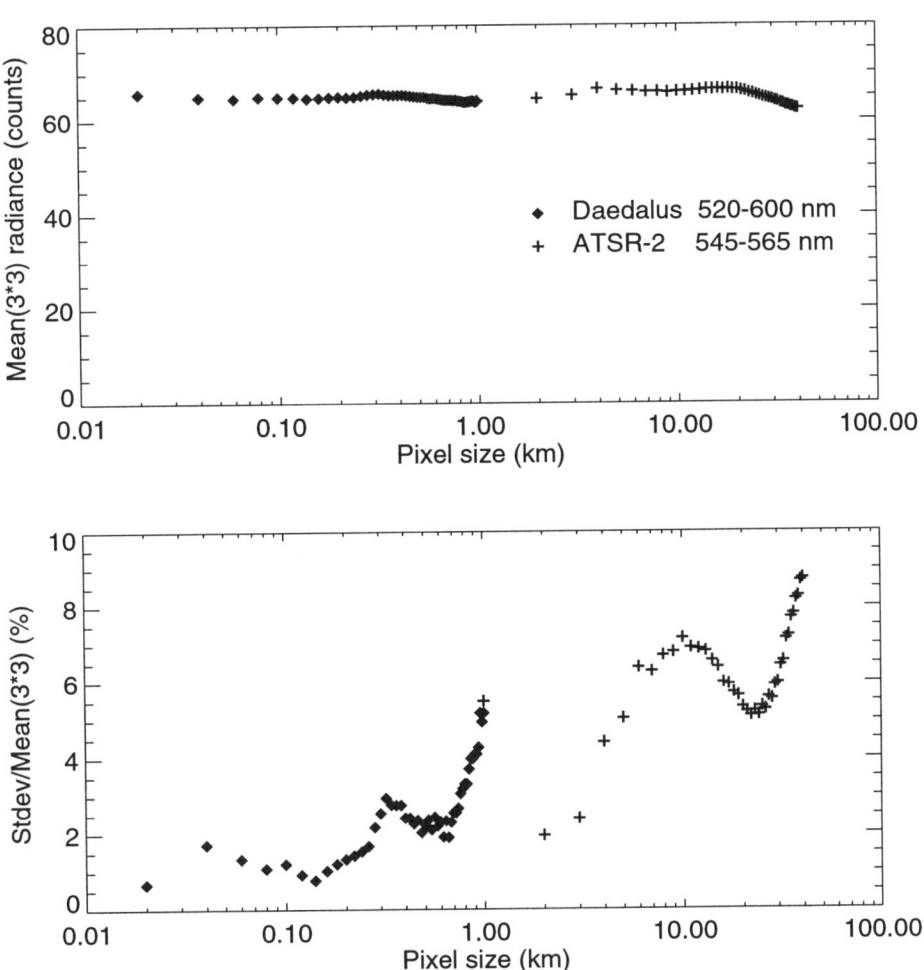

Figure 4. The pixel mean and variability as a function of scale (20 m - 100 km) for matching Daedalus and ATSR-2 channels

The suitability of the site for extrapolating ground-based surface measurements to larger spatial scales is demonstrated by the constancy of the mean to within a few percent for pixel sizes from 20 m to 1 km. At larger scales it is seen that the radiance of 1-km pixels measured at the site by

ATSR-2 is representative of the surrounding region to scales of tens of kilometers.

4. SPECTRAL BRDF MEASUREMENTS

CSIRO measures the surface-leaving spectral radiance with a triple grating CCD spectrometer. The spectrometers operate over three overlapping wavelength ranges: 380-540 nm, 480-870 nm and 500-1000 nm. Each has 1024 detector elements and is fed through a fiber-optic cable from input optics with a roughly 10° field of view. Single spectra are acquired over 200 ms and are logged onto a notebook computer on which they can be immediately displayed.

The spectrometer input optics are mounted on a computer-pointable dual-axis mount (the Remote AXis Controller - RAXC). For the results presented here, the upwelling radiance was sampled on one side of the principal plane, at zenith angle increments of 15° and azimuth increments of 30°, and then symmetry in the principal plane was assumed to fill the unmeasured azimuths. The measurements were repeated at several places across an area of a few km^2 at each site to assess and reduce the effect of spatial variability.

Figure 5 shows the hemispherical distribution of directional spectral reflectance measured at Amburla and integrated over the ATSR-2 555 nm band. The distribution has the characteristic bowl shape expected from vegetation, and a distinct hot-spot in the backscatter direction. Figure 6 shows that the Roujean three-parameter model (Roujean et al., 1992) of the bi-directional reflectance distribution function (BRDF) fits the measurements well.

24. AUSTRALIAN SITES FOR THE VALIDATION OF SATELLITE RETRIEVALS OF THE RADIATIVE PROPERTIES OF LAND SURFACES

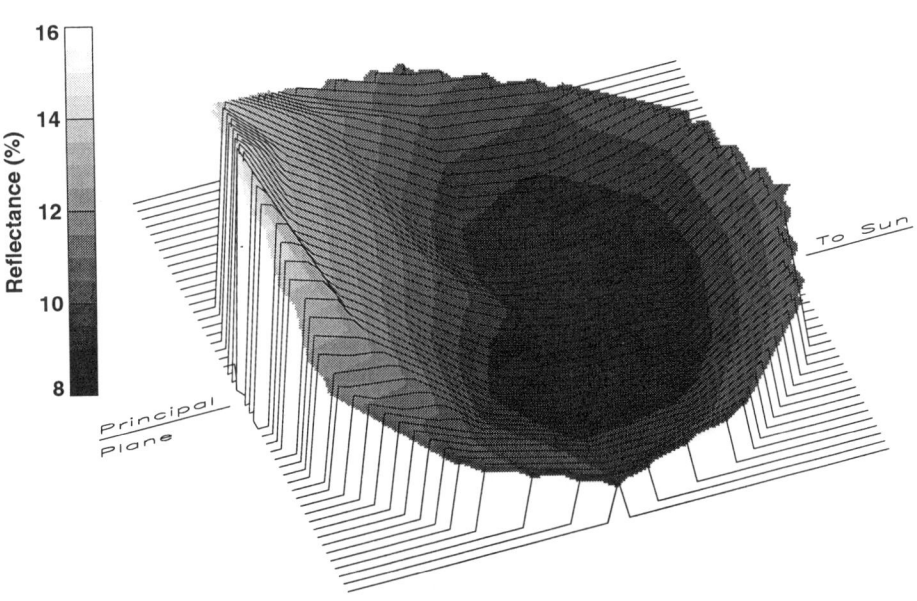

Figure 5. Directional reflectance measurements at Amburla in the ATSR-2 555 nm channel. The center of the plot represents the nadir view and the edges represent the horizon

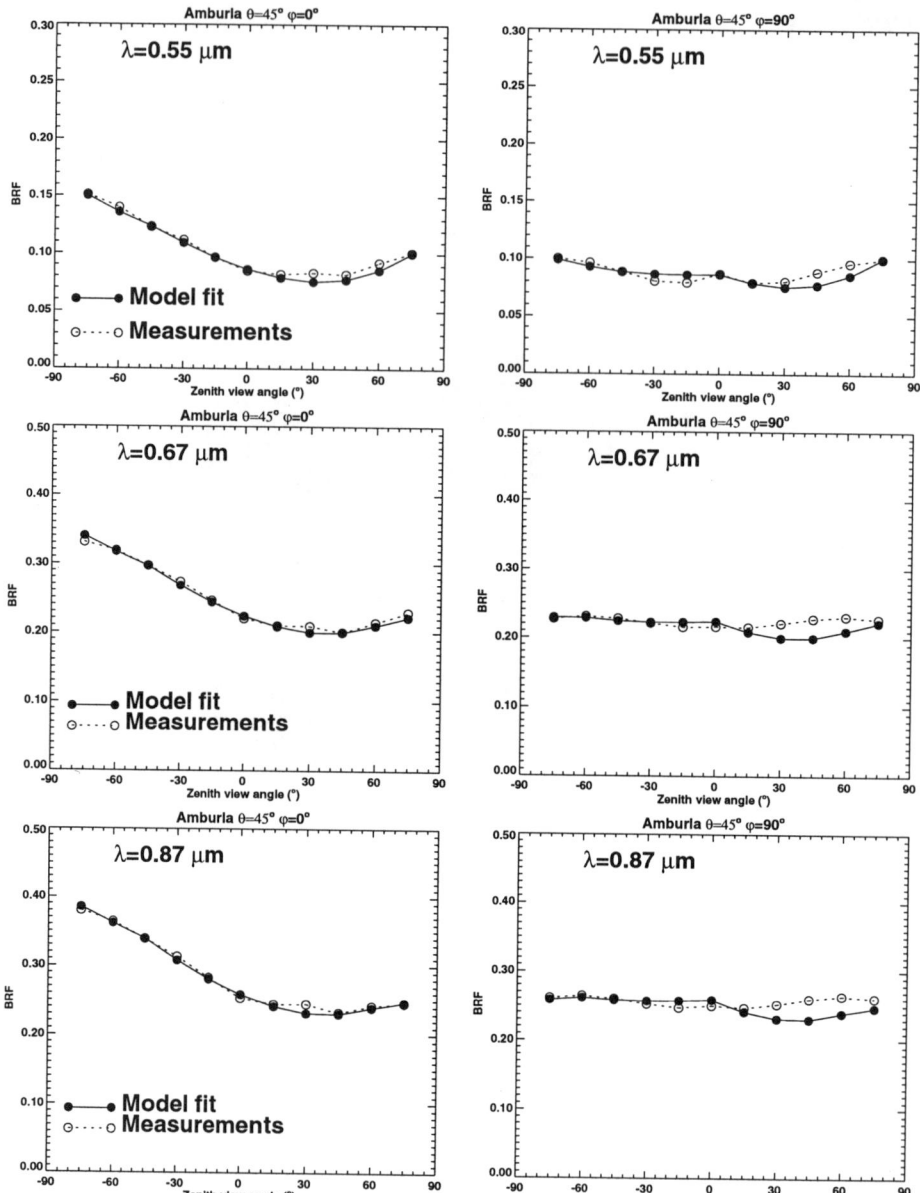

Figure 6. Sections in, and orthogonal to, the principal plane of the ACEX Amburla upwelling radiance measurements, and the corresponding values from the fitted Roujean model

5. VALIDATION OF ATSR-2 SHORTWAVE CHANNELS

The utility of the site for validating satellite measurements is illustrated by comparing measurements of the spectral radiance of the site made at the top of the atmosphere by ATSR-2 and at the ground during a collaborative campaign in April 1997 between CSIRO and the University of Nottingham. During the ATSR-2 atmospheric Correction EXperiment (ACEX) the directional surface reflectance was measured during one week (three ATSR-2 overpasses) at each site. The atmospheric water vapor was characterized by radiosonde profiles taken at the site at the overpass times and the aerosol optical depth was measured by a shadowband radiometer. The CSIRO and University of Nottingham teams each took measurements with different instruments and analyzed them with different radiative transfer codes.

The validation procedure consists of
1. measuring the surface directional spectral reflectance at the validation site: in the view direction used by ATSR-2 at the time of overpasses, and by interpolating measurements over the whole hemisphere made at other times;
2. applying an atmospheric correction to the surface measurements, using the measured water vapor profile and aerosol amount - CSIRO used the MODTRAN3B radiation code (Kneizys et al., 1988), and the University of Nottingham used a scheme (Mackay et al., 1994) built around the 5S code (Tanré et al., 1990);
3. integrating the top-of-atmosphere (TOA) spectral radiance with the filter response function over each ATSR-2 band;
4. comparing the result with the calibrated ATSR-2 TOA radiances.

Figure 7 shows preliminary validation results using the CSIRO/University of Nottingham measurements and MODTRAN3B modeling. The results suggest good agreement to within a few percent with the existing calibration of the ATSR-2. Understanding of the differences in detail requires further investigation.

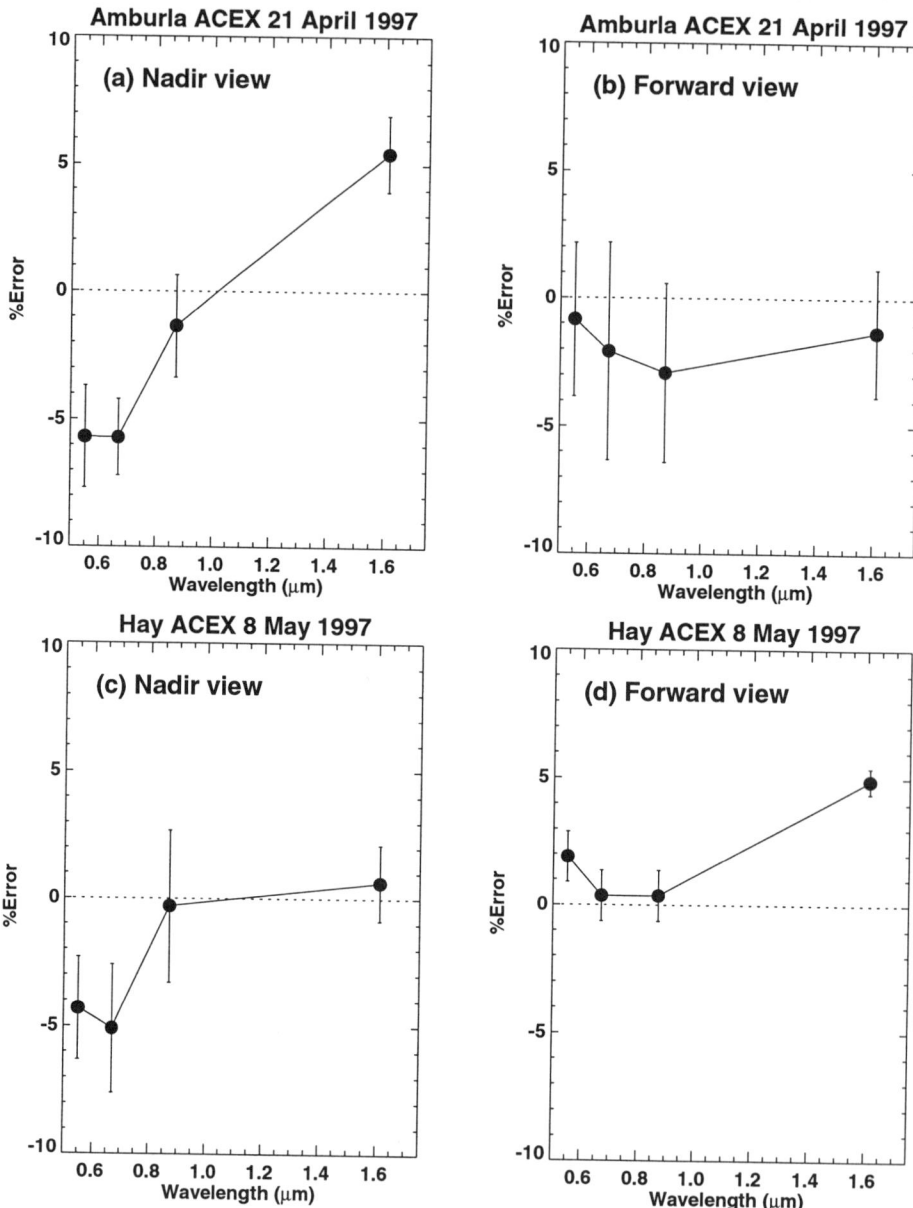

Figure 7. Preliminary validation results for ACEX. The ordinate is %Error = (O − C) / O, where O is the TOA radiance observed by ATSR-2, and C is the surface measurement corrected to the top of the atmosphere. The error bars are derived from the standard deviation of a 3×3 array of ATSR-2 pixels

6. CONCLUSION AND FUTURE

This paper has outlined the goals of the CIGSN network, described its first two sites, and presented some preliminary results from a campaign to validate the comparison of top-of-atmosphere radiances measured at the sites by the ATSR-2 with surface measurements. Similar campaigns will be planned to validate the data and products of several satellite sensors to be launched over the next few years. A third CIGSN site is planned for tropical Australia to extend the validation activities to conditions of high humidity.

7. ACKNOWLEDGMENTS

John Wright (RAL, UK) is thanked for his assistance with planning the ATSR-2 acquisitions during ACEX. The ATSR-2 data were processed and supplied by RAL and are used courtesy of ESA.

8. REFERENCES

Kneizys, F. X., E. P. Shettle, L. W. Abreu, J. H. Chetwynd, G. P. Anderson, W. O. Gallery, J. E. A. Selby, and S. A. Clough (1988) *Users guide to LOWTRAN-7*, Technical Report AFGL-TR-88-0177, Optical/Infrared Technology Division, U.S. Air Force Geophysics Laboratory, Hanscom Air Force Base, Massachusetts.

Mackay, G., M. D. Steven, and J. A. Clark (1994) A pseudo-5S code for the atmospheric correction of ATSR-2 visible and near-infrared land surface data, *Sixth International Symposium "Physical measurements and signatures in remote sensing"*, CNES, France.

Prata, A. J. (1994) Validation data for land surface temperature determination from satellites, *Technical Paper* No. 33, 36 pp., CSIRO, Division of Atmospheric Research, Aspendale, Victoria, Australia.

Roujean, J. L., M. Leroy, and P. Y. Deschamps (1992) A bidirectional reflectance model of the Earth's surface for the correction of remote sensing data, *Journal of Geophysical Research*, **97**, 20,455–20,468.

Tanré, D., C. Deroo, P. Duhaut, M. Herman, J. J. Morcrette, J. Perbos, and P. Y. Deschamps (1990) Description of a computer code to simulate the satellite signal in the solar spectrum - the 5S code, *International Journal of Remote Sensing*, **11**, 659–686.

Chapter 25

INSTRUMENTS AND METHODS FOR THE GROUND-LEVEL REFERENCE MEASUREMENT OF SOLAR RADIATION, ALBEDO AND NET RADIATION

M. Sulev, J. Ross and E.-M. Maasik
Tartu Observatory, Tõravere, Estonia.

1. INTRODUCTION

Regular actinometrical observations in Tartu Actinometric Station started in 1950. Since 1954, it has been regarded as one of the most effectively working actinometric station in the Soviet Union. In 1965, the station was moved, and it is now located at Tõravere, about 20 km southwest from Tartu, on the territory of Tartu Observatory (φ =58°16', λ =26°28'). The station currently belongs to the Estonian Meteorological and Hydrological Institute, but is working under the supervision and with the collaboration of scientists from the Tartu Observatory (former Institute of Astrophysics and Atmospheric Physics). Right from the start of scientific activities, methodical investigations to improve the accuracy of the measurements were initiated. Investigations have proceeded continuously, albeit with variable intensity, and encompassed a very wide range of problems: instrumental and methodical errors, comparison between different sensors, calibration of the instruments, designing new instruments, etc. In the former Soviet Union and Eastern European countries, instruments manufactured in the Soviet Union were in use as a rule. After the breaking up of the Soviet system, the modernization of the equipment to meet worldwide standards was started in these countries. This on-going process requires new additional methodical investigations. In Tõravere's Actinometric Station and Tartu Observatory, these problems became especially acute when the station was incorporated into the Baseline Surface Radiation Network, demanding measurements of all radiation fluxes with high quality as reference data for different tasks. Some conclusions and ideas from this experience are presented below.

2. RADIOMETRIC SCALE, REFERENCE INSTRUMENTS AND CALIBRATION METHODS

The radiometric scale in the Soviet Union was based on the national reference instrument—the Ångström pyrheliometer No. 212, which was periodically compared with the World Radiometric Reference (WRR). In Tõravere's Actinometric Station, an Ångström type pyrheliometer M-59-8 No. 1981 designed by Y. Yanishevsky is in use as the reference instrument. The pyrheliometer was periodically compared with the Soviet Union national reference. Direct comparison with the WRR in Davos in 1995 demonstrated good agreement with an error < 0.1%. Since 1997 a new reference is in use, namely the absolute cavity radiometer PMO-6 No. R850405. As a secondary reference instrument for the routine calibration of the working instruments the Yanishevsky actinometer AT-50 No. 596 is used. The results of the calibration of the secondary reference are presented in Fig. 1. In 1997, a new secondary reference, the normal incidence pyrheliometer (NIP) No. 16418, was also repeatedly calibrated using AT-50 No 596 and PMO-6 No R850405 as references (Fig. 2). The figures confirm that there are no essential differences between the M-59-S and PMO-6 and the radiometric scale is correctly presented at the station. Fig. 3 exhibits an example of the simultaneous record of the direct solar radiation in an almost cloudless day using NIP No. 16418 and AT-50 No. 331. In spite of the different angles of view of the instruments (NIP $-5.7°$, AT-50 $-10°$) the agreement is good.

25. INSTRUMENTS AND METHODS FOR THE GROUND-LEVEL REFERENCE MEASUREMENT OF SOLAR RADIATION, ALBEDO AND NET RADIATION

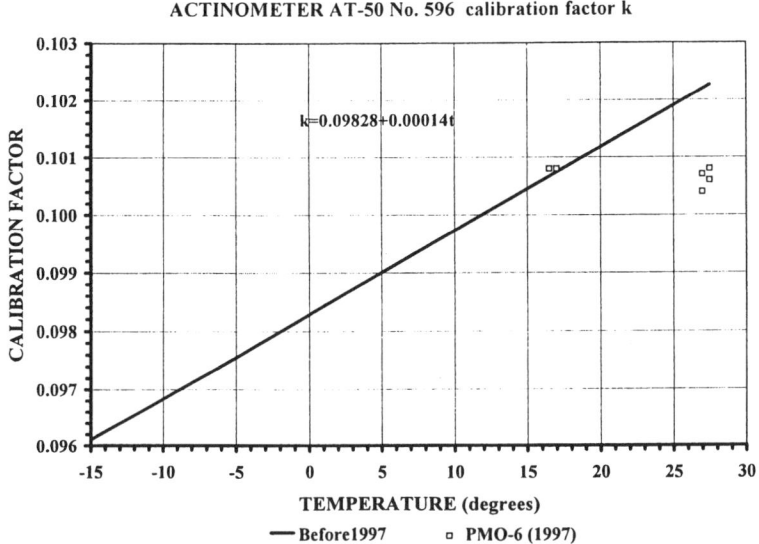

Figure 1. The calibration constant of the secondary reference actinometer AT-50 No. 596, compared with reference Ångström pyrheliometer No. 1981 (since 1997) and PMO-6 No. R850405 (1997)

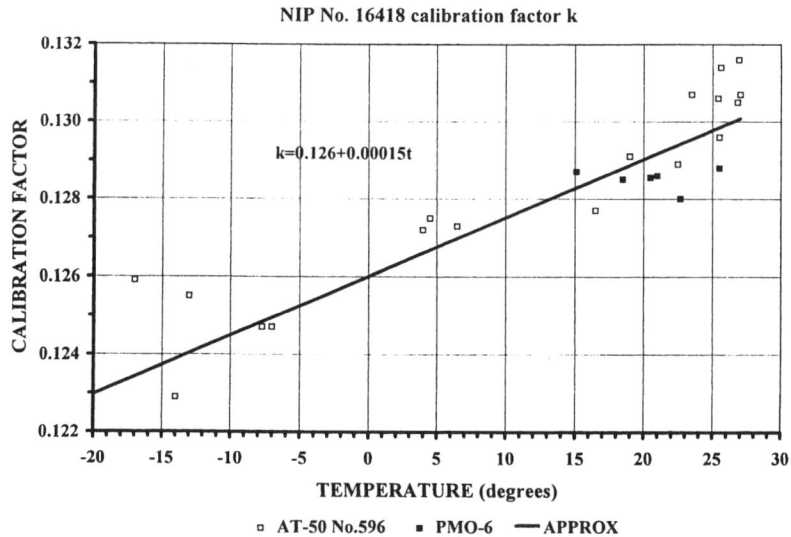

Figure 2. The calibration constant of the NIP pyrheliometer No. 16418.

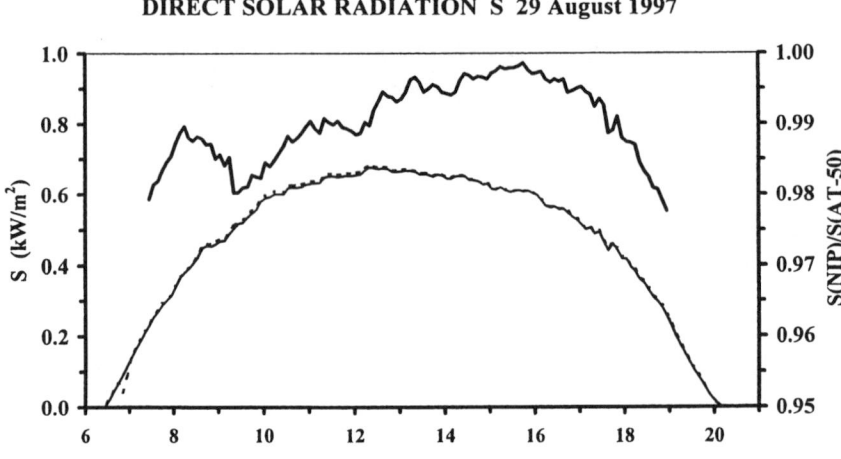

Figure 3. The flux density of direct solar radiation measured using actinometer AT-50 No. 331 and NIP No. 16418. August 29, 1997, clear sky with slight Ci clouds

The Sun-and-shadow method is commonly used for calibration of the pyranometers. According to our experience, better results can be derived using special tube (Fig. 4), geometrical proportions of which are in accordance with these of the reference pyrheliometer, i.e., $d/d_1 = D/D_1 = l/l_1$, where d is the diameter of the aperture diaphragm, D the diameter of the field diaphragm and l the distance between them of the reference pyrheliometer, d_1 is the diameter of the receiving surface of the pyranometer, D_1 the diameter of the field diaphragm of the tube and l_1, the distance from the sensitive surface of the pyranometer to the field diaphragm of the tube. When both the pyrheliometer and the tube are directed towards the Sun, the output of the pyranometer enclosed into the tube is proportional to the direct solar radiation measured by the reference pyrheliometer. To avoid overheating of the pyranometer in the bottom of the tube around pyranometer the openings must be made for air circulation. The temperature of the pyranometer has to be measured as the sensitivity (calibration factor) of all types more or less depends from the temperature.

25. INSTRUMENTS AND METHODS FOR THE GROUND-LEVEL REFERENCE MEASUREMENT OF SOLAR RADIATION, ALBEDO AND NET RADIATION

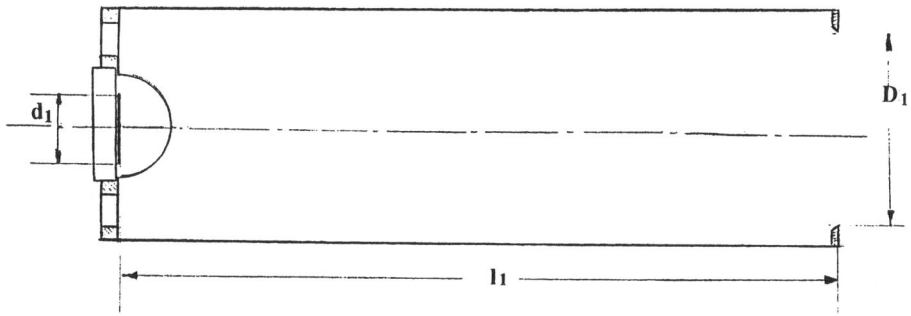

Figure 4. A scheme of the calibration tube for pyranometers

3. PYRANOMETERS

Two types of the pyranometers are in use: devices with black and white receiving surface and single glass dome, and with black receiving surface and double glass dome (Fröhlich, London, 1986). The first ones are more stable under variable conditions and their sensitivity depends less on the temperature, but they are more selective and the sensitivity may change due the aging of the white paint. Instruments with the black receiving surface do not have the shortcomings mentioned above, but the output of these may be affected by the thermal radiation of the domes when the temperature differs from the temperature of the sensor. It is why the moderate ventilation of the instrument is recommended to stabilize the thermal regime. Another possible reason of instrumental errors is less well-known. Usually the instruments are designed so that active thermocouples are in thermal contact with the receiving surface and reference ones with the instrument body, the thermal inertia of which is far larger. For that reason essential deviations of the zero point are possible under variable temperature regime of the instrument. To avoid these errors, a thermally symmetrical pyranometer TP-3 was designed by J. Reemann (Fig. 5). The thermocompensated hermetical pyranometer is a precision thermoelectrical sensor designed to measure global, diffuse or reflected solar radiation under field conditions. Fully symmetrical construction guarantees high stability of sensitivity and zero point at variable temperature regime. The instrument is hermetical to avoid inside condensation. The thin double glass domes and painted by NEXTEL 2010 Velvet Coating Black sensitive surface cover measurements in the spectral region of 0.3 to 4.8 µm. Sensor-elements are artificially aged and protected against corrosion and oxidation. The sensitivity of the instrument depends on temperature by about 0.2% per degree and must be taken into account (Fig. 6). The pyranometer is weatherproof and does not need special inspection. The instrument is appropriate for routine actinometrical and meteorological measurements, but is especially so for special investigations in atmospheric physics, ecology, bioactinometry, forestry etc.

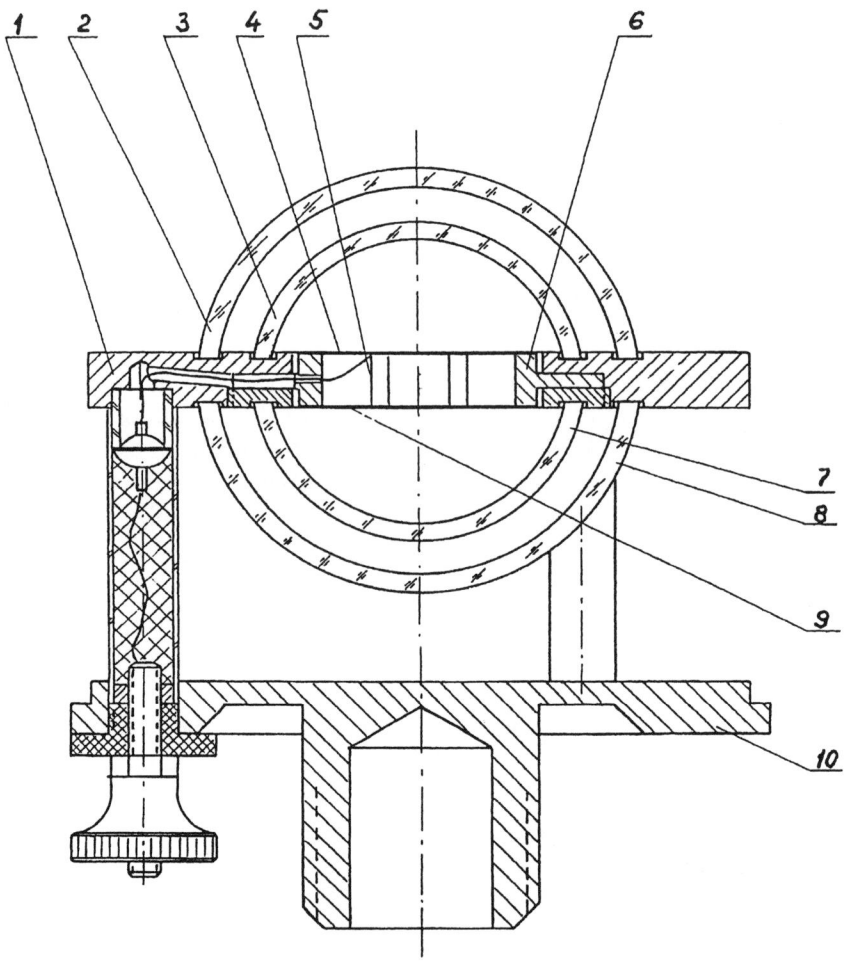

Figure 5. A schematic cross-cut of the minipyranometer TP-3 designed by J. Reemann

25. INSTRUMENTS AND METHODS FOR THE GROUND-LEVEL REFERENCE MEASUREMENT OF SOLAR RADIATION, ALBEDO AND NET RADIATION

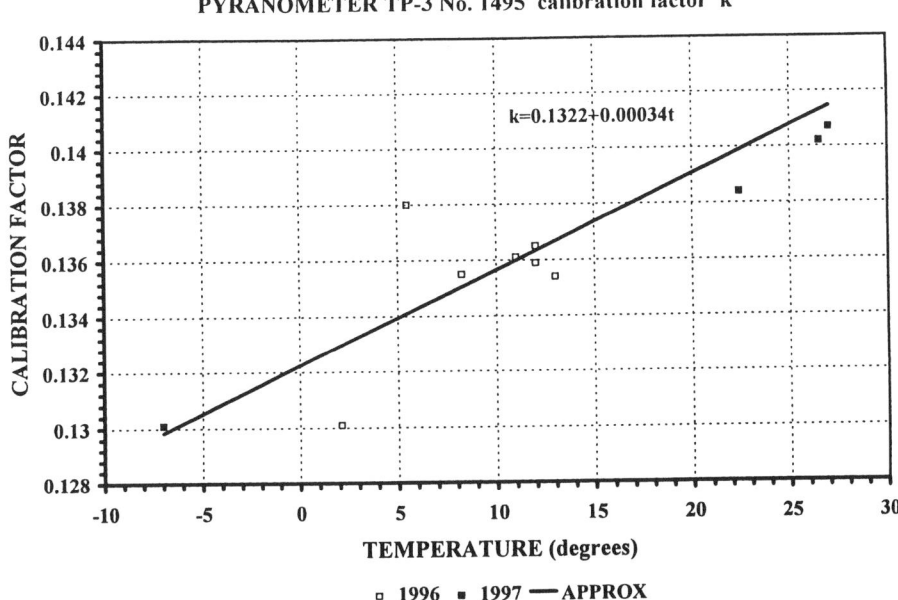

Figure 6. The calibration factor of the pyranometer TP-3 No. 1495 versus temperature

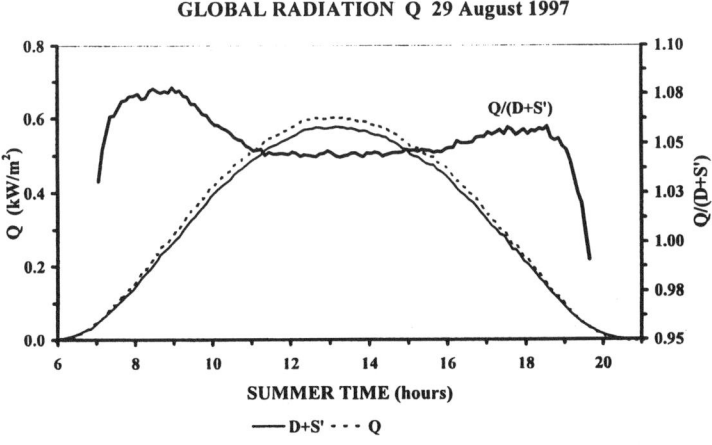

Figure 7. Global radiation flux density Q measured by mini-pyranometer TP-3 No. 1495 and as a sum of diffuse radiation D (measured by Janishevsky pyranometer M-1 15M No. 829 with shading ring) and direct solar radiation S' (actinometer AT-50 No. 554).

In Fig. 7 and Fig. 8 the values of global radiation Q measured using different instruments are presented. To measure diffuse radiation a shading ring was used but the corresponding correction is not taken into account, which explains the disagreement of the results.

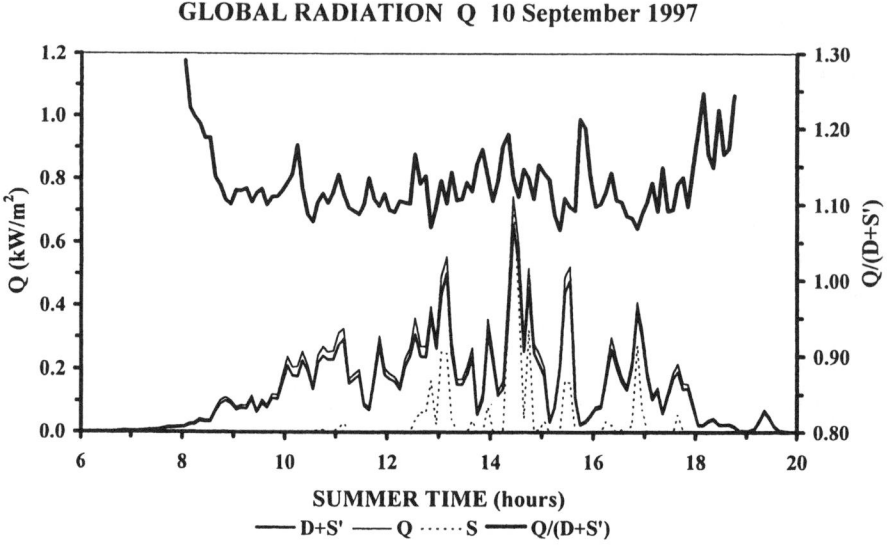

Figure 8. As in Fig. 7. September 10, 1997, mostly cloudy sky

4. NET RADIOMETERS

Net radiation, as a difference between all incoming and outgoing radiation fluxes, includes a wide spectral region containing both the solar radiation and thermal radiation of the atmosphere and the ground surface. Therefore, the correct measurement of net radiation requires instruments that are non-selective in the spectral region 0.3–40 μm. Moreover, it is quite complicated to eliminate the influence of other parameters (wind speed, temperature, etc.) other than radiation on the output signal. That is why net radiation measurements are much more complicated in comparison with short-wave (solar) radiation measurements, and why the comparison between different instruments is unsatisfactory (Dehne et al., 1993, Halldin, Lindroth, 1992).

An "ideal" net radiometer, as mentioned above, must be non-selective in the wide spectral region between 0.3–40 μm. Due to major technical difficulties in the long-wave calibration, net radiometers are, as a rule,

25. INSTRUMENTS AND METHODS FOR THE GROUND-LEVEL REFERENCE MEASUREMENT OF SOLAR RADIATION, ALBEDO AND NET RADIATION

calibrated only in the short-wave region, and an assumption about non selectivity within the limits about 1–2% is made.

Practically, the requirement of non-selectivity can be fulfilled only for instruments with open (unprotected by domes) receiving surfaces. On the other hand, the sensitivity of instruments with open receiving surfaces is affected by wind speed and air temperature, and the receiving surfaces are unprotected from precipitation and mechanical damages.

To minimize these disturbing factors, the receiving surfaces of net radiometers are often protected with domes or covers transparent to radiation. Such instruments are handy and comfortable to use, but the plastic domes have also undesirable perturbing effects, which are very often forgotten. The plastic domes may cause following difficulties:

- Selectivity. No optical material is known to be non-selectively transparent in the spectral region 0.3–40 µm. This means that net radiometers with plastic domes are much more selective than these with unprotected surfaces. Optical properties of plastics are quite variable and depend on the chemical composition, technology used, etc. so that the characteristics of each instrument must be determined individually.
- Since the domes are also sources of thermal radiation, its temperature regime may cause additional errors or zero-point displacement.
- Optical heterogeneity of the domes may cause additional cosine and azimuthal response.
- Optical properties of plastic materials may alter, especially if using under outdoor conditions. Therefore recurrent sensitivity control (calibration) is needed.
- Since thin plastic domes are mechanically not durable, the instruments need careful handling and periodical replacement of the domes.

The implication of these problems is that net radiometers with unprotected receiving surfaces must be preferred. Typically the influence of the wind speed is about 1–2% per 1 m/sec, and the necessary correction is not hard to estimate. At the same time the inaccuracy due to the spectral sensitivity of the instruments with plastic domes may reach 10% or more.

A miniature net radiometer MB-1 with open receiving surfaces was designed in Tartu Observatory by Jüri Reemann. The device was designed to measure net radiation inside vegetative cover but it can be used as well for routine measurements in meteorology, micrometeorology, environmental science etc. Its specific feature is the particular shape of receiving surfaces with a firm glossy black coating to protect the receiving surfaces from damage by contact with the leaves of plants.

Hollow receiving surfaces, made of copper and situated on an annular thermobattery are placed into a cylindrical copper case. The thermobattery is made of Constance wire, which has been coiled on an insulation ring and the outer part of which is galvanically covered by a copper coating. The diameter of the wire is 0.1 mm, the number of coils is about 200. A double layer of epoxy-impregnated paper with the thickness of 0.05 mm insulates the thermobattery from the receiving surfaces and the case. To reduce heat loss, the receiving surfaces are surrounded by foamed plastic insulation. The space between the details is filled with an epoxide component, which avoids the effects of humidity on the thermobattery. Equal sensitivity of receiving surfaces is adjusted individually for each instrument with the help of heat shunts, which increase thermal conductivity between the receiving surfaces and the case when the screws compress the plasticine column between the screws and the receiving surfaces. The accordance between the angular characteristic of the black glossy receiving surfaces and the cosine-law is achieved by shaping them in the form of cavities (elementary models of the black body).

The calibration of the net radiometer in the solar radiation spectral region has been provided in the pyrheliometric scale by the Sun and shadow method. To calibrate the net radiometer in the infrared (thermal) radiation region a special device—a double black body—has been designed and a suitable method developed.

Calibration of net radiometers by hemispherical low-temperature radiators is rather complicated due to considerable non-radiative heat transfer between the radiation receiver and the black body. As a result, sources of infrared radiation with a narrow aperture have been used lately for this purpose (Drummond, 1970, Faraponova, 1968). Since thermal radiation, measured with a net radiometer in field conditions, falls on the receiving surfaces from the hemisphere, it would be methodically correct to carry out the calibration of the sensors with the help of radiators with a wide aperture. For this kind of calibration, a radiation standard was designed. The standard consists of two low-temperature black bodies with an aperture of 2π. A method for the elimination of non-radiative heat transfer has also been devised.

25. INSTRUMENTS AND METHODS FOR THE GROUND-LEVEL REFERENCE MEASUREMENT OF SOLAR RADIATION, ALBEDO AND NET RADIATION

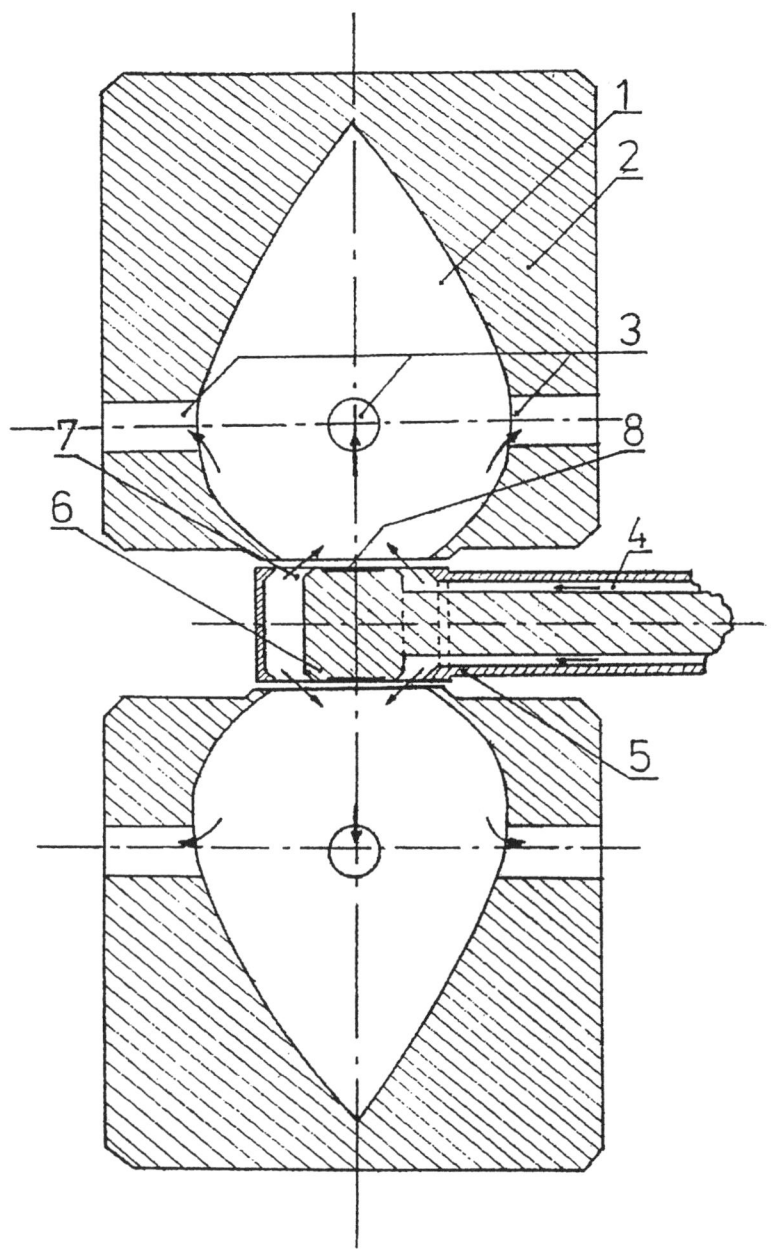

Figure 9. The scheme of the double black body model for calibrating net radiometers

The scheme of the device for the calibration of net radiometers is presented in Fig. 9. The net radiometer being compared (6), in a special copper casing (5), is placed between two massive copper cylinders (2) with blackened cavities (1). The upper block is heated by an electric current, the lower block is either kept at the temperature of the surrounding air, or cooled by a liquid with the help of a thermostat. The temperature of the radiators is determined by mercury thermometers located in the walls of the cylinders.

To eliminate molecular heat transfer from the walls of the cavities to the receiving surfaces, cooling by a faint airflow (arrows in Fig. 9) from a low-powered ventilator is used. The air, flowing through the passage (4) between the receiver and the casing, enters into the cavities of black bodies through annular slots (7), flowing past the receiving surfaces (8), and leaves the system through the openings (3). For the calibration of the sensor, it is necessary to determine the dependence of the output signal of the net radiometer on the difference of radiation fluxes of the radiators and on the intensity of cooling by the airflow.

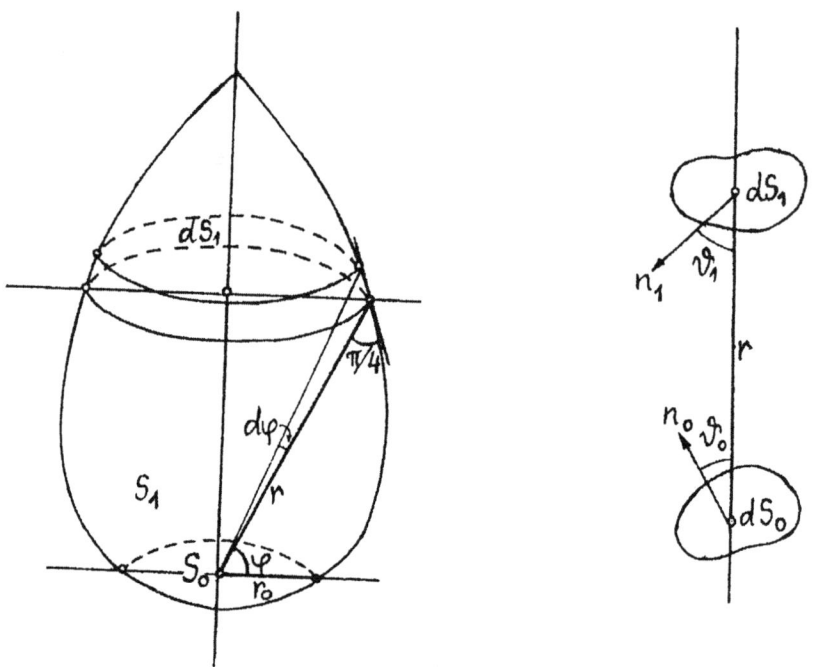

Figure 10. Geometry of black radiator and a scheme for calculation of radiative heat transfer

The black cavities, covered with the Parson varnish, have the configuration of bodies of rotation (Fig. 10), described by the formula $r = r_0 e^{\varphi}$, where r and φ are polar coordinates, r_0 is the radius of the opening of the radiator, while $0 \leq \varphi \leq \pi/2$, and $r_0 = 1.5$ cm. It can be shown that the angle between the surface of the cavity and the direction r is constant and equals $\pi/4$. We have chosen this shape of radiators as a result of the following considerations.

It is known (Blokh, 1962, Drummond, 1970) that consideration of the non-spherical indicatrix of radiation in the calculations of blackbody emissivity is rather complicated. To simplify the calculations, we have made an attempt to find a configuration of the cavity which allows the application of spherical indicatrixes without considerable decrease in the accuracy of calculations. This may be done if the emissivity of the surface element dS_1 in the direction of the opening S_0 does not differ essentially from the emissivity of dS_1, which is determined from the integral flux, radiated by the surface element into the hemisphere. In view of identity the indicatrixes of radiation and absorption, the experimental data, necessary for the determination of the shape of the black body, can be obtained from angular characteristics of radiation detectors, which are covered by the Parson varnish. These studies, as well as the results published in (Harris, 1967, Kmito, 1976) have shown that, for the geometry of the cavities presented in Fig. 8, the requirement mentioned above is fulfilled with sufficient accuracy, and that the use of spherical indicatrixes cannot cause considerable errors.

The resulting reflectivity of this radiator is calculated at $k = 0.00385$, and this result is expressed as $k = 0.0040 \pm 0.0005$ after considering the approximation of neglecting the reflections of higher orders, and estimating the errors. This value of k determines the self-radiation of the blackbody with a high enough accuracy, but it does not permit to consider the increase of the outgoing radiation flux caused by the reflection of external fluxes from an insufficiently black cavity. This phenomenon is more essential for low-temperature radiators, and calls for a correction of reflectivity according to the formula $k_k = (1 - \Phi/\sigma T^4 S_0) k$, where k_k is the corrected reflectivity, Φ is the radiation of external sources entering into the cavity of the blackbody, T the temperature of the blackbody. The external flux Φ must be estimated separately in every specific case. In our experiments, Φ was up to 50% of the flux of the radiator. The accuracy of the calculations of the radiation of the blackbody exceeds the accuracy of modern net radiometers several tens of times. When calibrating modern net radiometers, special attention should be paid to methodical errors, but errors in the determination of radiation fluxes can usually be neglected. To eliminate the reason for the main methodical error, the non-radiative heat transfer, a method of artificial ventilation is devised.

It can be concluded that this method of calibration of miniature net radiometers in the diffuse flux of infrared radiation enables to determine the conversion factors with the accuracy of ±5%. At the same time the accuracy of the short-wave calibration under field conditions is estimated about 3–5% The method can be applied to those devices whose design allows to carry out slight ventilation along the whole receiving surface. Otherwise it is possible to use a high quality reference net radiometer, calibrated using black body model.

5. REFERENCES

Blokh, A.G. (1962) *The Basis for Radiative Heat Transfer*, Gosenergoizdat, Moscow, Leningrad, 331 p. (in Russian).

Dehne, K., Bergholter, U., Kasten, F. (1993) *IEA Comparison of Longwave Radiometers*, Report No. IEA-SHCP-9F-3, Deutscher Wetterdienst, Hamburg, 72 p.

Drummond, A.G. (Editor) (1970) Precision Radiometry. *Advances in Geophysics*, **14**, Academic Press, New York & London, 272 p.

Faraponova, G.P. (1968) A Study of Radiation Thermoelements Designed for the Measurement of Radiation Fluxes in the Atmosphere, in: *Actinometry and Optics of the Atmosphere*, Valgus, Tallinn, pp. 202–218 (in Russian).

Frölich, C., London, J. (Editors) (1986) *Revised instruction manual on radiation instruments and measurements*, WCRP Publication Series No. 7, 139 p.

Halldin, S. and Lindroth, A. (1992) Errors in Net Radiometry, Comparison and Evaluation of Six Radiometer Designs, *Journal of Applied Meteorology*, **9**, 762–783.

Harris, L. (1967) The Optical Properties of Metal Blacks and Carbon Blacks, Massachusetts Institute of Technology, *Monograph Series* No. **1**, 116 p.

Kmito, A .A. (1976) Spectral Reflectivities of Blackened Surfaces, *Transactions of the Main Geophysical Observatory*, No. **370**, 39–44 (in Russian).

Chapter 26

THE NASA EARTH SYSTEM SCIENCE PROGRAM IN 21ST CENTURY

G. Asrar
NASA Headquarters, Washington D.C., USA.

1. INTRODUCTION

The Earth is a dynamic system of land, oceans, and atmosphere that has evolved as a result of the complex biological, chemical and physical processes. These processes are interconnected by an intricate and fragile web that helps modulate such things as weather, droughts and floods and can ultimately have profound impacts on global food production, natural resource management, commerce and the global economy. The recently experienced 1997–98 El Niño cycle illustrates this complex interaction that began when the normal state of the Pacific ocean was perturbed by a weakening of the westerly trade winds; a state that is necessary to maintain a "pool" of warm water within the western tropical Pacific. The weakening of trade winds allowed warm water to flow eastward along the equator to coastal Peru. The eastward-spreading water brought with it the associated tropical rainfall. As the warm pool reached the coast of Peru, it spread to the north and south progressively cutting off the nutrient-rich upwelling along the Peruvian coast, and causing a collapse of the fishery and impacting the seabird population. As the heat from the warm pool escaped into the atmosphere it impacted global atmospheric circulation including the global jet streams. This perturbed atmospheric state brought unusual weather around the world, where normally wet regions such as Indonesia and northern Australia, experienced drought and wildfires, and normally dry regions, such as southern California, experienced floods and mudslides. These changes, in turn, had a significant impact on plant, animal, and human life. The disruption to society was profound: the last cycle was associated with an estimated US$ 8 billion of economic losses.

Figure 1. El Niño Impacts: The 1997-1998 El Niño altered weather patterns all over the globe. Because people and animals pattern their lives on average weather, most of the impacts were negative. Some of the major impacts are listed below:
- Severe drought lowered crop yields in regions of Africa, Southeast Asia, Australia, and South America. Over 60% of Algeria's wheat crop was lost, drought sent food prices soaring in Uganda, 70% of North Korea's maize crop failed with more than 60 rain-free days with temperatures near 32°C.
- Flooding caused loss of life and road and property damage in central and southern South America.
- Atlantic tropical storms and hurricanes were dramatically reduced.
- The strength of tropical cyclones was increased in the eastern Pacific.
- Ice storms in eastern Canada and New England caused some areas to be without power for over a month as rain froze on contact with objects such as power pylons, telephone poles, wires, and trees.
- Sea lions and other marine mammals starved off the California coast because warm water blocked the normal upwelling of cold nutrient-rich coastal water.
- Forest fires in Indonesia, started by people clearing land in areas suffering from prolonged drought, blazed out of control because vegetation was so dry.
- Relatively warm temperatures across much of the southern Canada and northern USA was good news to golfers, but bad news to skiers.

The complex El Niño phenomena illustrate the need for the multidisciplinary, "systems" approach for studying Earth. NASA's Earth

Science Enterprise (ESE) endeavors to understand such complex phenomena related to the total Earth system and the effects that natural and human-induced changes have on the global environment. From the unique vantage point of space, ESE and its domestic and international partners are beginning to understand these dynamic and interacting processes. The fundamental knowledge gained from these efforts is also being used for practical societal needs—such as more accurate weather and climate forecasting, in precision farming and forestry, in natural resource management, in urban and regional planning, in disaster management, in human health and safety planning, and more.

NASA has implemented a complex program including science and applications research, observational systems and technology development, and information system management, in cooperation with domestic and international partners. This article provides a brief overview of the NASA Earth Science program and its implementation strategy. In short, NASA is poised to bring together the powerful combination of science, space-based remote sensing and information system technologies to enable policy and decision makers at all levels of the government, public and private sector practitioners, to establish sound, knowledge-based environmental decisions in the 21st Century.

2. SCIENCE OF THE EARTH SYSTEM

To study the Earth, NASA is taking full advantage of a strong and complex program of fundamental research in Earth system science. Satellite and airborne remote sensing observations, coupled with ground-based measurements, coordinated and multidisciplinary field experiments and powerful computers are used to achieve a basic understanding of the Earth's large-scale processes such as changes in the atmosphere, oceans and continents. The knowledge gained from these efforts allows us to mimic, or model, the behavior of the Earth system at the regional-to-global scale. These large-scale investigations continue as a critical part of the ESE scientific endeavor because they form the context within which to gain understanding of regional-to-local scale concerns. As the behavior of the Earth's past climate is successfully modeled, confidence is gained that future long- and short-term trends can someday be predicted. The information resulting from these efforts is necessary to address local-scale concerns such as suburban land use planning, natural resource management, or disaster mitigation. Both the growth in scientific understanding and the needs of economic and policy decision-makers are leading ESE to formulate more focused questions such as:

Is climate changing in ways we can understand and predict? NASA is attempting to uncover the basic mechanics of climate, and then distinguish natural from human-induced impacts on the climate system leading to more accurate predictive models of ocean-atmosphere interactions, cloud formation, radiative balance, and chemical transport from land to atmosphere. This predictive knowledge can then be translated to support decisions based on improved understanding short-term weather and seasonal climate variations. With the aid of satellite observations, scientist were able to predict the 1997-98 El Niño approximately six months in advance and monitor its progression which provided decision-makers time to plan mitigation strategies including planting alternate crop types and modifying harvest schedules, implement flood control activities and better prepare for its impacts.

Can we understand and predict how terrestrial and marine ecosystems are changing? NASA will attempt to distinguish natural from human-induced changes in ecosystem characteristics towards addressing uncertainties in understanding the carbon and nitrogen cycles in terrestrial and marine ecosystems. More reliable knowledge is needed of the ability of ecosystems to recover from disturbances, human-induced or natural, and of implications regarding the long-term sustainability of the planet's biological productivity. This information is directly applicable, for example, to natural resource management, and can support decisions makers and practitioners in agriculture and the food production and fishery industries.

How is the chemical composition of the atmosphere changing? NASA has made great strides in understanding the concentrations and distributions of ozone and ozone-depleting chemicals in the stratosphere. Validation of the new substitutes for the banned chlorofluorocarbons (CFCs) is needed to find out if they have adverse impacts themselves. A better understanding of the chemistry in the troposphere, where humans live, is needed to determine the consequences of human-produced ozone and other pollutants, and is necessary for urban and city planning in the future to sustain and protect the quality of the environment.

Can we improve our understanding of the processes and dynamics of the Earth's surface and interior, and use this knowledge to prepare for and respond to natural hazards such as volcanoes and earthquakes? NASA seeks to understand the dynamics of the solid earth and its interaction with the atmosphere, oceans, and biosphere. Fundamental scientific questions about the rates and magnitudes, and the spatial and temporal variability of these dynamics remain unanswered. Such information also provides critical knowledge to improve assessments of the vulnerability to natural hazards, and to predict and mitigate the consequences of natural disasters.

3. OBSERVATIONS AND MEASUREMENT STRATEGIES

For the purposes of this article, ESE missions are described below in phases. It is beyond the scope of this article to describe and discuss each ESE mission, however, most are listed and described on the ESE World Wide Web home page at http://www.earth.nasa.gov/.

Table 1. Earth system enterprise: Mission description

Mission (with planned launch date)	Area of Study
Phase I: Primary Missions Through 1997	
Total Ozone Mapping Spectrometer (TOMS) 1978, 1991, 1996, 2000	Ozone measurements *(co-operative efforts with Russia & Japan)*
Earth Radiation Budget Satellite (ERBS) – *1984*	Radiation budget, aerosols, ozone
Upper Atmosphere Research Satellite (UARS) – *1991*	Upper atmospheric chemistry *(with UK & Canada)*
ATLAS series - *1992-94*	Shuttle experiments on the atmosphere and effects of the Sun *(with Germany, Belgium, & France)*
TOPEX/Poseidon - *1992*	Ocean circulation *(with France)*
LAGEOS-2 - *1992*	Crustal motion & Earth rotation *(with Italy)*
Space Radar Laboratory 1&2 missions - *1994*	Synthetic aperture radar scans of the Earth's surface to classify vegetation and penetrate loose ground cover *(with Germany and Italy)*
Optical Transient Detector (OTD) – *1995*	Lightning tracking experiment *(with commercial firm)*
NASA Scatterometer (NSCAT) - *1996*	Ocean surface wind speed & direction *(with Japan)*
ORBCOMM-2 (SeaWiFS) - *1997*	Ocean biological productivity *(data purchase)*
Tropical Rainfall Measuring Mission (TRMM) - *1997*	Tropical rainfall and storms *(with Japan)*
Phase II: The Earth Observing System (measurement groupings)	
EOS AM Series (including MODIS, MISR, ASTER) - *1999*	Atmospheric/surface/solar processes
MOPITT, CERES - *1999*	controlling fresh water resources and ecological processes affecting global climate *(with Japan and Canada)*
Landsat-7 - *1999*	Land surface features & changes *(high resolution)(with USGS and NOAA)*

Mission (with planned launch date)	Area of Study
EOS ACRIM Flights - *1999*	Changes in total solar output *(partner TBD)*
EOS PM Series - *2000*	Causes of climate variations and basis for improvements in long-term weather & climate prediction *(with Japan and Brazil)*
QuikScat / SeaWinds - *1999, 2000*	Role of ocean winds in climate system and interaction with atmosphere *(with Japan)*
Jason 1 Radar Altimetry Mission - *2000*	Role of Ocean in Climate System *(with France)*
EOS Laser Altimetry Mission (ICESAT) - *2001*	Ice sheet topography mapping
EOS Chemistry Series - *2001* / SAGE - *1999, 2002*	Behavior of ozone, other greenhouse gases; aerosols and their impact on global climate; as well as regional and global studies of pollution *(with Japan, U.K. and Russia)*
EOS SOLSTICE Flights - *2002*	Changes in ultraviolet radiation output from the Sun
Phase III: Complementary Missions Through 2002	
Shuttle Radar Topography Mission - *1999*	Shuttle-based synthetic aperture radar flight to produce precise digital elevation models of most of the Earth's surface; partnership with DoD and USGS
New Millennium Program missions - *1999/2001/2002*	Technology demonstration missions for advanced instruments to reduce the cost of the next EOS series.
EO-1 including ALI, Hyperion	Advanced land surface imaging and hyper-spectral observations
Earth System Science Pathfinder missions - *2000/2001/2002*	Small satellites for new science measurements; low- cost, rapid development missions using innovative academia/industry partnerships
VCL	Vegetation height and distribution
Grace	Time variable gravity and climate
PICASSO-CENA	The roles of clouds and aerosols in the atmosphere

Phase I of the ESE has been comprised of focused, free-flying satellites, Space Shuttle missions, and various airborne and ground-based studies. Phase one has provided global measurements of major Earth system components. Their purpose was to improve our knowledge of basic processes prior to the Earth Observing System (EOS) era. These include the currently operating

1. Upper Atmosphere Research Satellite (UARS), which confirmed the anthropogenic origin of ozone-depleting substances
   (http://umpgal.gsfc/nasa.gov/),
2. TOPEX/Poseidon, which is detecting variations in sea level worldwide
   (http://topex-www.jpl.nasa.gov/),

26. THE NASA EARTH SYSTEM SCIENCE PROGRAM IN 21ST CENTURY

3. NASA Scatterometer (NSCAT), which detected ocean surface wind vectors from Japan's ADEOS satellite (http://winds.www.jpl.nasa.gov/),
4. SeaWiFS, a mission which is providing ocean color information useful to understand biological productivity in the oceans (http://seawifs.gsfc.nasa.gov/)
5. and the Tropical Rainfall Measuring Mission (TRMM), also conducted jointly with Japan, which for the first time is providing an accurate assessment of the structure of storm systems, rainfall and associated energy release over the oceans at low latitudes (http://trmm.gsfc.nasa.gov/).

Phase II of the ESE program began with the launch of Landsat-7 and the Earth Observing System (EOS) satellites in late 1999. EOS is the first observing system to offer integrated measurements of the Earth's processes and will generate a long-term environmental database focusing on climate change. EOS will usher in an unprecedented observational capability for understanding the planet. EOS, the largest element of NASA's ESE observational program, is a program of multiple spacecraft designed to provide measurements of the key, multi-disciplinary parameters needed to understand global climate change. The first spacecraft—"Terra" (formerly called EOS AM) and Landsat-7—have been launched in 1999. The instruments aboard Terra will enable scientific studies of the physical and radiative properties of clouds; air-land and air-sea exchanges of energy, carbon and water cycles; measurements of trace atmospheric gases; and other land processes. Landsat-7 will make important land-use and land processes measurements, complementing and improving upon those made by previous Landsat spacecraft in building the largest database of medium resolution land surface images of Earth's continents. The EOS PM mission will be launched in year 2000, and Chemistry-1 in 2002. Both will help achieve an understanding of short-term weather, long-term climate records and atmospheric chemistry. The EOS program also includes several small spacecraft such as a joint venture with France known as Jason-1 (a follow-on to the highly successful TOPEX/Poseidon mission), QuikScat providing ocean surface wind vector, ICESat focused on study of the polar caps and their contributions to climate and sea level variations, and Solstice and ACRIM to continue the past records of total and spectral solar irradiance and variability.

Phase III of the ESE program includes missions that are complementary in nature to EOS and address unique, specific, and highly focused mission requirements in Earth science research, called Earth Probes. The program was designed to have the flexibility to take advantage of unique opportunities presented by domestic/international cooperative efforts and/or technical innovation. It complements the EOS by providing the ability to investigate specific Earth processes that require special orbits or have unique observation requirements. Currently approved Earth Probes include the Shuttle Radar Topography Mission (planned for launch in January 2000) which will provide the first global (within ~60 degrees N and S latitude) moderate-resolution digital topographic characterization of the Earth land surface. The Earth System Science Pathfinder missions including Vegetation Canopy Lidar (VCL), to measure the height of global forests and their contributions to global carbon cycle, the Gravity Recovery and Climate Experiment (GRACE), to document long baseline Earth's gravity field, and PICASSO-CENA to measure the vertical structure of the clouds and aerosols in the atmosphere and their contributions to the variability in Earth's climate. A combination of the EOS and Earth Probes satellites will provide some unprecedented capabilities for Earth scientists to study and understand some complex aspects of the Earth's atmosphere and terrestrial ecosystems which have not been possible before. For example, a combination of Terra and VCL satellites will enable understanding the role of terrestrial ecosystems in sequestering the atmospheric carbon dioxide a major greenhouse gas. We do not have a reliable estimate of the capacity of these ecosystems for removing carbon dioxide from the atmosphere. The observations resulting from Terra and VCL will help reducing the uncertainties associated with our current knowledge at least by 30%. Similarly, observations resulting from a combination of EOS PM and PICASSO-CENA will address the current ambiguity surrounding the role of clouds and aerosols in absorbing the solar radiation which contributes to the heating of atmosphere.

Embracing the NASA "faster, better, cheaper" philosophy and pursuing advanced technological developments, ESE intends to drastically shrink the size, cost and development time for missions in the next decade while maintaining capabilities and performance of these systems. Such missions will be more sharply focused on addressing a specific science or application questions rather than conducting broad surveys. An integrated technology program is paving the way for a "faster, better, cheaper" mission set. The Instrument Incubator Program, for example, supports the development of new instruments and measurement techniques from concept to laboratory development and ground or air validation. The New Millennium Program focuses on identifying and demonstrating advanced technologies or engineering capabilities requiring space-based validations that reduce cost or improve performance of spacecraft instruments for future. The first New Millennium missions include the Earth Observing mission (EO-1), which will carry an Advanced Land Imager which is one-half the size, one-tenth the mass and requires one-seventh the power required by the current

Enhanced Thematic Mapper instrument on Landsat, with an overall performance 4 to 10 times better than Landsat system. The second New Millennium mission will test a Space-Readiness Coherent Lidar Experiment aboard the Space for measuring the atmospheric wind speed and direction for the first time.

4. THE APPLICATIONS IMPERATIVE

As stated previously, ESE has recognized that its knowledge and data have significant practical value to society and therefore is fostering increased access to, and use of, the information to make better, more informed decisions on a daily basis. Decisions regarding important issues such as those related to weather and climate forecasting, flood monitoring and mitigation, agriculture productivity, natural resource management, urban and regional planning, drought impact assessment, and transportation planning. To this end, the ESE formulated and implemented a new program aimed at applications research and commercial developments including communication of ESE data and science to the non-science communities.

The content of this new initiative is defined based on six applications themes. These applications themes are similar in concept to the high-level science themes and questions that drive the scientific research in ESE. The major difference is that the applications themes are defined by public and private sector markets and necessities, and not scientific curiosity. The current applications themes include food and fiber, natural resources, disaster management, environmental quality, urban and infrastructure, and human health and safety.

> ESE Applications and Commercialization Themes
> - Food and Fiber
>   – e.g., Precision Agriculture; Pest control; Forestry; Rangelands
> - Natural Resources
>   – e.g., Land Use/Land Cover; Wetlands; Geology; Mineral/Energy Exploration & Extraction; Recreation; Water Resources; Wildlife Management; Bio-diversity & Habitat Analysis; Coastal & Ocean Systems(Fisheries, Human Impact on Marine Systems)
> - Disaster Management
>   – e.g., Earthquakes; Volcanic Eruptions & Ash Clouds; Landslides; Coastal Hazards; Wildfires; Flooding; Severe Storms; Short-term Climate Change Effects
> - Environmental Quality
>   – e.g., Air Quality; Tropospheric Ozone; Water Quality; Soils; Abandoned Mines; Brownfields; Electromagnetic Energy; Contingency Spill Events; Urban Heat Islands
> - Urban & Infrastructure
>   – e.g., Growth Management; Urban & Regional Planning; Infrastructure Planning (Transportation, Communication and Utilities)
> - Human Health & Safety
>   – e.g., Public health (Water; Air; Carcinogens (aerosols), Ozone); Vector-borne & Infectious Diseases
> - Cross-Cutting Themes: Land Use/Land Cover, Weather And Climate

5. COMMUNICATION OF INFORMATION AND KNOWLEDGE

The EOS Data Information System (EOSDIS) is the backbone of ESE's ability to convert raw remote sensing data to information and communicate that to science and applications users. In addition to serving thousands of current users, EOSDIS will operate the EOS spacecraft and acquire and distribute the basic data gathered by the instruments. It provides the basis for both the government and its commercial and academic partners to generate the higher-level data products that will make the measurements more easily understandable and usable by researchers, educators, policy makers and the public.

The EOSDIS will process and archive nearly all ESE data, making it one of the largest civilian information systems ever conceived. The enormous variety and utility of these data, and consequently the wide variety of potential users, make creation and distribution of final data products a challenge. ESE is also establishing a "federation" through Earth Science Information Partners who will participate in producing and distributing selected data products for targeted users. If the EOSDIS federation is successful, eventually all higher level data products will be produced by competitively selected Earth Science Information Partners, bound together by common standards and protocols to make the accommodate a seamless exchange of information among them. ESE will continue to utilize new and creative mechanisms to communicate its data and information to the science and applications communities, as well as the general public.

6. PARTNERSHIPS

It is virtually impossible to implement a program as diverse and complex as ESE without cooperation and collaboration from external sources. ESE employs a multifaceted approach to making and using Earth science information by significant cooperation and partnerships with other Federal agencies within United States, the commercial sector, and the international community. EOS, for example, is already integrated with climate-related research and capabilities conducted by 17 other Federal agencies through the U.S. Global Change Research Program (USGCRP). ESE will continue increasing its partnerships with Federal Agencies as it continues expanding, particularly through its applications efforts. Recognizing the continued emergence and expansion of many commercial remote sensing capabilities, ESE is also attempting to stimulate the development of this sector, both as a provider and a user of ESE data. Developing commercial capabilities can enhance efficiency in the ESE science mission and promote the long-term stability required by science research and commerce.

In the international arena, ESE is leveraging its investments through partnerships with other nations. Foreign governments are providing instruments to fly on ESE satellites, and the United States is providing instruments for other nations satellites. ESE also negotiates for access to data from non-US missions. This not only stretches the dollars invested in ESE by American taxpayers but also builds worldwide confidence in scientific results and environmental assessments. The outcome and payoff of this international involvement extends to the development of a framework for nations to commit to fulfilling requirements of international efforts fostered by the World Climate Research Program (WCRP), International Geosphere-Biosphere Program (IGBP), the Intergovernmental Panel on Climate Change (IPCC), and the ultimate realization of an International Global Observing Strategy.

7. SUMMARY

NASA and its partners in the U.S. Government and around the world have responded to the challenge of Earth system science by developing complementary and interconnected observations, data, and analysis capabilities designed to provide the most cost-effective means of conducting the research and conveying that information to societal uses. The resulting science and applications will provide global leadership in the development of an international consensus on the state of the Earth today and in the future. This approach will yield long term predictions of the state of the Earth system as well as provide practical information that will benefit practitioners, policy makers, educational institutions, the business community, and the public at large. Just as the weather and communications satellites changed the way we view and perceive those fields, the elements of ESE will change the way we perceive the environment and climate in the 21st Century.

Chapter 27

ESA'S PLANS AND STRATEGY FOR OPTICAL REMOTE SENSING OF TERRESTRIAL SURFACES IN THE NEXT DECADE

M. Rast
ESA ESTEC, Noordwijk, The Netherlands.

1. OBJECTIVES OF ESA'S EARTH OBSERVATION ACTIVITIES

In the past, ESA has focused mainly on the development of Earth Observation techniques addressing oceanographic and atmospheric applications. Microwave technology was the main technology developed in line with the views of the ESA Member States as demonstrated by the payloads of ERS-1 and ERS-2. Out of more than 6000 scientific ERS-1/ERS-2 publications during the past years, about 70% address land applications (ESA, 1995, 1998a). Considering the large number of SAR land applications, and realizing the need for complementary optical data, the development of new optical sensor systems will play an important role in addressing ESA's long-term Earth Observation objectives, as outlined in detail in the 'Living Planet Program' (ESA, 1998b).

Future Earth Observation missions will contribute to the four 'Living Planet' Themes. These are:
- Theme 1 – Earth Interior:
  Advancing understanding of the structure and dynamics of the Earth's crust and interior.
- Theme 2 – Physical Climate:
  The study and monitoring of the Earth's climate as well as the

continuation and improvement of the service provided to the worldwide operational meteorological community.
- Theme 3 – Geosphere/Biosphere:
 The study and monitoring of the environment and the resources both renewable and non-renewable on various scales from local, to regional and to global.
- Theme 4 – Atmosphere and Marine Environment:
 The study and monitoring of anthropogenic impacts on the atmosphere and marine environment (change of atmospheric composition, change of water level, change of water quality).

It should be noted that optical remote sensing instruments have the potential to contribute to all four of these Themes.

2. CURRENT AND APPROVED ESA EARTH OBSERVATION MISSIONS

The following table lists all current and approved ESA Earth Observation missions.

Table 1. Current operational and approved ESA EO missions

Mission	Current Status	Launch Date
ERS-1	in orbit: reserve satellite	launched 1991
ERS-2	in orbit: operational	launched 1995
ENVISAT	Phase C/D	2001
METOP (3 satellites) *	Phase B completed	first launch 2002
METEOSAT operational and transitional programs (MOP, MTP)*	in orbit: operational	MOP-1,2,3: 1989, 1991, 1993 MTP: 1997
MSG (3 satellites)*	approved (Phase B + C/D of MSG-1)	first launch 2000

*: In cooperation with EUMETSAT

METEOSAT and METEOSAT Second Generation (MSG), as well as METOP, are missions undertaken in cooperation with EUMETSAT and dedicated to operational meteorology. ERS-1 and its follow-on ERS-2 are dedicated mainly to ocean/ice observations, though ERS-2 includes capabilities for land and atmospheric chemistry investigations. ENVISAT focuses on the observation of environment, ocean-ice and the atmosphere.

None of these are specific land observation missions but all, to varying extents, contribute to it. Four optical terrestrial observations contributions are made by:
- ATSR on ERS-1
- ATSR-2 on ERS-2

- SEVIRI on Meteosat Second Generation
- MERIS and AATSR on Envisat.

MERIS and AATSR are the main sensors from which strong contributions to optical terrestrial remote sensing are expected in the near future. Through its spectral bands in the visible and near-infrared, MERIS will mainly allow assessments to be made of the status and health of vegetation by observing the vegetation "red-edge". AATSR will, by observing land-surface temperatures, contribute to the bio-geophysical modeling of land surface processes and energy balance investigations. For both sensors, algorithms supporting the establishment of data products are being developed for implementation in the Envisat ground segment. For MERIS, this will include an improved vegetation index and for AATSR a land surface temperature product.

Considering this, and taking into account the objectives of the new 'Living Planet' program, there is a clear and pressing need for additional missions focussing on land observations.

The future strategy for Earth Observation, beyond the currently operational missions and those presently being developed, is being elaborated with the user community, scientists, industry and the Delegations of the ESA Member States. In general two classes of missions are proposed, which are intended to complement each another:

1. The EARTH EXPLORER Missions, these are research/demonstration missions concerned with advancing understanding of the different Earth System processes and/or the demonstration of new observing techniques. Each mission will focus on a limited set of objectives. No two missions would be the same.
2. The EARTH WATCH missions, these are pre-operational missions concerned with the operational needs of user communities. There is a need to ensure the continuous provision of data so many of the individual missions will be identical. A key element of these is the identification of a partner who ultimately will assume total responsibility for these missions.

3. THE EARTH EXPLORER MISSIONS

The Earth Explorer component of the 'Living Planet' program is the implementation of research and demonstration missions for advancing understanding of the different Earth System processes. It will build on the heritage of research missions such as ERS, TOPEX/POSEIDON and

Envisat, set within the general forum of Europe's operational missions, namely Meteosat, Metop and SPOT.

Europe has the ability to design and implement, in a coordinated way, scientific space missions addressing each of the four 'Living Planet' themes. Only by defining such a long-term program will Europe be in a position to play a key role in the coordination of international activities and in the implementation of fruitful collaboration. The requirement for continuity is associated automatically with the need for a long-term view of the program. However, this must be complemented by an appropriate flexibility to take advantage of shorter-term opportunities and evolving requirements.

This need is covered by the implementation of two types of Earth Explorer missions, namely, the Earth Explorer Core Missions, which are larger research/demonstration missions lead by ESA and the Earth Explorer Opportunity Missions, which are smaller research/demonstration missions not necessarily ESA lead. Earth Explorer Opportunity Missions will be implemented on shorter intervals providing the scientific flexibility required.

The consultation of the scientific community led to the ESA consultation meeting in Granada in 1996. During this meeting the wide span of interests of Europe's scientists in the Earth Explorer component of the Living Planet Program was highlighted by the set of nine Explorer Core mission, which was reviewed at that meeting. See ESA (1996) for a detailed description of these nine missions.

Out of the nine proposed Explorer Core missions, four were selected for Phase A studies, which are:
- The Gravity Field and Steady State Ocean Circulation Mission - GOCE
- The Atmospheric Dynamics Mission - ADM
- The Earth Radiation Mission - ERM
- The Land Surface Processes and Interactions Mission - LSPIM

The first two are currently selected for further studies. In addition, the Agency released a Call for Earth Explorer Opportunity Missions. The high level of response (27 proposals) to this call serves to highlight the high interest of the European scientific community in the Earth Explorer missions. Of these 27 proposals a scientific evaluation resulted in three missions being retained for further consideration. These missions are:
- Cryosat, a mission addressing the study of ice sheets and their influence in sea level change.
- SMOS, a mission addressing the quantification of soil moisture and sea surface salinity and
- ACE, a mission studying temperature and Ozone profiles in the upper troposphere.

Of all missions currently studied and envisaged by the Agency the Land-Surface Processes and Interactions Earth Explorer Core Mission is the Earth

Observation program component most focussed upon optical terrestrial observations.

4. THE LAND SURFACE PROCESSES AND INTERACTIONS MISSION

This proposed mission is dedicated to advancing the understanding of our ecosystem processes and its interactions with the atmosphere (including scaling issues). A hyper-spectral imager, covering the VNIR/SWIR spectral range in narrow bands, complemented by a thermal IR imaging radiometer is proposed as the payload. The required accessibility, different viewing geometries (BRDF capability) and revisit will be achieved by along-track and cross-track sensor steering capabilities.

The international scientific community has devoted considerable effort over the past decade to addressing questions related to land surface processes and their role in climate change. Thus, the World Climate Research Program (WCRP) and the International Geosphere Biosphere Program (IGBP) have provided the first insights into the relevant processes, on local, regional and global scales. However, even if the processes occurring at the interfaces of the biosphere/geosphere system with the atmosphere and the ocean are well understood and parameterized on the very local scale, extending that knowledge to regional and global scales is still difficult. The strong coupling between these processes and the extreme heterogeneity of the geosphere/biosphere system render spatial aggregation very complicated.

The primary focus of a high spatial and spectral resolution space-borne mission is the land surface processes and their interactions with the atmosphere, concentrating on the measurement of surface characteristics such as albedo, hyper-spectral reflectance, BRDF (bi-directional reflectance distribution function) and surface temperature which are linked to geophysical variables involved in biogenic processes such as productivity, evapotranspiration and nutrient cycles. The space segment of the mission would proceed in parallel with a measurement program involving a set of carefully instrumented ground sites. The most important processes to investigate and understand more quantitatively (on different spatial and temporal scales) have been identified as a result of experience gained over the last few decades in modeling the evolution of the geosphere/biosphere system and its links to the climate. These processes involve:
- the energy and water cycles

- interactions between terrestrial surfaces and the atmosphere, in particular through the exchange of biogeochemicals, including trace gases such as carbon dioxide and methane
- the productivity of the different ecosystems
- the state and dynamics of land cover/land use.

The energy and water fluxes are determined by the biophysics of the surface, the carbon and nitrogen fluxes are controlled by its biochemistry. The interrelations are shown in Figure 1.

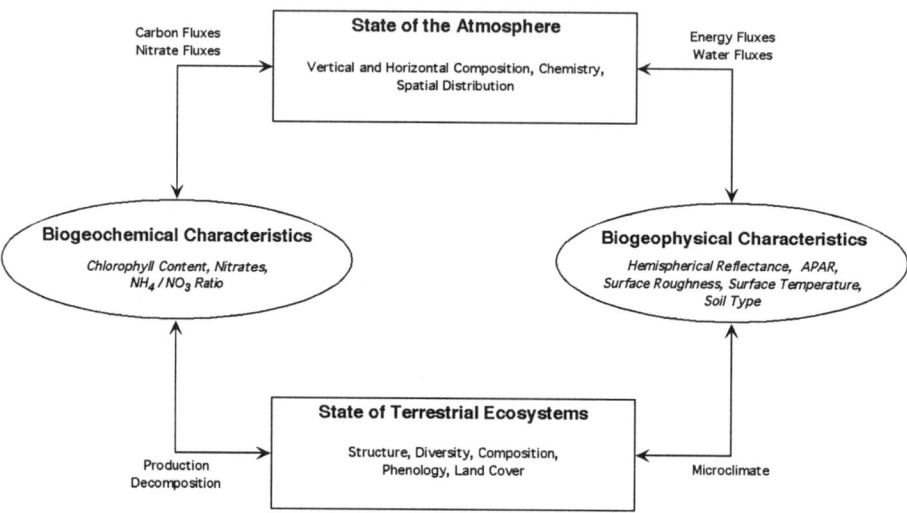

Figure 1. Simplified representation of biospheric interactions, processes and variables

In summary, without entering into the requirements of all land-related applications, it is possible to identify a set of basic land features and characteristics, which must represent the focus of attention in a space observation strategy for the next decade. They are related to:
- surface characteristics and conditions (type and condition of land cover, terrain characteristics, soil, surface – and subsurface water, and changes in those conditions, including environmental degradation and pollution);
- surface processes such as primary productivity, biochemical cycles, energy and matter interactions (e.g., evapotranspiration, radiation balance).

Knowledge of the surface/atmospheric exchange processes for trace gases, together with the exchange processes for energy and water, are determining factors for furthering our understanding and the modeling of ecosystems. The links between the biospheric interactions in the time domain are displayed in Figure 2.

27. ESA'S PLANS AND STRATEGY FOR OPTICAL REMOTE SENSING OF TERRESTRIAL SURFACES IN THE NEXT DECADE

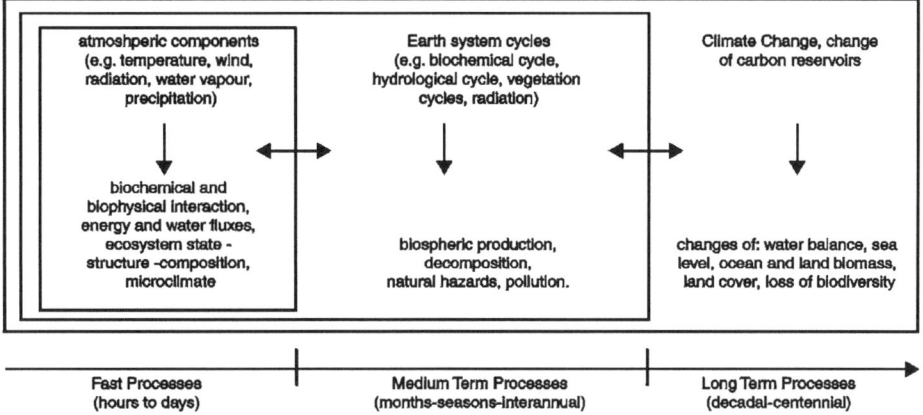

Figure 2. Examples of geo-biospheric interaction: the links between the components

Within this context, the mission objectives of the Land-Surface Processes and Interactions Mission could be summarized as follows:
- The modeling and monitoring of land surface processes—mainly concerned with change detection and furthering understanding of geophysical and biological interactions.
- The study of interactions—addressing interactions of the land surface with both atmosphere and ocean contributing to the development of land / hydrosphere / cryosphere / atmosphere interaction models (including model initialization, model validation and data assimilation for forecasting tasks).
- The analysis of climate impacts on the biosphere and on the water cycle—emphasis would be placed on the observation of key vegetation, soil and hydrological parameters in order to assess possible influences of natural climate variations and human activities.

A hyper-spectral imager, covering both the visible/near infrared (imaging spectrometer) and the thermal infrared (imaging radiometer), is the core of this mission. It would be flown in a near polar orbit to ensure access to all land areas. The mission would exploit a considerable technical and scientific European heritage in the observation of the land surface and its interpretation.

The preliminary ranges for the performance parameters of a candidate space-borne mission are:

Spectral range: 450 nm–2350 nm at < 15 nm spectral sampling interval (spectral absorption features of vegetation and soils). 8–8.5 and 8.6–9.1 microns surface temperature (vegetation and soils).
Spatial sampling: around 50 m

Spectral range: 450 nm–2350 nm at < 15 nm spectral sampling interval (spectral absorption features of vegetation and soils). 8–8.5 and 8.6–9.1 microns surface temperature (vegetation and soils).
Swath width: around 50 km
Pointing: across track depointing capability of about ± 35° for global access and BRDF measurements.
along track pointing: ±70, ±60, +45, one programmable angle and 0 degrees for BRDF measurements
Data transmission: Possibility of selection of spectral bands (up to 60 bands per scene, to be confirmed), also for data reduction

5. LAND SURFACE PROCESSES AND MODEL VARIABLES

To derive observational requirements from the conceptual schematization of processes and interactions, the latter have to be linked with observations of spectral radiances. It should be noted that linking model with radiometric variables implies the availability of algorithms. In some cases such as with inverse modeling of directional radiances this link is based on the expected developments of algorithms described by recent literature and the studies carried out to define specific aspects of the Land Surface Processes and Interactions Mission.

First, more specific categories of land surface processes have to be identified. The LSPIM will focus on the following land surface processes:

- <u>Heat and momentum exchange</u> at the land-atmosphere interface; radiation balance; turbulent heat transfer; control of heterogeneous land surfaces on processes in the atmospheric boundary layer;
- <u>Regional processes of photosynthesis</u>, including regulation of canopy conductance through environmental factors (water stress, temperature);
- <u>Regional hydrological processes</u> including evapotranspiration, run-off generation, ground water recharge, soil water flow, lateral water and pollutant transports, infiltration, generation of floods and droughts, accumulation and melt of snow and ice, erosion and sedimentation.
- <u>Biogeochemical processes</u> which control the cycle of carbon and nitrogen through the water cycle; biogeochemicals solution and transport in surface and ground-water; soil organic matter turnover; waterborne diseases;
- <u>Ecosystem processes</u> determining structure and functioning of ecosystems, including biodiversity, succession dynamics of plant associations; interrelation of vegetation patches; soil development, erosion and degradation. The LSPIM will focus on the determination of

the spatial structure of ecosystems and on the relation of ecosystem structure and functioning with the processes (1) through (4).

To address these needs, two parallel sets of activities must be pursued:

1. A scientific program leading to the derivation of more appropriate and accurate models of geosphere/biosphere processes and surface atmosphere interaction processes, formulated in terms of parameters and variables, which are well identified at each scale and unambiguously related to the quantities measurable from space. Coupled with this must be an improvement in assimilation procedures and processing algorithms.
2. The development and provision of instruments capable of ensuring observations of key geo-biophysical parameters characterizing the state of the biosphere/geosphere system and its evolution. These data would be used in the validation and improvement of the corresponding models, as well as in advancing our understanding of scaling from local to global.

Such a program implies two complementary systems of observation, namely:

1. an observing system with high spatial and spectral resolution, directional sampling capability and, if possible, high temporal repetitivity, but not necessarily with a global coverage, for studying processes and scaling-up procedures
2. one or more instruments with lower spatial and spectral resolution, but suitable for long-term, global observations and complementing the above observing system.. Examples include AVHRR, MERIS, VEGETATION and MODIS.

However, a major concern is the scaling up of local processes to regional and global scales. By advancing the understanding and characterization of these processes at small to medium scales, the mission should enhance their realistic representation in spatially distributed models. This improvement would immediately result in major advances in environmental monitoring and management.

6. THE EARTH WATCH MISSIONS

Earth Watch missions are operational/pre-operational missions driven by the operational needs of the user communities. Thus it is intended that the need the for continuous provision of data (and the accessibility and revisit for monitoring purposes) will be ensured by a series of individual but identical missions. A key element of an Earth Watch mission is the identification of a partner who will be responsible for and ultimately take over full responsibility for the mission, and who will guarantee the data exploitation. A possible candidate as an Earth Watch mission is one dedicated to land applications (crop forecast, damage assessment, forestry, land-use) which includes optical instruments as the payload.

The Earth Explorer missions can be viewed as both, research and demonstration missions (technology demonstrator) so they pave the way for the Earth Watch missions. The strong link lies in their complementarity and its implementation from research (Earth Explorer) to operation (Earth Watch). The main difference is in their funding and selection mechanism.

7. CONCLUSIONS

A satisfactory pursuit to respond to the ENAMORS questions and the advancement of EO capabilities is tried to be achieved by ESA with the competent support of the European User/Science community. This will allow ESA to develop optical remote sensing instruments providing the required spatial, radiometric, spectral and temporal resolution for land observations and advancing retrieval algorithms, which bears the opportunity for novel applications. Surface/atmospheric exchange processes are being studied to advance our understanding of ecosystems. A more accurate knowledge of these processes is needed to improve climate predictions. Spatially distributed information on the bio-geophysical properties and temporal changes of the terrestrial ecosystems, as provided by satellites, are the requisite inputs for work in this area. The Land-Surface Processes and Interactions Mission will enable for the first time and with detailed spectral and directional coverage the observation of important geophysical variables at the appropriate scale, this responding to the ENAMORS objective.

Within the frame of ESA's planned post 2000 mission programs, the Earth Explorer and Earth Watch missions, optical remote sensing system, particularly designed for land observation are proposed for implementation.

8. REFERENCES

ESA (1995) *New Views of the Earth, Scientific Achievements of ERS-1*, **ESA-SP 1176/I**, Noordwijk.
ESA (1998a) *Further achievements of the ERS missions*, **ESA-SP 1228**, Noordwijk.
ESA (1998b) *The Science and Research Elements of ESA's Living Planet Programme*, **ESA SP-1227**, Noordwijk.
ESA (1996) *The Nine Candidate Earth Explorer Missions*, **ESA SP-1196(1-9)**, Noordwijk.

Chapter 28

EARLY RESULTS FROM ADEOS AND FUTURE EARTH OBSERVATION MISSIONS

T. Igarashi
Earth Observation Research Center, National Space Development Agency, Tokyo, Japan.

1. OVERVIEW OF ADEOS

The Advanced Earth Observing Satellite (ADEOS), also known as the Earth Observation Platform Technology Satellite "Midori" in Japan, was launched on 17 August 1996. It stopped operating on 30 June 1997 after an operational phase of about eight months, due to a malfunction of its flexible solar paddle. The mission of ADEOS is to obtain global Earth Observation data using eight onboard sensors to contribute to the international cooperative monitoring of the global change, such as the global warming, the atmospheric ozone depletion over the Antarctic and the Arctic circles, the decrease of tropical rain forest, the abnormal weather and so on.

This space mission is addressing a number of global issues and will prove useful in a variety of practical and operational applications, including

- the frequent observation of the sea surface temperature and ocean color (an indicator of the chlorophyll-a concentration) by the Ocean Color and Temperature Scanner (OCTS),
- the improvement of weather forecast accuracy using NASA's Scatterometer (NSCAT),
- the continuous monitoring of global total ozone content in the atmosphere with the Total Ozone Mapping Spectrometer (TOMS),
- land use and land cover classification, the generation of digital elevation models, and the observation of disaster areas using the Advanced Visible and Near-Infrared Radiometer (AVNIR).

ADEOS' premature halt in operation affects the long term Earth science research on seasonal or annual change data, and those practical applications relying on near real-time data.

On the other hand, so far, the development of algorithms, calibration and validation, as well as the generation of data sets have been well under way and OCTS and AVNIR data are available for distribution to general users (Please consult http://www.eorc.nasda.go.jp/ADEOS/). In the archived data acquired during the eight months long operational period, a large amount of Earth observations have been accumulated from these new sensors; they need to be processed and analyzed. The National Space Development Agency of Japan (NASDA) is now intensively processing these data and reviewing the operation of the ground system, the research using such data and the promotion of data use to get early results from the Earth science or the demonstration of applications.

2. ADEOS DATA ACQUISITION, PROCESSING AND DISTRIBUTION

2.1 Result of operation and data acquisition

ADEOS was launched on 17 August 1996. After an initial verification of its function, the operational phase started on 26 November 1996. Until the end of its operation on 30 June 1997, sensors on-board ADEOS worked almost normally, except for an abnormal tilting mechanism of OCTS which was recovered soon, so that observational data were regularly acquired. The total amount of data archived or planned to be archived as of 8 July were 33,014 scenes of AVNIR-MU, 19,300 scenes of AVNIR-PAN, 7408 paths of OCTS, 4423 paths of IMG, 3433 paths of POLDER, 3340 paths of NSCAT, 3343 paths of TOMS and 6545 paths of ILAS.

2.2 Status of data processing

The data received in NASDA's Earth Observation Center (EOC) at Hatoyama Japan as well as at NASA's ground stations in Fairbanks and Wallops, have been processed into Level 0 data at EOC and have been distributed to each sensor provider. As of beginning of July 1997, EOC processed OCTS, POLDER, ILAS Level 0 data of middle of April 1997. Each sensor provider has done the Level 1 and higher level processing. At EOC, as of 18 August 1997, the following data have been processed: AVNIR level 0, OCTS level 0 and OCTS Real Time Coverage (RTC), and

OCTS Global Area Coverage (GAC) data processing was under way. The standard products of each sensor up to June 1997 are quality checked and archived at EOC. All OCTS Level 1 processing using version 2 algorithm was completed by the end of August 1997 and products are available for users. Moreover, after version 3 of the algorithm becomes available by the end of September 1997, all data will be re-processed and distributed to users.

2.3 Data distribution

2.3.1 Data distribution to PIs, NASDA internal users and general users

Data distribution to about 150 Principal Investigators (PIs), as well as for internal uses in NASDA Earth Observation Research Center (EORC) such as calibration, validation and data set generation, started at the beginning of the operational phase. Data distribution of AVNIR and OCTS to general users became available from 7 May and 30 June 1997, respectively. By 18 August 1997, 547 scenes of AVNIR in digital media, 4563 scenes of AVNIR photograph, 5130 scenes of OCTS in digital media and 3111 scenes of AVNIR photograph were distributed to users. In January 1997, when the oil tanker "Nakhodka" accident happened in the Sea of Japan and the oil spill flow from the sunken tanker reached the beach of Hokuriku district of Japan, ADEOS was used to detect the oil spill with other satellites, JERS-1, SPOT, LANDSAT, ERS-2, RADARSAT.

2.3.2 Real-time and near real-time distribution

NASDA had been providing data via network directly to users who need near real-time data. OCTS, NSCAT, TOMS data were delivered to the Japan Meteorological Agency (JMA), and OCTS data were similarly provided to the Fishery Information Center (JAFIC) in near real-time. For its part, NOAA received OCTS, NSCAT, TOMS data from EOC, Fairbanks and Wallops ground stations via network on a near real-time basis.

ADEOS and OCTS browse data were transmitted in real-time to thirteen research institutes under the Fishery Agency, and to two fishery research institutes under the authority of local governments, fishery unions and so on, through onboard Direct Transmission for Local Users (DTL).

2.3.3 Distribution via Internet

NASDA is delivering data, images and ADEOS-related information to the public via the World Wide Web. Users can browse such public information on the Internet directly and can download processed images using visualizing tools. The information contents include the ADEOS science program, information on events, user's guide, near real-time data, cities in Asia observed by AVNIR, ozone data measured by TOMS, sea surface temperature and chlorophyll-a concentration maps retrieved from OCTS data, sea surface wind maps retrieved from NSCAT data, the first images of each sensor, ADEOS data sets and so on.

The URLs of contact are the followings:
- NASDA Hqs   : http://www.nasda.go.jp/
- NASDA/EOC   : http://www.eoc.nasda.go.jp/
- NASDA/EORC   : http://www.eorc.nasda.go.jp/

3. EARLY RESULTS FROM ADEOS OBSERVATION

3.1 Calibration and validation

NASDA carried out image quality estimation, calibration and validation activities to confirm the expected performance of the data generated by OCTS and AVNIR. Some line noise remains in some images, but NASDA is continuously improving the image quality. For the other sensors, each instrument provider has been carrying out similar calibration and validation tasks.

3.1.1 OCTS

NASDA has estimated the characteristics of processed products by measuring the spatial resolution, the geometric accuracy and the variability of sensitivity. The spatial resolution of 700 m and the signal to noise ratio from 700 to more than 900 were confirmed and met the specifications. The line noise was verified as well reduced level for the scientific research and application use. After version 3 of the algorithm is developed, the improvement of the sensitivity and the geometric accuracy will be re-evaluated. The reprocessing of OCTS using version 2 algorithm will be accomplished by the end of September 1997.

The sensor stability was estimated by electrical calibration, internal light source calibration, and nighttime observation image, and it is confirmed that OCTS is stable after the beginning of October 1996.

NASDA has calibrated the absolute radiance using the NASA airborne sensor AVIRIS. As a result, an accuracy of 10% was achieved. Then, using calibrated sensor input radiance and ground truth data, the algorithm was verified and its parameters were tuned up. Moreover, NASDA is conducting improvement of the accuracy using additional observation data.

So far, as a result of such verification, an accuracy of 50% in the estimation of chlorophyll-a has been attained and the next version of the algorithm is expected to improve the accuracy further. On the sea surface temperature, the tuning of the algorithm is under way to improve the accuracy using about 530 points from NOAA truth data. So far, OCTS SST data is about 1.8 degrees lower than NOAA AVHRR SST data, and the standard deviation is 0.8.

In addition, a global vegetation index map using OCTS band 6 (660-680 nm), 7 (745-785 nm) and 8 (845-885 nm) is planned to be generated.

3.1.2 AVNIR

NASDA/EORC had estimated the spatial resolution, the signal to noise ratio and the geometric accuracy. The performance is almost meeting the specification. On the geometric accuracy, further improvement will be made. The calibration was made using AVIRIS data, and the parameters of algorithm were confirmed.

3.2 Algorithm development and data sets generation

OCTS algorithm has been improved gradually. The first version was used from the beginning of operation to 9 April 1997, and then after 10 April, version 2 was used to process the data. Version 3 will be available by the end of September. The image quality is also improved by the same date.

So far, 10 CD-ROMs including OCTS and AVNIR data sets were made by EORC as a sample data for users. After this, EORC is planning to produce data sets such as match-up data set s of satellite data and truth data dedicated to scientific users.

3.3 Results from ADEOS observations

Using data already acquired, sensor providers have processed and distributed data such as the chlorophyll-a concentration in the ocean, atmospheric ozone content, and so on. In these data, some new scientific data were obtained such as the ozone depletion over the Arctic region. On the other hand, the validity of some applications using AVNIR and OCTS were verified. Some examples are introduced below.

1. Ozone depletion over the Arctic region (TOMS)
   In the Arctic region, the ozone depletion was found in the springtime like the Antarctic ozone hole, in small scale. However, in the Northern Hemisphere, the stable circumpolar vortex is really generated, in this year it is stable and the ozone depletion had happened. On 25 March 1997, the total ozone content decreased to less than 225 Dobson units.
2. Monitoring of ozone hole over the Antarctic region (TOMS, ILAS)
   ADEOS has continuously monitored of the ozone hole in the Antarctic region, of which the largest scale was observed in September 1996 and it has disappeared in December 1997.
3. Vertical temperature profile observation (IMG)
   Using IMG the first space-borne Michelson interferometer observing in the infrared band with very high spectral resolution, the atmospheric temperature was retrieved from the observed atmospheric radiance spectral. The variations in content of greenhouse gas such as CO_2, O_3, H_2O, CH_4 were also observed by IMG.
4. Detection of abnormal flow of the Kuroshio Current (OCTS)
   An abnormal flow of the Kuroshio Current was detected using OCTS the SST and the chlorophyll-a simultaneous mapped data. In the SST, information of the thin skin surface of the sea is reflected and in the chlorophyll-a, information of water color down to the depth of 30m below sea surface is reflected, therefore the ocean current or the water mass were detected correctly using these simultaneous information.
5. Global map of chlorophyll-a and SST (OCTS)
   These data has been processed and distributed as the standard products.
6. Global aerosol map (OCTS)
   A global map of atmospheric aerosol distribution over the ocean was made using OCTS.
7. Global chlorophyll, aerosol and cloud mapping (POLDER)
   The optical thickness, the reflectance of land surface, the atmospheric radiance and the cloud classification was made by CNES.
8. Sea surface wind data improving weather forecast (NSCAT)

In April 1997, JMA initiated the research use of NSCAT data for the improvement of weather forecast. NASDA also analyzed the relationship between convergence and geographical features on the winter seasonal wind.

4. OPERATIONAL SATELLITES

4.1 JERS-1

JERS-1 is still operational since the launch in 1992, but in 1997 the bit error rate of onboard Mission Data Recorder increased and the onboard data storage function is not working reliably. However, the real-time operation over the coverage area of ground stations is still in good condition. The onboard sensors OPS and SAR are operational, and a large amount of data have been archived covering world wide land area.

Recently, NASDA/EORC has been conducting research project such as the application of SAR interferometry applied to the detection of displacement by earthquakes, the Global Rainfall Forest Mapping and the Global Boreal Forest Mapping.

4.2 GMS-5

The Geostationary Meteorological Satellite-5 (GMS-5) is operated by JMA for the weather forecast application, since the launch of the first satellite in 1977. A follow-on satellite MTSAT was planned but failed at launch in FY 1999.

5. LONG TERM MISSION SCENARIO

The latest draft of Japanese scenario of long term Earth Observation mission is the version as of 25 June 1997. However the draft will be under discussion in FY 1997, so far, the strategy and the mission was summarized.

The basic strategy is broken into eight categories corresponding to Earth Observation missions as the following.
1. Mission continuity/research oriented
2. Mission demonstration for application
3. Balanced international cooperation/research oriented
4. Cooperation with foreign mission
5. Feasibility study
6. Small mission for technical demonstration
7. Large sensor technical demonstration
8. Algorithm development

The following chapters briefly described particular missions.

6. APPROVED MISSIONS

6.1 TRMM

Tropical Rainfall Measuring Mission (TRMM) is a US-Japan joint project to measure tropical rainfall and its variations, scheduled to be launched on 1 November 1997. TRMM will carry five sensors: a Precipitation Radar (PR) of 13.8 GHz, the five-frequency radiometer TRMM Microwave Imager (TMI), the Visible Infrared Scanner similar to AVHRR (VIRS), the Clouds and Earth's Radiant Energy System (CERES) and the Lightning Imaging Sensor (LIS).

The microwave radiometer TMI has a wide swath width (as much as 700 km) and can make quantitative rain measurement over oceans, PR can measure the rain both over land and ocean and can provide vertical rainfall profile, although its swath width is about 220 km and is thus narrower than TMI.

The expected result from TRMM is to provide the three dimensional rain rate distribution, the detailed vertical structure of rainfall, the vertically integrated rain rate and the cloud height/properties in wide swath, and finally the time/space averaged rainfall and the storm structure to users.

6.2 ADEOS-II

The mission objectives of ADEOS-II, as the successor of ADEOS, is to acquire data contributing to the international global change research program, as well as useful for applications such as meteorology and fishery. Among various research areas, ADEOS-II is particularly dedicated to the water, energy and carbon cycling. The data from ADEOS-II are expected to be used by international research programs such as the Global Energy and Water Cycling Research Experiment (GEWEX), the Climate Variability Research (CLIVAR) Program and the International Geosphere and Biosphere Research Program (IGBP).

SeaWinds, POLDER and ILAS-II data will be provided to both international and domestic partners in addition to data from the Advanced Microwave Scanning Radiometer (AMSR) and the Global Imager (GLI).

ADEOS-II satellite and related ground segments are under development, with launch currently planned in 2001.

6.3 MDS-2

Mission Demonstration Satellite (MDS) is a small satellite program to demonstrate mission payloads. The second mission concerns the demonstration of a Mie Scattering Lidar, which is planned to be launched in FY 2000. The scientific objectives include the development of a high altitude cloud (cirrus) climatology, a multiply-layered clouds climatology, and stratospheric and tropospheric aerosols studies. This experiment would be a precursor of ATMOS-B.

6.4 AMSR-E on EOS-PM1

AMSR-E is a microwave radiometer provided to the EOS-PM1 platform and is similar to AMSR on ADEOS-II. The combination of ADEOS-II and EOS-PM1 will make twice as much frequent observation possible and it will improve the diurnal variation observation and the observation of fast evolving phenomena such as the cloud liquid content or the water vapor in typhoon. EOS-PM1 is planned to be launched in 2000.

7. PROPOSED MISSIONS

7.1 ALOS

The mission objectives of the Advanced Land Observing Satellite (ALOS) include technology development, mapping, regional observation, hazard monitoring and Earth resources survey. The sensors required for this mission are a Panchromatic Remote-sensing Instrument for Stereo Mapping (PRISM), the Advanced Visible and Near Infrared Radiometer type 2 (AVNIR-2) and Phased Array type L-band Synthetic Aperture Radar (PALSAR). ALSO is planned to be launched in FY 2002. Phase B is on going in FY 1997 and Phase C/D is anticipated to start in FY 1998.

7.2 Experiment on ISS/JEM

The Superconducting Sub-Millimeter Wave Limb Emission Sounder (SMILES), which works at a frequency of 640 GHz, was selected for the first experiment of new remote sensing technologies on the International Space Station/Japan Experiment Module. The objectives of SMILES are to make three-dimensional observations of trace gas species such as ClO, BrO,

H_2O, distributed vertically from the upper troposphere to the stratosphere. These observations should prove useful to understand chemical process, the dynamics of the atmosphere, and the interaction between these processes and atmospheric trace gas species to study the ozone depletion or the global warming mechanism. This proposal was accepted in March 1997 and SMILES is under phase A study, with a view towards a launch in 2002.

8. MISSIONS UNDER CONCEPTUAL STUDY

8.1 ADEOS-III

ADEOS-III is the ADEOS-II follow-on mission. Since 1997, a user requirements definition study has been conducted within the science community, the ADEOS-III Science Team of the Earth Science and Technology Forum. Proposed sensors include AMSR-2, GLI-2, ODUS, ILAS-3, SeaWinds-2, IMG-2, TERSE and GPSR. Feasibility studies are planned for some of these sensors in 1997. Phase A is expected in FY 1998 while launch is scheduled in 2004.

8.2 ATMOS-A, B, C

ATMOS stands for Atmosphere Observing Satellite and the mission concept is categorized into three types, which are A: Precipitation Mission, B: Cloud Radiation Mission, C: Atmospheric (Chemistry, Wind) Missions.

ATMOS-A is a TRMM follow-on precipitation mission. The technical feasibility study for 2.5t class mission had been done by NASDA and the first iteration for the mission definition had been completed in April 1997. In this study, a dual frequency precipitation radar PR-2, TMI and VIRS-2 are considered as the core sensors. At present, a mission definition study is being carried out in collaboration with NASA/GSFC.

ATMOS-B is a cloud/radiation mission. A feasibility study is planned in the latter half of JFY 1997, and some of the results of ATMOS-A F/S are applicable to ATMOS-B F/S. There are possible collaborative missions with JPL's CloudSat, if a joint proposal to NASDA's MDS selected and possible collaborative mission with ESA's Earth Radiation Earth Explorer mission. For these collaborations, joint/collaborative phase A is planned in 1998.

ATMOS-C is an atmospheric chemistry mission. The user requirements definition study by ATMOS-C Science Team is on-going.

9. CONCLUSION

The sudden failure of ADEOS early in its three-year expected mission lifetime affected the researchers and users who hoped for continuous and operational data provision, however a lot of results using existing data can be derived.

The implications of this accident should be taken into account in future plans for the continuing contribution to the Earth Observation users.

Chapter 29

VEGETATION: AN EARTH OBSERVATION SYSTEM TO MONITOR THE BIOSPHERE

G. Saint
CNES, Toulouse, France.

1. INTRODUCTION

The VEGETATION Program is co-funded by the European Commission, Belgium, France, Italy and Sweden. It will span a period of at least 10 years, as two instruments are now approved to be flown on the SPOT 4 and SPOT 5 satellites. The first one was launched in March 1998, SPOT 5 being planned for launch around 2001.

The overall objectives of the VEGETATION system are to provide accurate measurements of basic characteristics of vegetation canopies on an operational basis, either for scientific studies involving both regional and global scales experiments over long time periods (for example development of models of the biosphere dynamics interacting with climate models), or for systems designed to monitor important vegetation resources, like crops, pastures and forests.

The VEGETATION system, consisting of a satellite-borne sensor and of its associated ground segment, will provide long term basic measurements appropriate for biosphere studies. Opportunities for scale integration are provided by the combination with the main SPOT instruments (HRVIR) which allow high spatial resolution for detailed modeling activities or multilevel sampling procedures. Availability of data to different types of users is facilitated through the centralization of reception and archiving global data sets.

Clearly this system will benefit from detailed studies based on other systems that are dedicated to specific studies of the characteristics of remote sensing measurements or to their relationships with surface or processes' parameters. It must be envisaged that the evolution of the mission specifications will have to take into account results of such studies to provide improved characterization of the biosphere state and dynamics.

2. MISSION OBJECTIVES

2.1 Surface parameters mapping

This is the basic requirement, especially for climate and meteorological studies where boundary conditions have to be prescribed as in the case of General Circulation Models or forecasting models. Factors such as albedo, surface roughness, or resistances to heat exchanges (sensible and latent) are important variables for these models, and they can be either determined directly from the measurements or inferred from identification of land cover. The seasonal and long-term variations of such variables are related to vegetation dynamics. The capability to identify, through these variations, physical characteristics of land cover is a key to accurate prescription of these variables. Scales addressed in GCM or forecasting models (typically about 100 km) require that land cover and its variability must be determined with a sampling of about 8 to 10 km: the basic spatial resolution needed for identification of land cover and its variability is 1 km.

2.2 Agricultural, pastoral and forest production

Since the beginning of the land surface satellite remote sensing era (1972), important projects (for example, LACIE, AGRISTARS for USDA, MARS for CEC, TREES for JRC/ESA) have been set up to develop methodologies and strategies to use remote sensing data either for mapping of land use in anthropogenized or natural ecosystems, or for estimation of production potential. Their specific objective was to determine the evolution of productions. This objective had to be adapted to the management of crop production for agricultural exporting countries, to the monitoring of pastoral resources and their dependence from meteorological evolution, to the evaluation of possible global impacts of deforestation and more generally to the need for information related to political or social orientations and decisions.

2.3 Terrestrial biosphere monitoring and modeling

The contribution of the continental biosphere to the biogeochemical cycles (exchanges of carbon and other trace gases) and to water and energy exchanges is one of the objectives of the development of global models. Interaction with human activities is also one of the main points to be studied, because the effect of human pressure on the biosphere might be one of the means by which man is acting on climate in the long term. Biosphere processes and land cover characterization are the basis for quantification: estimations of land cover variables as well as the dynamics of these variables have to be made in order to obtain a good understanding of these processes upon which models may be built. Predictions of impact of climate change on the biosphere and of interactions of the biosphere with the climate (either due to natural factors or to human pressure) can only be inferred from quantification and formalization of the mechanisms by which vegetation cover and ecosystems are functioning. Multilevel series of models have to be developed and linked, ranging from ground studies, local parameterization and exchange models to regional or global dynamics and interaction models. Remote sensing of the vegetation as shown above offers a unique tool for these developments, providing the specification of the systems is adapted to each particular need.

3. VEGETATION SYSTEM DESCRIPTION

Designed to provide elaborated products for thematic studies, the overall system is based on two major components: the payload onboard the satellites and a ground segment which archives the data, process images to generate products and distribute them to various users. Detailed description and updates on the quality of the system and products can be found at the URL: http://www-vegetation.cst.cnes.fr:8050/ (Saint et al., 1998).

3.1 The payload

It is composed of a wide field of view instrument which allows a coverage on the Earth surface of 2250 kilometers, a solid state memory to store about 90 minutes of acquisition, image telemetry systems and an onboard calibration device to monitor radiometric performance of the instrument.

The instrument allows measurements in 'basic' spectral bands where vegetation cover can be characterized with sufficient accuracy for most applications today, as shown in the following table.

Table 1. Spectral bands of the VEGETATION instrument

Band	Spectral range, in nm
Blue	440–475
Red	620–690
Near-infrared	780–880
Short wave infrared	1570–1690

Red, Near Infrared and Short Wave Infrared bands give information about the structure of the canopy (density), the absorption by the 'green' component and water content: these measurements are actually used by more and more models to infer the biophysical state of the vegetation. A Blue band was added to get some indication of the atmospheric effects on radiation coming to and leaving from the surface, so that the final products might be corrected and give as accurate as possible equivalent ground measurements. This feature is especially important as the system is designed to provide time series, which are recognized to be the way to get valuable information on seasonal and inter-annual variations.

Radiometric quality is expressed in terms of noise equivalent reflectance ($NE\Delta\rho$) and by the relative and absolute calibrations. The measured $NE\Delta\rho$ appears to be within specifications. Estimates of the calibration accuracy will need some months before they are validated, although it appears already that calibration based on Rayleigh diffusion is very stable and consistent with the onboard calibration system, while the use of ground sites shows a large variability.

A second important feature of the instrument is the geometric quality: the spatial resolution is 1 km in any part of the field of view (due to the unique instrument concept based on the use of telecentric optics and CCD line detectors) and, using a limited number of Ground Control Points, all images acquired at different dates during a year can be registered with an accuracy of about 300 meters. Again, this specific feature is of particular importance for time series analysis: while previous systems could not provide the same accuracy and allowed analysis on individual areas which could not be less than between 5 and 10 kilometers in diameter, VEGETATION will decrease this size to less than 2 kilometers.

The field of view of the instrument, associated to characteristics of the orbit of the SPOT series, provides a daily coverage of all terrestrial areas. For areas located at latitudes higher than 35°, they may be seen several times a day.

Finally, the simultaneity between the VEGETATION acquisition and the High Resolution Instrument acquisition, giving the same measurements (same spectral bands) allow 'zooming' on some areas where models for interpretation of VEGETATION data can benefit from a detailed knowledge of the ground cover. Use of both types of acquisition give full access to the high frequency (once a day) and high spatial resolution (20 m).

3.2 The ground segment

Data are received from the satellite in a station located in Sweden and transmitted to the Archiving and Processing Center located in Belgium, where all products are generated. Satellite Control and Data Quality are done from France.

From the beginning of the Program, users were involved to define its characteristics and the products it should deliver: they are adapted to the particular missions described above and coherent as much as possible with the needs of existing projects. Two general categories of users could be identified:

1. research teams which are developing methodologies for the use of VEGETATION data or scientific biosphere studies: they generally have a study site (about 500 x 500 km2) and need long time series (one year of daily or weekly data),
2. projects which are based on the use of both VEGETATION and other data sets, for which the data delivery has to be fully operational and for long periods, for comparison or historical studies (one continent every day). Typically, these projects include MARS, TREES, IGBP, etc.

To illustrate the special characteristics of the instrument, high priority was given to design products that would allow direct multi-temporal registration as well as simple superposition with simultaneously acquired high-resolution data.

VGT-P products: They are adapted for the first type of users for which physical quality of data is important. They correspond to data which would have been acquired by an ideal instrument: they are corrected for system errors (mis-registration of the different channels, calibration of all the detectors along the line-array detectors for each spectral bands) and re-sampled to geographic projections for multi-temporal analysis as well as for comparison with high resolution data. Annotations giving full information on applied corrections (calibration information, geometric parameters taking into account attitude and position on the orbit), or for further non-system corrections ("standard" atmosphere parameters) are attached to the data sets.

VGT-S products: They are most probably the data sets which will be frequently used operationally: they correspond to VGT-P data to which corrections have been applied using the annotations and for which some syntheses are provided:
- a daily synthesis using all available measurements on one day for a specific location,
- a 10-day synthesis, based on the selection of the "best" measurement of the entire period. The selection is based on the maximum NDVI value, as it is commonly accepted today, even if many problems associated to that selection are identified.

To adapt to the evolution of users needs as well as to the validation of new algorithms, a procedure to regularly update the processing system is requested: it should provide capabilities to include new methods for data correction, synthesis, etc, as soon as they are commonly accepted by the user community.

Support to users is provided to facilitate the use of VEGETATION data: a catalogue with browsing capability on the data quality (cloudiness) is accessible through Internet. Validated software templates for the common operations for data handling and standard correction will be made widely available.

3.3 The distribution component

The design of the entire system was adapted to provide measurements which are mostly fitted to thematic developments: it was considered that, either for scientific studies or application projects, the lack of operational sets of data, with the insurance that they would be available in a long term was a blocking factor. While numerous scientific or experimental satellites do exist for a long time for the benefit of physicists, none of them is providing data sets adapted to the use in either scientific development on the study of the biosphere at different scales or application projects where continuous provision of basic and robust information is essential. For these reasons, the distribution of VEGETATION products is being done through the existing network which is providing the SPOT high resolution images, on a 'commercial-like' basis. Innovative solutions for distribution through modern ways (e.g., Internet on line catalog on World Wide Web, access to data through electronic transmission) have been implemented to facilitate the contact with users and limit the delay between acquisition and product delivery to about 2 days.

29. VEGETATION: AN EARTH OBSERVATION SYSTEM TO MONITOR THE BIOSPHERE

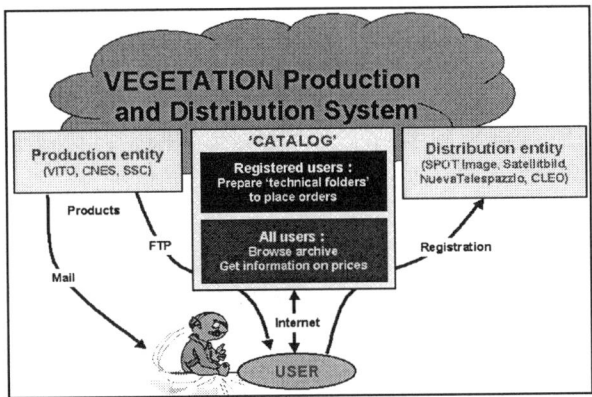

Figure 1. The VEGETATION production and distribution system.

The following diagram identifies the principal components of the distribution system and the relationships between the user and the entire VEGETATION program.

4. SOME APPLICATIONS

While the operational distribution of products began before spring 1999, some results were already obtained from previous studies. A Preparatory Program has been running for four years, 33 international investigations being conducted on various projects and types of applications. These investigations dealt first with methodological development on simulated data sets as long as real data were not available and will continue validating their methods on real products. Two examples are given here.

4.1 Monitoring agricultural areas

The following example (from Husson et al., 1998) is related to the preparation for the integration of VEGETATION products into the European system for monitoring agriculture through remote sensing (MARS). During 1998, the following images show the difference measured between beginning of May and end of August on France (Fig 2 and 3). While residual atmospheric effects remain on the Northern part in August, on the Southern part, the difference between the two images is related to the regional evolution on regions where homogeneous agricultural practices are used (winter/spring crops, permanent prairies, vineyards...).

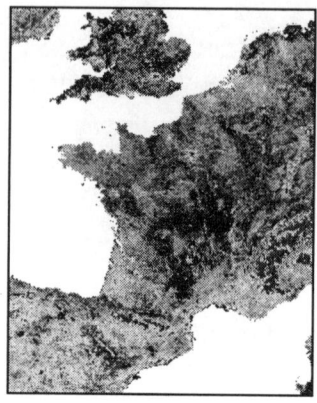

Figure 2. First decade of May 1998.

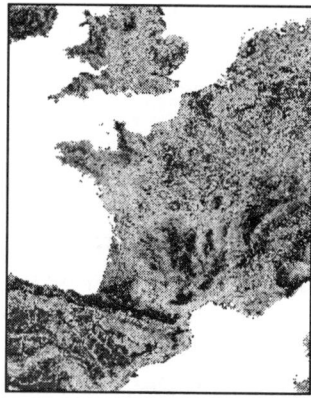

Figure 3. Last decade of August 1998.

More details can be obtained following the evolution of an index related to red and near infrared spectral bands. On simulated data from 1994 and 1995, the difference between wheat dominant and corn dominant areas (Fig. 4 and 5). A corn dominant area, as south of the Landes forest shows a much higher index (dark gray) in July-August (decades 18-20) than wheat dominant areas (southeast from Toulouse) where crops are harvested.

29. VEGETATION: AN EARTH OBSERVATION SYSTEM TO MONITOR THE BIOSPHERE

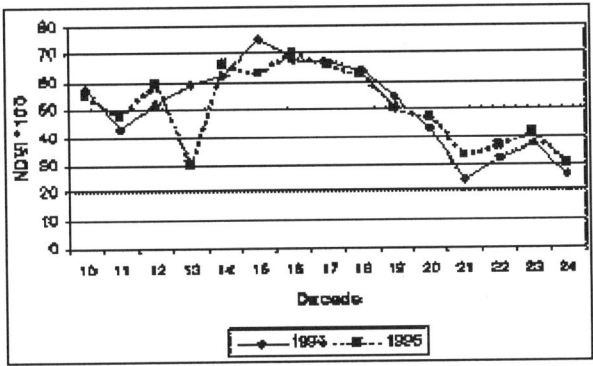

Figure 4. NDVI for wheat dominated area

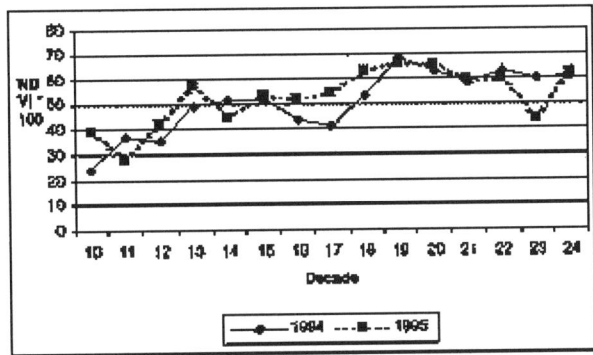

Figure 5. NDVI for corn dominated area

In the same corn dominant area, comparing (on Fig. 6) two successive years (1994 and 1995) shows the different evolution of corn in 1995 where the density became higher very soon (decade 16). This is an example of a way to monitor crop growth and get information which can help for prediction of yield and production.

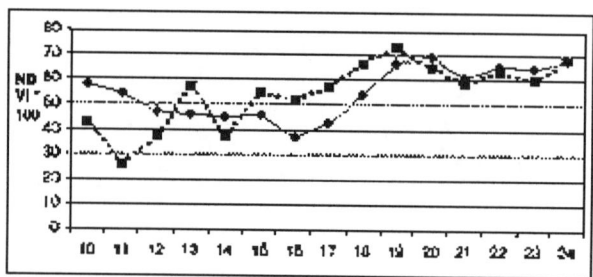

Figure 6. NDVI for successive years on a corn dominated area

4.2 Modeling the biosphere

To acquire knowledge on ecosystems, numerous models are being refined, new ones being based on the basic eco-physiological processes. While they do not generally need satellite data to be run, improvements can be obtained from the comparison of their outputs with satellite measurements of surface parameters. For example Kergoat (1998) and Dedieu (1998) adapted rooting depth of forest biomes and introduced a carbon budget constraint to limit leaves production so that the outputs of the model became coherent with satellite measurements (Fig 7). To get comparable measurements, important changes similar to some values suggested in the literature had to be introduced: the continuous and accurate measurements provided by VEGETATION will allow validation of these changes on a global scale and will then provide better hypothesis for interactions between the biosphere and atmosphere for global climate change studies.

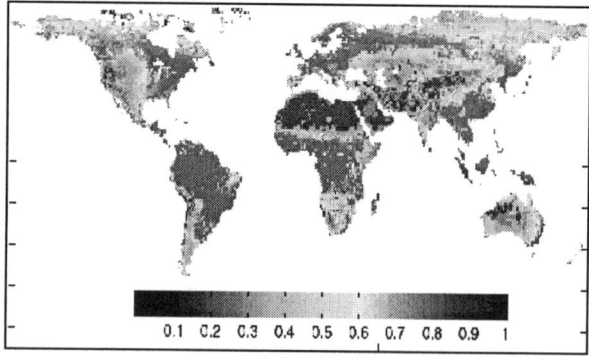

Figure 7. Using satellite measurements to adapt biosphere model parameters: Indicator derived from the adapted model

29. VEGETATION: AN EARTH OBSERVATION SYSTEM TO MONITOR THE BIOSPHERE

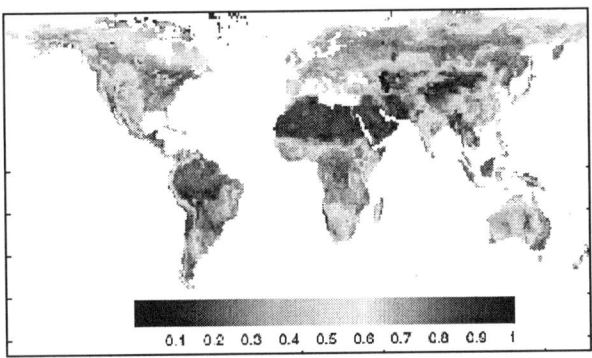

Figure 8. Using satellite measurements to adapt bopsphere model parameters: Indicator derived from satellite measurements

5. CONCLUSION

While VEGETATION products will begin to be progressively used, methods to infer information will be derived and validated. Due to previous studies, many programs are waiting for them to really begin operational management of ecosystems as presented in the above examples.

Concurrently, research scientists are already developing more accurate indicators of vegetation canopy state both for a refined use of these products and for future generation systems.

6. REFERENCES

Husson, A. et al. (1998) Integration of VEGETATION and HRVIR data into yield estimation.
Kergoat, L. (1998) A model of hydrological equilibrium of Leaf Area Index at global scale, *Journal of Hydrology* (in print).
Dedieu, G. (1998) STEM-VEGETATION: satellite measurements and terrestrial ecosystem modelling using VEGETATION instrument, Final Report 07/98.
Saint, G. et al (1998) The VEGETATION Information Web server, http://www-vegetation.cst.cnes.fr:8050/

Chapter 30

HYPERSPECTRAL IMAGER SURVEY AND DEVELOPMENTS FOR SCIENTIFIC AND OPERATIONAL LAND PROCESSES MONITORING APPLICATIONS

B. Kunkel (1), J. Harms (1), U. Kummer (1), E. Schmidt (1), U. Del Bello (2), B. Harnisch (2) and R. Meynart (2)
(1) Dornier Satellitensysteme GmbH, Ottobrunn, Germany and (2) ESA ESTEC, Noordwijk, The Netherlands.

1. APPLICATION SURVEY FOR IMAGING SPECTROMETERS

Table 1 below presents a schematic survey of the typical and proven imaging spectrometer (IS) application categories, and a preliminary allocation of the useful spectral bands per application. This survey has been generated under DSS in-house funding, and does neither claim to be complete, nor to be correct in every aspect. This is the simplified view of an industrial team, although we attempted to take into account as much as possible scientific publications and personal contacts with scientists and operational users.

This survey is kept rather schematic, and does not show major parameters such as spectral sampling interval, IFOV, swath width etc., which are summarized below as typical space-borne IS mission requirements.

The main message for us as an industrial team is: A useful IS must cover the visible (VIS), NIR (near-infrared) and SWIR (short wavelengths IR) spectrum.

Of the three main groups of optical remote sensing instruments—multispectral / high-resolution imagers, imaging spectrometers and atmospheric sounders—the IS represents the group with the widest range of applications. Its geometric and spectral resolution is about half way between the other two optical instrument categories (a rather crude categorization).

Table 1. (Part 1) Survey of Imaging Spectrometer applications. The larger the digit, the more important the spectral band for the indicated application. Bold figures refer to primary applications, while figures in italics indicate relevance for supporting measurements.

Application	Parameter	VIS/NIR											TIR
Spectral range (VIS/NIR: nm, TIR: µm)		400–450	450–500	500–600	625–700	700–730	730–800	800–1000	1000–1100	1100–1300	1500–1650	1650–1750	1900–2000	2050–2400	10–12
Vegetation, Agriculture, Forestry														
Status, health	Chlorophyll, carotine	*3*			**3**	1								
N content	Cellulose, lignin				2						**4**	**4**		
Maturity	Cellulose, lignin, color			*3*			*3*				**4**	**4**		
Species discrimination	λ reflectance, MRDF			*3*	**4**						**4**	2		
Trees & bush health	Chlorophyll, blue shift	2		2 2							**4**			
Deforestation, desertification	λ reflectance				**4**	1		*1* 2	2 2		2	2		
Timber production	λ reflectance								2	2			**2**	
Burning, erosion	Aerosols, smoke		1	2 2				**4**					*2*	**3**

There is one fundamental difference between these three main groups of instruments: while countless data archives from space-borne sensors are available for high-resolution imager and sounder type instruments, there are, until now, only airborne IS data, at least of the Earth.

30. HYPERSPECTRAL IMAGER SURVEY AND DEVELOPMENTS FOR SCIENTIFIC AND OPERATIONAL LAND PROCESSES MONITORING APPLICATIONS

Table 1. (Part 2) Survey of Imaging Spectrometer applications. The larger the digit, the more important the spectral band for the indicated application. Bold figures refer to primary applications, while figures in italics indicate relevance for supporting measurements.

Application	Parameter	VIS/NIR													TIR
Spectral range (VIS/NIR: nm, TIR: µm)		400–450	450–500	500–600	625–700	700–730	730–800	800–1000	1000–1100	1100–1300	1500–1650	1650–1750	1900–2000	2050–2400	10–12
Water quality, Environmental Monitoring															
Quality	Chlorophyll, yellow substance	**4**	**2**	*1*											
Eutrophication	λ reflectance, color	**4**	**3**		**1**										
Bathymetry	Ground reflectance	**4**	**2**												
Micro–climate	Aerosols, snow/ice				*3*			*3*			*3*	*2*			*3*
Glacier melting, snow/ice	Reflectance change, moisture		*2*					*3*			*3*	*3*		*3*	*4*
Coastal zone changes	Chlorophyll, yellow substance, erosion	**4**	**2**												
Acid deposits detection	Overtone absorption bands				**4**	*1*						**2**		*1*	
Atmospheric measurements, Climate															
Aerosol load	Backscatter			**4**	**1**										
CH$_4$ load	Absorption band @ 1.13µm									**1**					
CO$_2$	Absorption band @ 1.9–2.0µm												**3**		
Disaster Management															
Flood	Reflectance, color	**4**	*1*	**3**			**3**	**3**							
Volcanoes	Aerosols, temperature			**3**	**1**									*1*	**4**
Land Use Planning															
Settlements	λ reflectance, color					**4**	**3**	**3**	**2**			**2**		**2**	
Soil	λ reflectance, moisture	*1*			*1*	**3**	**2**	**3**	**3**	**3**				**4**	

Space-borne hyper-spectral imagers are under development at present for various missions. These include MERIS on ENVISAT, HSI on the Lewis Test Satellite, ARIES on an Australian platform (with the support of CSIRO and commercial mining user groups), or MODIS on the NASA Terra platform. Additional information on these instruments is provided in Tables 2A and 2B below.

Particularly, IS allows the distinction of species in the water, in the vegetation canopy (biochemistry), as well as applications in surface mineralogy and petrography. Table 1 provides an example survey of such applications and the corresponding useful spectral regions, though with inadequate spectral resolution. While the early push-broom type IS developments (FLI in Canada, ROSIS at DSS) covered the visible/NIR spectrum (VNIR) only, more recent developments typically include short wavelengths IR channels (SWIR)—both operating in the reflected sunlight spectral range. Somewhat less emphasis has been given to the thermal infrared (IR: 3–12 µm).

The primary mission emphasis is widely dependent on the IS IFOV and TFOV, and will be mainly focused on:
- energy and water fluxes
- water quality, organic and inorganic inlets (on global scale with medium resolution IS, on regional scale high resolution IS)
- land surface processes and changes (see Fig. 1 as example), particularly for:
  – vegetation status, health/stress, multi-temporal BRDF measurements along- and across-track
  – soil condition, surface minerals
  – coastal and inland water bodies for narrow FOV IS, global ocean monitoring for wide FOV IS
  – carbon cycle detailing
  – detection of indicators for the effect of traces gases and aerosols (see Fig. 2 for the Etna) on the environment (e.g., vegetation, glacier melting).

30. HYPERSPECTRAL IMAGER SURVEY AND DEVELOPMENTS FOR SCIENTIFIC AND OPERATIONAL LAND PROCESSES MONITORING APPLICATIONS

Figure 1. Example of DAIS Imaging Spectrometer Image Cube (DLR) VNIR/SWIR & TIR Composite for Land Use Classification (web page)

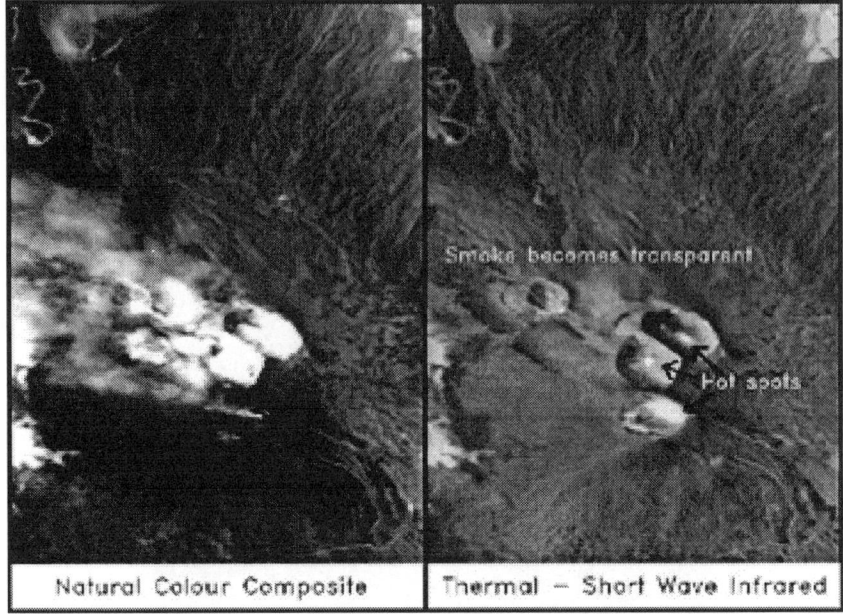

Figure 2. Example of DAIS (DLR) –Etna in VNIR/NIR versus SWIR/TIR Composite (web page)

Secondary mission objectives refer to topics such as:
- geological mapping (particularly in the NIR and SWIR channels)
- land use planning, typically for higher geometric resolution IS.

2. SURVEY OF EXISTING AND PLANNED IMAGING SPECTROMETERS

A survey of planned and existing IS is given in Tables 2A and 2B below. It may not be complete, but refers to the instrument developments most often quoted in literature (Rast, 1991; Staenz, 1992; Kunkel et al., 1994; Blechinger et al., 1995; Baudin et al., 1993; Kunkel et al., 1996; Pagano et al., 1995; Basedow, 1995; Kunkel et al., 1997).

There are two principal types of IS, with two main sub-groups:
1. *Whiskbroom IS* are instruments with mechanical across-track scan and a linear detector array for the spectrum representation, e.g., AVIRIS (Rast, 1991; Staenz, 1992), MODIS Pagano et al., 1995), the "classical" instruments dating back to the times when 2-d detector arrays were not mature enough; radiometrically, this principle leads to very short pixel dwell times at given

IFOV, thus, either large apertures or reduced radiometric or geometric performance compared to the other category:
2. *Pushbroom IS,* which are sensors with electronic across-track and spectral scan, thus, the inverse of the line frequency being equal the pixel dwell time, requiring 2-d detector arrays per wavelength range; this group will form the emphasis of this paper—and here we have two basic sub-groups again:
a) *Wide-field IS,* such as MERIS (Baudin et al., 1993) or ROSIS (Blechinger et al., 1995) for global mapping, often by assembling multiple optics module aside, and
b) *Narrow-field IS,* for spot-mode observations of discrete targets, typically addressed by an additional front optics-pointing mirror.

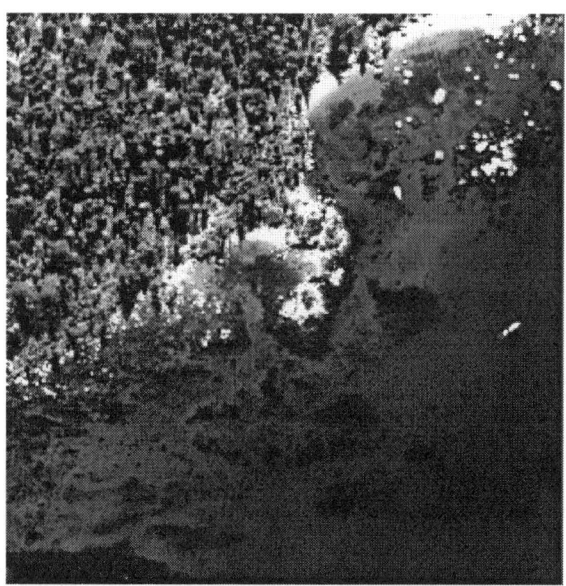

Figure 3. HYDICE Image Example: Lake Tahoe. Bands: 476/542/677 nm (Web page)

This categorization has the following basic technical reasons:
- The main limitation arises from the 2-d detector arrays which, at least for the spectral range $\lambda \geq 1$ µm, typically are limited to less than 1024 detector elements in the spatial scanning direction
- Correspondingly, the front and spectrometer optics have to provide a planar focal plane of a certain physical extension, typically 30–50 mm width, which excludes a priori several optics concepts

- In addition, the spectrometer entrance slit in the front optics focus tends to suggest a telecentric beam, which further limits the optics choice.

Table 2A. (Part 1) Survey of existing airborne Imaging Spectrometers

Name	AVIRIS	AIS-II	CASI
Type	VIS/IR IS	IS	Compact IS
Manufacturer/ Developer	JPL, USA	JPL, USA	ITRES / Canada
Operational	Since 1987	Since 1987	Since 1989
Scanning Principle	Whiskbroom	Whiskbroom.	Pushbroom
Spectral Range (µm):	0.4–2.45	0.8–2.4	
* VNIR	0.4–1.0		0.43–0.87
* SWIR	1.0–2.45	0.8–2.4	
# Spectral channels	224		≤ 288
Spectral bandwidth (nm):			
* VNIR	9.6–14		2.9
* SWIR	16–25		
Encoding (bit)	10		12
Total FOV (deg)	30		35.5
IFOV (mrad)	1		1.2
F-number			2
Detector array format used	Linear 224 elements	Linear 224 elements	256 × 612
Detector array type		CCD+CMT	CCD
# Recording channels @ full resolution	All	All	≤ 39
Disperser type	Transm. grating		Reflection grating
Specials		4 cameras	
Remarks	Extensive use for spectral feature recovery		

As a schematic example, the HRIS all-reflective optics and FPA concept are shown in Fig. 7. This 3-mirror (aspheric, elliptic) optics concept is currently realized by Carl Zeiss. Its peculiarity is the spatial and spectral slit curvature correction by the elliptic primary and secondary optics mirrors.

As shown in Table 2A, a relatively large variety of IS has been developed for airborne missions (a few more are left out here). Quite clearly, these have a high merit even when advanced space-borne hyper-spectral imagers will be in orbit because of their ability to show the smallest details such as individual bushes, ponds, fields or mineral lodes.

Both surveys (Tables 2A and 2B) show that the future oriented instruments and several existing sensors place emphasis on the provision of both VNIR and SWIR spectral ranges; only dedicated water/ocean instruments can afford the limitation to VNIR. The main references for these tables were Rast (1991), Staenz (1992) and Pagano et al. (1995), however, at least 50 more have been partly used.

30. HYPERSPECTRAL IMAGER SURVEY AND DEVELOPMENTS FOR SCIENTIFIC AND OPERATIONAL LAND PROCESSES MONITORING APPLICATIONS

Table 2A. (Part 2) Survey of existing airborne Imaging Spectrometers

Name	FLI/PMI	GERII/DAIS	IRIS
Type	Fluorescence line Imager/Program. Multispectral Imager	Geophysical Environmental Research Imager	IR Imaging Spectrometer
Manufacturer/ Developer	Moniteq, Canada	GER, USA	ERIM, USA
Operational	Since 1984	Since 1986	Under development
Scanning Principle	Pushbroom	Whiskbroom	Pushbroom
Spectral Range (µm):		0.4–10.6	2.0–5.0
* VNIR	0.43–0.805	0.4–1.0	
* SWIR		1.1–2.5	2.0–5.0 MIR
# Spectral channels	≤ 288	63	256
Spectral bandwidth (nm):			
* VNIR	2.5	25.4	
* SWIR		16.5–120	≥ 20 MIR
Encoding (bit)		10	12
Total FOV (deg)		90	0.92 × 0.8
IFOV (mrad)		2.5/3.3/4.5	0.9–1.5
F-number			
Detector array format used	4 × 288 × 385	2 × 32-elements + 2 discrete CMT	256 × 256
Detector array type			InSb
# Recording channels @ full resolution	≤ 32	all	
Disperser type	Transm. grating	Prism/grating	Grating
Specials			
Remarks	First operational Imaging Spectrometer		

The emphasis in this paper is on the two space-borne IS which are under study or development for ESA (of which finally only one will go into the flight hardware phase). These are HRIS (High Resolution Imaging Spectrometer, currently an airborne demo model is under development under DSS prime contract), and PRISM (Processes Research by an Imaging Space Mission), currently in Pre-Phase A study level in two teams, one again lead by DSS.

Table 2A. (Part 3) Survey of existing airborne Imaging Spectrometers

Name	ROSIS	HYMAP	HYDICE
Type	Reflective Optics System Imaging Spectrometer	Hyperspectral Mapper	Hyperspectral Digital Imagery Collection Experiment
Manufacturer/ Developer	MBB (DSS)/ DLR/ GKSS	DSS/ESTEC	Hughes, USA
Operational	Since 1992	Early 1999	Prototype
Scanning Principle	Pushbroom	Pushbroom	Pushbroom
Spectral Range (µm):	0.425–0.85	0.44–2.4	0.4–2.5
* VNIR	0.425–0.85	0.44–1.03	0.4–1.0
* SWIR		1.0–2.4	1.0–2.5
# Spectral channels	90	205	210
Spectral bandwidth (nm):			
* VNIR	4.3	4.9/9.8 average	4.7
* SWIR		9.2 average	11.7
Encoding (bit)	10	12	12
Total FOV (deg)	32	± 4	7.2
IFOV (mrad)	0.5	0.9	0.5
F-number	3.5		
Detector array format used	128 × 512		220 × 308 sp. monol. array
Detector array type	CCD	CCD + CMT	hybrid InSb
# Recording channels @ full resolution	≤ 28	30	
Disperser type	Reflecting grating	Dual prisms	Prism dual-p.
Specials	All-reflecting optics		Baker Paul
Remarks	First European operational IS	Currently flown by DSS	Near completion by mid 1995

30. HYPERSPECTRAL IMAGER SURVEY AND DEVELOPMENTS FOR SCIENTIFIC AND OPERATIONAL LAND PROCESSES MONITORING APPLICATIONS

Table 2B. Planned spaceborne Imaging Spectrometers (Part 1): Australian Research Imaging Spectrometer (ARIES), Hyperspectral Spaceborne Imager (HSI), and High Resolution Imaging Spectrometer (HRIS)

Name:	ARIES	HSI	HRIS
Manufacturer/ Developer	CSIRO, Australia	TRW, USA	ESTEC/DSS
Status/Launch	Under development	Launch Aug. 97 Lewis launch failed	Demo model under development
Scanning type	Pushbroom	Pushbroom	Pushbroom
Spectral range (µm)	0.4–2.4	0.43–2.5	0.45–2.35
* VNIR	0.4–1.1	0.43–1.0	0.45–1.02
* SWIR (MIR, TIR)	2.0–2.45	1.0–2.5	1.0–2.35
# Spectral channels	380	220	205
Spectral Bandw. (nm)	12.5 average	10	8.9–12
Encoding (bit)	12	12	12
Tot. FOV (deg)	14.7	0.9	2.2
Swath variation (± deg)		± 25	± 30
IFOV (mrad) - Spectr.	1	0.06	0.05
High-resol.?	no	yes	no
HR Channel IFOV		0.01	
SSP @ 500 km altitude (m)	500	30	25
Swath @ 500 km altitude (km)	128	7.7	19.2
Array format	256 × 256 VNIR 256 × 256 SWIR	256 × 256	128 × 768 VNIR 140 × 768 SWIR
Array type(s)	CCD/CMT	CCD/ NiCMOS	CCD/CMT CMOS
Specials		Technology test mission only, 2 Mbps downlink	
Raw data rate (Mbps)	148	110	367
Transmiss. Rate (Mbps)		2	100

PRISM is conceived as a candidate for a dedicated "Earth Explorer" mission and platform (among other candidates). Both instruments represent the "spot mode" IS category.

Table 2B. Planned spaceborne Imaging Spectrometers (Part 2): Medium Resolution Imaging Spectrometer (MERIS), MODerate Resolution Imaging Spectrometer (MODIS), and Processes Research by an Imaging Space Mission (PRISM)

Name:	MERIS	MODIS	PRISM
Manufacturer/ Developer	ESTEC / Aérospatiale	NASA GSFC/Hughes	ESTEC/ DSS
Status/Launch	Launch on ENVISAT 2001	Launch on Terra December 1999	Explorer Mission post 2005
Scanning type	Pushbroom	Whiskbroom	Pushbroom
Spectral range (μm)	0.40–1.05	0.415–14.3	0.45–2.35
* VNIR	0.40–1.05	0.4–1.0	0.45–1.02
* SWIR (MIR, TIR)		1.1–2.33	1.0–2.35
# Spectral channels	15 selectable	36	185
Spectral Bandw. (nm)	≥ 1.25	10–500	3 min., 9 average
Encoding (bit)	12	12	12
Tot. FOV (deg)	68.5	± 52	3.7
Swath variation (± deg)			± 32
IFOV (mrad) - Spectr.	0.375	0.3–1.3	0.064
High-resol.? HR Channel IFOV	no	no	Possible
SSP @ 500 km altitude (m)	187	220–850	32
Swath @ 500 km altitude (km)	810	2330	33
Array format		Linear 64 elements	200 × 1024 VNIR 140 × 1024 SWIR
Array type(s)	CCD	CCD + InSb + CMT	CCD/CMT CMOS
Specials		14 MIR/TIR channels	2 TIR channels, onboard storage
Raw data rate (Mbps)	18,3	26	320
Transmiss. Rate (Mbps)			≈ 100

3. TYPICAL REQUIREMENTS AND INSTRUMENT IMPACT DISCUSSION

The typical mission requirements for IS—see Table 3—reflect the unequalled wide range of applications of these relatively new instrument types. The combination of fairly high geometric resolution, a wide spectral range (more recent developments) and fine spectral sampling allow to detect, in a quasi 3-dimensional imaging mode, features which are not detectable with any other optical or microwave sensor type. The designs presented here are specific to planet Earth. However, this kind of instrument will be part of all future planetary (as well as comet) missions. Of all celestial bodies in our so-

30. HYPERSPECTRAL IMAGER SURVEY AND DEVELOPMENTS FOR SCIENTIFIC AND OPERATIONAL LAND PROCESSES MONITORING APPLICATIONS

lar system, planet Earth is the only one carrying higher developed forms of life in a fragile environment. This makes the permanent monitoring of the thin biosphere layer such an important task—this is where imaging spectrometers play an increasingly dominant role for environmental change processes, essentially changes of the biosphere, climate and cryosphere.

Of the requirements presented in Table 3 below, the ones defining the instrument layout and dimensions are best expressed in the typically required radiometric resolution calculation equation:

$$NE\Delta L = \int_{\Delta\lambda} \{E(\lambda) \times \tau_{opt}^{-1}(\lambda) \times \cos^4\phi \times i_n(\lambda) \times d\lambda\} \times F$$

where

$$F = 4h^2 \times \tau_p^{-1} \times R(\lambda) \times P^2 \times \sqrt{n} \times \pi D^2$$

where

- $E(\lambda)$ = Solar reflectance irradiance at platform level, given by spectral range and interval
- $i_n(\lambda)$ = integrated detector and readout noise, *subject to optimized selection*
- $R(\lambda)$ = Detector responsivity at given wavelength, *subject to optimized selection*
- $\tau_{opt}(\lambda)$ = optics total transmission per wavelength interval, *subject to optimized design*
- $\cos^4\phi$ = Sun zenith angle at selected orbit and nodal transmission time, typically given
- $d\lambda$ = spectral bandwidth or sampling interval, mostly given
- n = number of detectors available for scanning, mostly deleted for pushbroom imagers
- D = Optics effective aperture, main parameter to play with (impact on instrument size, mass)
- h = orbit (or aircraft) altitude, mostly given
- $D^2 h^2/P^2$ = defines, with detector pixel size, the effective system focal length, instrument dimension
- P = Ground pixel size (m)
- τ_p = pixel dwell time, given by orbit altitude dependent relative velocity and P

These indications demonstrate the small degree of freedom in the instrument's design approach in terms of selectable or scalable parameters. A high radiometric resolution (either expressed in terms of NEΔL, with low values being $\leq 10^{-3}$, or in terms of encoding bits) at a given orbit altitude and Sun zenith angle more or less automatically results in large aperture requirements.

The main design-driving parameters and requirements are briefly discussed below. We should underline that we have, based on frequent contacts with various science and application users, a rather fair understanding of these requirements:

- *Orbit altitude:* Impact on radiometric performance, see above, but also relative velocity (decreasing towards higher altitudes); impact on analog and digital readout electronics frequency
- *Ground pixel size (or IFoV):* Together with radiometric resolution most decisive parameter for radiometric aperture, effective focal length definition. Hence this affects instrument size and mass, as well as analog and digital processing electronics. Example: a 30 m SSP compared to 50 m brings either the radiometric performance down by a factor of 3.6, or the aperture needs to be increased by factor of 1.9 to maintain the performance, with a nearly doubled (ratio of $P^2 / \sqrt{\tau_p}$)
- *FoV or Swath width:* At least for the pushbroom type of instrument, a two-fold critical parameter is relevant. In terms of choice of optics, mostly introducing aspheric elements to yield FPA planarity and slit curvature correction, plus limited number of spatial pixels for 2-d detector arrays, for wavelengths ≥ 1.1 µm typically to ≤ 1024 spatial pixels unless optical butting is applied. For IS, the large number of spectral channels combined with a medium to high geometric resolution, also a design driver for the onboard storage or telemetry data rate
- *Swath variation or pointing (across-track for more rapid target access, and/or along-track, e.g., for BRDF measurements):* Particularly 2-axis mechanisms are technically cumbersome in space. A single axis pointing leaves the choice of rotation in the optical axis with the negative effect of image distortion at increasing angles, and rotation of the mirror surface causes incidence angle depending transmission and polarization changes, which need to be calibrated out
- *Radiometric resolution (in NEΔL or NE$\Delta\rho$ or bit, ρ = target albedo):* Impact on radiometric aperture, see above, optical throughput, i.e., choice of optics, particularly the disperser, ADC, signal processing electronics, and, most of all, calibration accuracy
- *Radiometric accuracy (mostly expressed in % or dNEΔL):* Impact on calibration unit and choice, thermal stability particularly of the FPA, widely dependent on detector noise characteristics

30. HYPERSPECTRAL IMAGER SURVEY AND DEVELOPMENTS FOR SCIENTIFIC AND OPERATIONAL LAND PROCESSES MONITORING APPLICATIONS

- *MTF, PSF:* Typically, high optical image quality is mandatory; however, too high demands will narrow down the choice of optics concept (this has been the case, e.g., for HRIS, for the TMA front optics as well as the spectrometer optics)
- *Spectral range:* Advanced IS typically call for wide band ranges such as 0.4–2.4 µm; this suggests a preference for all-reflective optics plus dedicated coatings. Some IS also provide discrete MIR and TIR channels. Technically, a pushbroom IS covering the region ≥ 1.1 µm could also be tuned for MIR and TIR ranges, yet the majority of applications tends to require VNIR and SWIR

Table 3. Typical Imaging Spectrometer Mission and Performance Requirements

Parameter	Instrument / Mission Level	
	High-resolution Imaging Spectrometer	Low-resolution Imaging Spectrometer
Orbit Height (km)	500–800	same
Relative Velocity (m/s)	7070–6625	same
TFOV (degree)	typically: 2–5	18–82
Across- /along-track Point, (BRDF)	typically: ± 30°	
Number of Pixels/Line	500–1000 (or: 512–1024 detector array formats)	same
Ground Pixel Size (m)	30–60	100–500
Result. Line Freq. (Hz)	125–200	25–60
IFOV (µrad)	50–150	200–1000
Req. EFL (mm @ 25 µm detector pixel, at 600 km orbit)	200–800	40–120
Overall Instrument Mass (kg)	< 400	< 200
Max. Data Rate (Mbps)	≤ 100–250	≤ 25–60
	Spectral and radiometric requirements		
	VNIR	SWIR	MIR/TIR
Spectral Range (µm)	0.40–0.105	1.0–2.40	3–5 / 10–12
Spectral sampling Interval (nm)	2–15	10–20	> 100
Radiometric resolution	≥ 12 bit	10–12 bit	10–12 bit
Radiometric accuracy	≤ 2% goal	≤ 2% goal	0.1–1 K

- *Spectral resolution/spectral sampling interval and its variation:* Typical values for pushbroom IS are 5–10 nm. For wide band IS from VNIR to SWIR, uniform dλ would require grating dispersers. These, on the other hand, should be limited to one optical octave (e.g., 450–900 nm), and

yield low optical throughput. The typical solution consists in using achromatized prism doublets. Wide band single optics may necessitate a spectral over-sampling in the 550–780 nm range to avoid saturation, since the radiometric aperture will be defined by the lower and higher wavelengths
- *Spectral and spatial co-registration accuracy (typically 0.1 to 0.2 pixels):* This is another incisive parameter for the optics and FPA choice, optical bench mechanical and thermal stability, OGSE. The preference is for single common optics for all channels.

4. WORKING PRINCIPLES OF IMAGING SPECTROMETERS, EXAMPLES

Fig. 7 shows, as a typical example for a pushbroom IS, the optical scheme of the HRIS instrument. For simplicity reasons, only the SWIR path of the spectrometer is presented.

The main sub-units, following the optical and electronics path, are briefly summarized below:
- A *pointing unit* (or scan unit for whiskbroom IS) to provide a variable swath (except large FOV instruments such as MERIS, MODIS, ROSIS, where an along-track tilt of the swath is partly introduced to avoid Sun glint over water bodies) or along-track pointing for BRDF measurements (e.g., PRISM), mostly used also to access in-flight and external calibration references; critical for two-axis pointing in terms of mechanisms
- A *common front optics* or telescope or a set of modular optics (e.g., MERIS), providing the TFOV, the required geometrical (EFL) and radiometric resolution via the performance compliant effective aperture, focal length and/or or f-number. The front optics can be a lens system for limited spectral ranges (see Fig. 4 for MERIS), or a reflective mirror optics for wider spectral ranges and high throughput (see Fig. 7 for HRIS), mostly combined with the need of some aspheric corrector elements, or mixed catadioptric systems (e.g., HYDICE)
- A *spectral separation unit* (when covering more than one spectral region), e.g., by dichroic beam splitters or in-field separation. A dedicated example for the latter is the HRIS in-field separation concept which is unique in providing a single pixel separation in along-track direction
- Following the *spectrometer entrance slit*, the spectrometer optics, single or dual, consisting of a collimating system for a collimated beam onto the spectral disperser (grating or prism(s)), and an imager optics, in a Littrow configuration (e.g., ROSIS), or a Non-Littrow arrangement (e.g.,

HRIS) following the disperser, and focusing the dispersed beams onto the detector array, often containing some corrective elements for focal plane curvature correction, aberration or slit curvature correction—a rather crucial sub-unit; as indicated above, the spectral disperser could be transmissive, reflective or holographic gratings (all with spectral range limitations and/or low optical throughput), or single or dual prisms

- The *Focal Plane Asembly (FPA)*, consisting of the linear (whiskbroom) or 2-d (pushbroom) detector array(s), its mounting, cooling for IR, and analogue control and readout electronics, plus the HRIS SWIR Array. For wavelengths ≥ 1.1 µm, silicon detectors can no longer be used, this is where IR detector array technology has to be introduced. For linear arrays of a whiskbroom IS, there is quite a choice. For 2-d arrays of pushbroom instruments, either narrow-band InGaAs type lattice materials with GaAs multiplexers can be used, or low-responsivitiy Si:Pt (or-:Pd) Schottky Barrier arrays with Si MUXes, or hybridized InSb arrays (HYDICE) or CMT (HgCdTe) ir diode arrays mated to a Si MuUX, as for HRIS. For technical reasons (CTE matching between MUX and IR detector material), the number of spatial elements is limited to 256 or at the most 512 pixels for a monolithic array, thus, optical or mechanical butting techniques have to be applied, which makes these arrays a costly item. Another problem is for MUX layout the low photon number in SWIR on one hand, asking for very small injection capacitances, and the high dynamic range and charge handling capacity.

Note that IR detector arrays (including SWIR) require cooling, either by a passive radiator cooler (MODIS, HRIS) or by active Stirling coolers (PRISM). Furthermore, their location is not indifferent: they need cold space viewing, which must be taken into account in the optics design. See Fig. 5 (PRISM) as an example.

- All these units are typically accommodated on a common optical bench in the *optics module* housing, including thermal control and harness
- The *digital processing electronics* also represents a rather critical sub-unit, at least for the higher resolution instruments with often over 200 spectral channels. Together with some additional control electronics, these systems provide for line frequencies above 100 Hz and pixel rates (at the FPA output channels) of the order of 5 Mpixels/s. The corresponding internal data rates often exceed 300 Mbps, requiring on-line data correction (calibration), channel and resolution selection and data compression—for space versions rad-hard. Fig. 4-4 schematically shows a generic digital signal processing electronics.

In addition to a real-time transmission mode, a *data storage system* is often provided, with advance solutions typically based on solid-state memory devices with capacities now reaching > 50 Gbit
- The *instrument control electronics* which provides the command interface to the platform, as well as mode and thermal control, plus some fair amount of processing software
- The *in-flight calibration unit*, for the VNIR and SWIR typically includes a diffuser against Sun, or an aperture plate with small holes for direct Sun viewing, and/or calibration lamps, integrating spheres (radiation sensitive), plus possibly active sources such as laser diodes for wavelength calibration, if atmospheric absorption bands and the solar Fraunhofer bands are not precise enough
- In addition to the instrument space flight hardware, there is a considerable demand for on-ground pre- and post-launch equipment for the instrument's assembly, alignment, performance and functional testing and, after launch, for calibration and data processing.
  The latter is still an area of basic research at least in Europe, as far as algorithms are concerned

All these sub-units have to be designed, verified by finite element and performance analyses harmonized on their interfaces, assembled and verified, where necessary, by breadboarding and space qualification testing prior to the submission to the platform and its integration. This is a long and thoroughly planned activity, whose complexity may often exceed that of designing the satellite itself.

Some examples of representative IS concepts are shown in Fig. 4 (MERIS optical scheme), Fig. 5 (PRISM optical scheme and instrument layout), Fig. 6 (HYDICE), Fig. 7 (HRIS with high-resolution TMA telescope), and Fig. 8 (HRIS with medium FoV front optics).

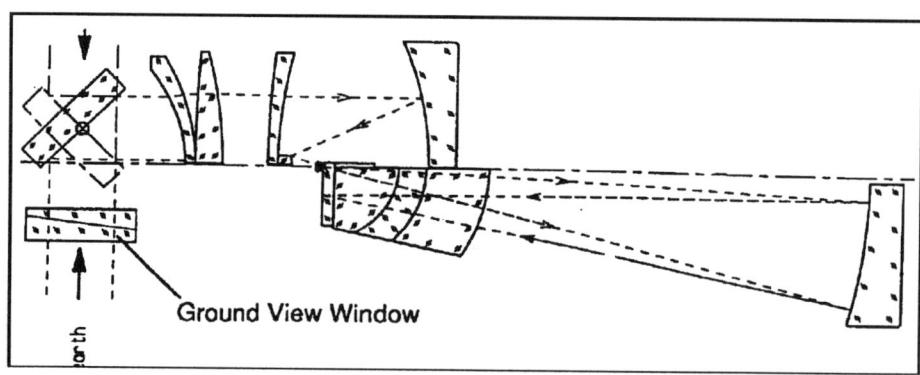

Figure 4. MERIS optical scheme

30. HYPERSPECTRAL IMAGER SURVEY AND DEVELOPMENTS FOR SCIENTIFIC AND OPERATIONAL LAND PROCESSES MONITORING APPLICATIONS

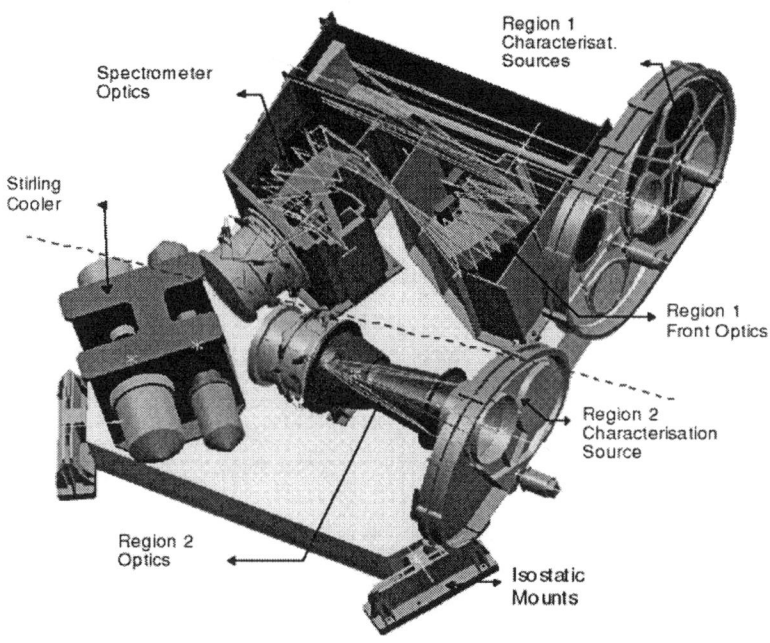

Figure 5. PRISM Optical and Instrument Scheme

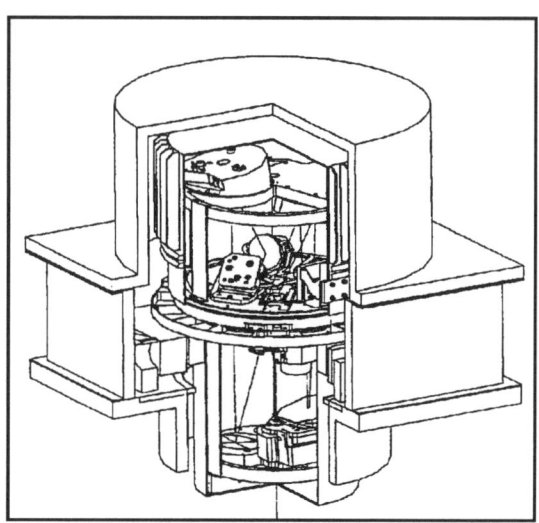

Figure 6. Airborne HYDICE (Hyperspectral Digital Imagery Collection Experiment), NRL (Kunkel et al., 1997), with single InSb Detector Array for 400 - 2500 nm spectral range

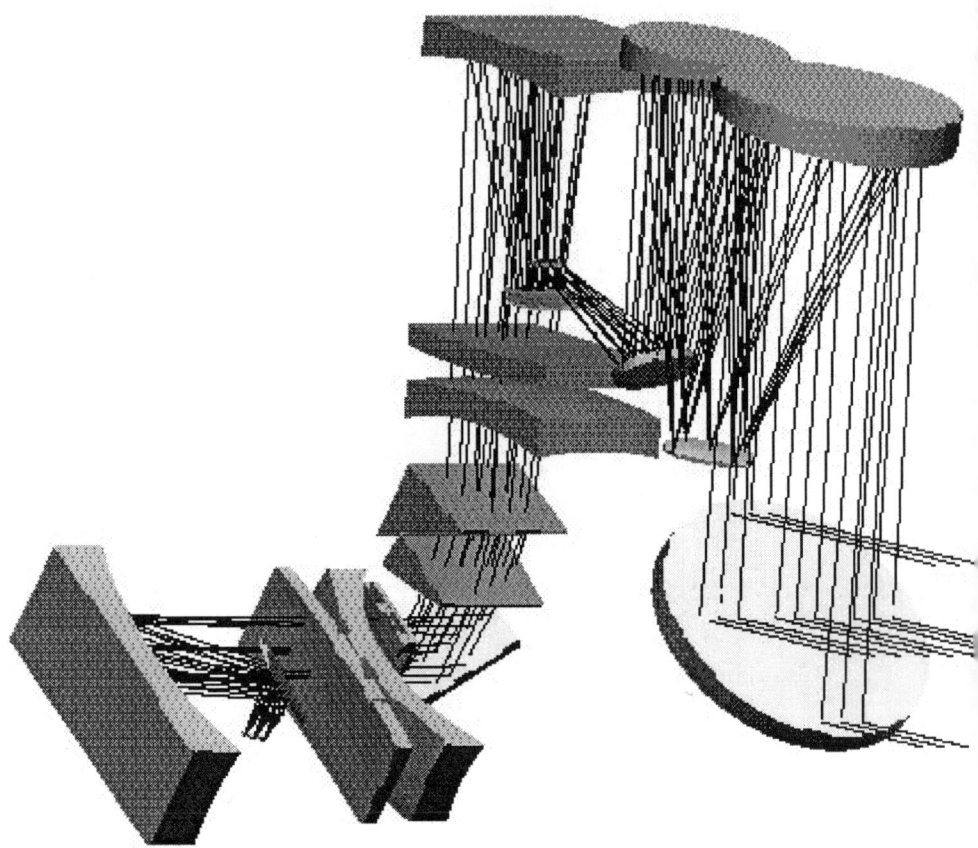

Figure 7. Nominal HRIS Design with TMA Front Optics for spaceborne Version (currently deleted in favour OF PRISM); for simplification only SWIR optical path in spectrometer (Kunkel et al., 1994)

30. HYPERSPECTRAL IMAGER SURVEY AND DEVELOPMENTS FOR SCIENTIFIC AND OPERATIONAL LAND PROCESSES MONITORING APPLICATIONS

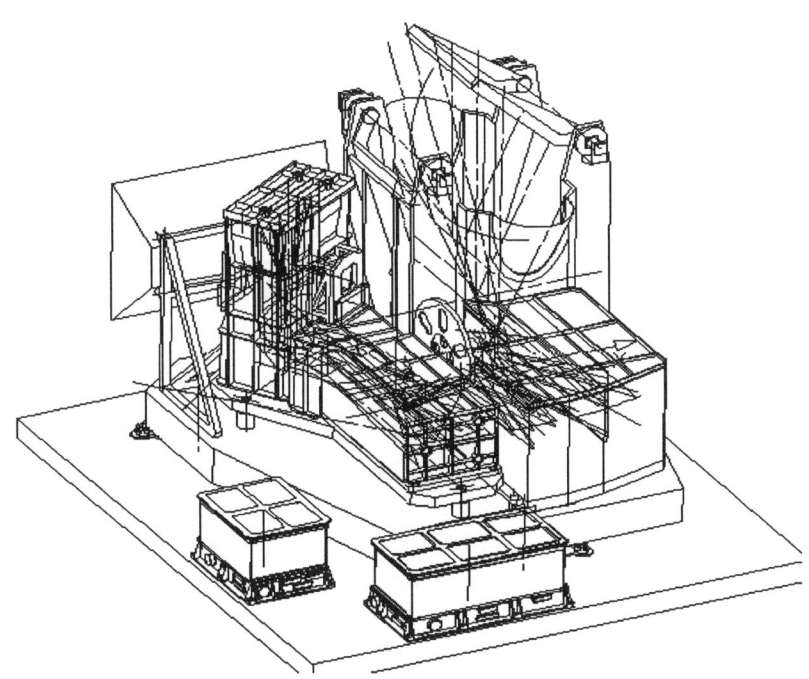

Figure 8 Modified HRIS optical Scheme with Medium FoV Front Optics for airborne Demo Model Version

Figs. 7 and 8 illustrate the IS optical scheme complexity, compared to "simple" pushbroom imagers. In the course of the ESA-ESTEC funded HRIS development program (see space-borne version design with its huge baffle and pointing unit), both optics are under development as hardware units. The spectrometer has been completed and fully tested by our subcontractor Officine Galileo, and is currently subject of some improvements. The objective now is to integrate all HRIS breadboards—

optics, FPA, electronics—to an airborne demonstrator. In the course of this program, it turned out that the high-resolution TMA (Three Mirror Anastigmat) telescope, in its final assembling stage at Carl Zeiss, does not make sense for aircraft missions since it would yield much too high spatial resolution, and at least in SWIR this would be sacrificed by a too small swath width. Thus, our co-author E. Schmidt designed a new medium FoV optics schematically shown in Fig. 8, to fit to the same spectrometer optics. As can be seen, the new telescope is comparatively small, but in order to get the telescope focus via a 200 mm back focal length into the existing spectrometer entrance, a massy Offner relay optics had to be introduced. Here, the wide spectral range was a design driver in view of the necessity to design all-reflective optics.

A rather peculiar and interesting IS represents the airborne HYDICE, see Fig. 6. It utilizes a single hybrid InSb detector array for the total spectral range from 400 to 2450 nm, with two different dotation (passivation) zones. This advanced array is currently not available outside the USA. A butting to about 1000 spatial pixels is currently not planned.

The PRISM instrument for the ESA Earth Explorer Programme, as sketched in Fig. 5, is peculiar with respect to its ambitious calibration scheme, induced by the rather tight ESA mission and performance requirements in terms of radiometric accuracy. The in-flight calibration devices are accessed by a pointing mirror rotated in the optical axis for across-track targets access, which also allows to point the optical entrance beam to cold space for the TIR channels, the Sun (seen through an aperture plate with a set of tiny holes, illuminating some 50 pixels each, and also occasionally to the moon (its value for on-board calibration is subject of an on-going debate among the experts). The calibration subsystems include:

- the mentioned aperture plate with a pattern of thin holes for absolute calibration of the VNIR and SWIR channels, used against the Sun over the equator, also providing a section for dark current measurements
- a protected rotatable diffuser against Sun for relative calibration of the total FoV in relation to the apertures
- a cavity IR blackbody for the two TIR channels, floating temperature, but precisely measured
- optional internal laser sources for calibration of the wavelength location on the arrays.

In earlier phases of the PRISM studies, a two-axis pointing unit had been included for along-track pointing as well as cross-track to enable multidirectional reflectance function (MDRF) measurements. In current concepts, the along-track pointing is left to the platform attitude control. The MDRF measurements have to be paid by long passes of more than 2500 km (angle dependent) without other target areas access, thus, it is primarily considered for scientific mission such as PRISM, while operational missions such as the Australian ARIES most likely can not afford such data take losses.

Not shown because of lack of suitable figures is ARIES, for which CSIRO claims that it shall be launched by the end of 1999. This is a rather dedicated and purely commercial IS approach, primarily aiming at providing data for Australian mining companies who also pay for the mission. Its SWIR part is therefore limited to 2.0 to 2.4 µm range, see Table 2B, giving up the tremendous use of the 1.5–1.7 µm range, e.g., for vegetation status monitoring. Since the HSI IS test mission on the Lewis small-sat just failed due to loss of the satellite control, it may be come the first space-borne IS.

5. OUTLOOK, DEVELOPMENT TRENDS

This paper can give a rather general survey of the status of IS developments and achievements. The main messages and conclusions are:
1. We have established through wisely defined, though not generously funded contracts, particularly by ESA, the necessary instrument technology towards advanced pushbroom IS in terms of:
   - smart optical design and manufacturing capabilities
   - sophisticated detector array technology, with high demands on the VNIR CCD´s, but emphasis on the SWIR arrays; the HRIS-based hybrid 2-d CMT technology appears to be promising despite some problems yet, and still some cost effort is needed to provide a 2-d butted array with 1024 spatial pixels
   - the rather demanding analog and digital processing electronics is just at hand. It is useful up to about 200 Hz line frequencies, or pixel sizes close to 30 m
   - sophisticated but elaborate calibration devices have been designed and partly successfully tested.
2. The least explored ground seems to be the on-ground data processing of IS data in terms of geometric and essentially radiometric correction algorithms compared to "classical" multi-spectral imager algorithms. There are a variety of activities, but a major lack is suitable data: DAIS and AVIRIS are not representative as whiskbroom instruments, ROSIS and CASI do not provide SWIR bands – "HyMaS" is our attempt to close such gaps.
3. DSS thus invites the attendant user representatives and organizations to support and join our just initiated HyMaS program, a proposal for the EU

is in preparation for the Area 3.2(2) program calling for "a system approach to understanding the operational requirements of EO systems". Support is welcome in the sense of target site provision, preparation and attendance during over-flights, resulting costs are to be paid by corresponding local/regional authorities or research groups. This would help to close the remaining funding gap for the HyMaS instrument and mission development.

While the IS envisaged herein and in associated programs are rather well covered technically, there is a variety of further mission and technical routes, some examples are briefly addressed here, without any priority assignment:

- The currently demanded *spectral regimes* cover always the same regions, i.e., the VNIR, typically 400–1050 nm, and SWIR (1000–2500 nm), occasionally discrete MIR and/or TIR bands are provided. However, technically, the *spectral range can be extended towards lower wavelengths* (e.g., down to 200 nm, for atmospheric measurements), as well as *longer wavelengths,* with a limit for push-broom type instruments at about 12 µm for detector array reasons, e.g., for geological/ petrological applications (NB: the CMT SWIR arrays are derivatives from longer wavelength detector arrays!)
- Though not demanded by agencies such as ESA, but by distinct users, it would be technically rather *easy to add a high-resolution panchromatic channel or 2–3 wide band spectral channels* to an IS for the a.m. optics developments (actually: HSI had it), simply by adding a linear array with detector pixel sizes a factor of 3–4 smaller, e.g., 10 m SSP compared to IS hyperspectral bands 40 m
- *Higher spatial and/or spectral resolution* is technology-limited to about 30 m at ≤ 800 km altitude (electronics); even 30 m would yield a rather large IS which cannot be accommodated on smaller satellites. A simplified "PRISM" is about the limit for "small-satellites", i.e., the requirements particularly for radiometric accuracy need to be relaxed to arrive at compact instruments
- American companies started to develop *Fourier transform imaging interferometers* instead of dispersing spectrometers; we do not see the drive for this trend unless spectral resolutions below 1 nm are demanded. However, the European industry is preparing to also address this topic, so far on in-house funds
- Another point to be addressed by users with experience in the processing of IS data is the general *reluctance to accept "quasi loss-free" data compression algorithms* instead of perfectly loss-free. We define "quasi loss-free" as losses in the noise of the instrument. The gain in data rate for on-board storage and transmission would be tremendous, typically a factor ≥ 4. We see this as essential for later operational space-borne IS,

requiring some basic research work to define the tolerance limits. In the same context, the burden of extreme redundancy in the spectral channel data could be addressed.

6. ACKNOWLEDGEMENTS:

The authors would like to thank all their colleagues at DSS as well as all subcontractors of DLR, GEC Marconi IR, MATRA MSF, Officine Galileo for their contributions. Particular thanks are directed to the three ESTEC co-authors and several of their colleagues. Finally, we would like to thank the conference organizers for giving us the opportunity for this presentation.

7. REFERENCES

Anonymous (1994) Final Report on the High Resolution Imaging Spectrometer HRIS, Doc. Ref. HS-DAS-FR-01, May 1994, plus Draft Final Report of "HRIS CCN-06 Follow-on Activities" (unpublished report by ESA).

Basedow, R.W. (1995) HYDICE System Performance- an Update, *SPIE* Vol. **2821**, p. 76 f.

Baudin, G. et al. (1993) Medium Resolution Imaging Spectrometer (MERIS), *SPIE* Vol. **2209**, 115.

Blechinger, F. et al. (1995) Optical Concepts for High Resolution Imaging Spectrometers, *SPIE Conference*, Orlando, USA.

Kunkel, B. et al. (1993) High Resolution Imaging Spectrometer "HRIS" - Optics, Focal Plane and Calibration, *SPIE Conference*, Orlando, USA.

Kunkel, B. et al. (1994) HRIS High Resolution Spectrometer - A challenging Instrument in terms of Photonics, *ICAPT 94*, Toronto, Canada.

Kunkel, B. et al. (1996) High-Resolution Hyperspectral Imagers for Land Processes in ESA Earth Observation Programmes (HRIS and PRISM), *IAF Congress*, Paper No. IAF-B 03.02.

Kunkel, B. et al. (1997) Hyperspectral Imager Survey and Developments for Scientific and operational Land Processes Monitoring Applications, *ENVIROSENSE*, Munich, Germany.

Pagano, T.S. et al. (1995) Development of the Moderate Resolution Imaging Spectrometer (MODIS) Proto-flight Model, *SPIE* Vol. **2820**, p. 10f.

Rast, M. (1991) *Imaging Spectroscopy*, **ESA SP-1144**, Noordwijk.

Staenz, K. (1992) A Decade of Imaging Spectrometry in Canada, *Canadian Journal of Remote Sensing*, **18**.

Vogel, P. (1997) DLR-IWS: Bestimmung des Blattflächenindexes und der absorbierten, photosynthetisch aktiven Strahlung der Vegetation aus bidirektionalen Reflexionsfaktoren am Oberrand der Pflanzendecke, *DLR-Forschungsbericht*, **97-25**, Juli 97.

Chapter 31

SUMMARY AND CONCLUSIONS

M. M. Verstraete (1) and M. Menenti (2, 3)
(1) Space Applications Institute, Ispra, Italy, (2) The Winand Staring Centre for Integrated Land, Soil and Water Research, Wageningen, The Netherlands and (3) Université Louis Pasteur, Illkirch, France.

This first ENAMORS conference provided a stimulating environment to discuss the challenges and opportunities provided by the upcoming generation of space sensors. This book constitutes the formal proceedings of that workshop. Additional information on the activities and achievement of this Concerted Action is available from the ENAMORS World Wide Web site (http://www.enamors.org/). The main conclusions of the meeting were as follows:

1. The participants expressed concern about the state of preparations in Europe to effectively address the scientific and technical issues arising from the upcoming new generation of space sensors to be operated by national and international agencies. While considerable efforts have been made to design and launch hardware in space, it was felt that insufficient attention and resources have been dedicated to the technical issues of acquiring, archiving, and distributing data and information in Europe, to the thorny issue of effectively translating radiance measurements made in space into useful information usable in practical applications, and to stimulating basic and applied research in support of the long term development of this economic sector in the coming decade.

2. Specifically, the participants endorsed and amplified a point eloquently stressed by John MacDonald concerning the large knowledge gap between the scientists and engineers developing the space platforms and sensors on one side, and the actual 'end users' of Earth Observation

information on the other. Much effort must urgently be directed to fill this gap by promoting methodological research and development, and demonstrating the feasibility and worthiness of remote sensing data gathered in space. This is particularly important if the field of Earth Observation is to stimulate, and sustain in the long run, commercially viable operational activities.
3. Another major concern of the participants was related to the access to European and non-European remote sensing data. It was widely recognized that the distribution of data in Europe would remain the duty and prerogative of the space agencies and other data providers, both commercial and non-commercial. The role of the CEO program of the Space Applications Institute in promoting standards, formats, and catalog interoperability as well as in organizing the providers and users of remote sensing data in a large decentralized network was well recognized and should be pursued.
4. Nevertheless, many participants suggested that a European institution should take the lead in providing the R&D necessary to stimulate new and better products and services, in collaboration with European universities, national research institutions, space agencies and industry. Such a research and coordination institution could also play a leading role in providing or encouraging the provision of ancillary data sets required for the proper interpretation of the data, in developing and evaluating new or better tools, techniques and approaches to convert data into useful information, and in offering calibration, validation and testing facilities to establish the accuracy and reliability of the information, products and services.

In his introductory talk, Alan Cross outlined the state of preparation of the European 5th Framework Programme of Research and Technological Development, and suggested that specific points be addressed at the conference. The following points were made in response to these questions:
- *Methodological research:* A much stronger effort should be made in this category than has been done in the past. The quantitative use of remote sensing data to address, accurately and reliably, significant climate and environmental questions has barely been started. Many of the techniques available so far rely on empirical qualitative approaches. Indeed, the advent of a new generation of advanced space sensors within the coming months and years will necessitate urgent developments to optimally exploit the new spectral, directional and spatial resolutions afforded by these sensors, as few of the existing algorithms, tools and techniques will even be appropriate to analyze these data. Similarly, taking advantage of the synergy between these instruments will also

require substantial new developments. The relevance of robust algorithms stretches beyond being scientifically correct. It is unlikely that an efficient market will select the most effective applications when the latter rely on empirical, qualitative approaches which mean different thing to different individuals.

The long term maintenance and growth of this economic sector can only be sustained if the information generated is really worth the efforts and costs. This implies carefully calibrated data, availability of extensive ancillary data sets, as well as availability of a panoply of efficient, accurate and reliable tools to analyze these data and generate information. A high priority should thus be given to the design, implementation, evaluation and delivery of accurate and reliable algorithms to derive useful information from EO data. The scope and difficulty of the tasks, as well as the cost of the required research and development programs necessitate a significant involvement of the European Commission, to stimulate creative initiatives, to maintain and enhance the position of European industries in this highly competitive field, and to guarantee the performance and quality of the products and services to the ultimate users of the information.

By contrast, those fields of application where the objectives and methodology have been established and properly validated should be encouraged to develop into more operational schemes. Thus, some 40% or more of the total 5FP budget devoted to EO could be allocated to methodological research to ensure the optimal exploitation of the new sensors and the exploration of new ideas and applications, while the remaining 60% could be used to promote the operationalization of existing applications.

- *European dimension:* Many of the critical R&D issues that affect the long term prospects for a commercial exploitation of Earth Observation from space require sizeable human, data processing and financial resources and can only be entertained at the European level. These include the creation, maintenance and continuous improvement of large scale facilities such as R&D institutions to generate and evaluate algorithms, to conduct research on synergy, or to implement approaches requiring the merging of multiple data streams from both space and non space data. The role of national initiatives was recognized as valuable, especially for specific applications mostly relevant to particular geographical areas. Nevertheless, integrated European programs of focused methodological research are highly desirable to permit an optimal use of available resources, to avoid duplication of investigations, and to ensure the wide distribution of advanced tools and techniques of data interpretation.

- *Role of users in the process:* Users should be involved in the specification of the ultimate outcome, the type and characteristics of the desired products, and the evaluation of the adequacy of the generated products to these requirements. On the other hand, true end users should be indifferent to the actual source of data exploited (which instrument is used) or to the processing (which algorithm is applied). The participants also noted that true end users cannot be the sole or perhaps even the main driver in a technological field so remote from their own experience. Hence, the scientific and engineering communities must strive to communicate the potential of new approaches and the feasibility of deriving new or better information from existing or future sensors.
- *Pre-operational pilot and demonstration projects:* The participants recognized that there might be value in a diversified approach towards the selection of such projects. Specifically, the projects selected for funding should be identified by a panel wide enough to represent the user and scientific community, industry, and the Commission. The latter, for its part, should be encouraged to publish its specific requirements for information to fulfill EU policies and mandates.
- *Technology transfers:* Workshop participants were uneasy and divided regarding the relations between the research world and the commercial world. All recognized and agreed that research should ultimately lead to applications in line with the requirements of the community of users. Initiatives in this direction are currently emerging, for instance through the activities of the Centre for Earth Observation. However, many participants insisted the worth and long term benefits of the research endeavor cannot be measured exclusively in financial terms. Public support for exploratory research not driven by immediate applications remains necessary. Indeed, past experience shows that many discoveries that turned out to be critical did not and probably could not have been obtained through a program of directed actions. In the world as we know it today, information bestows power. As a result, serious ethical issues and conflicts of interest arise when the open-ended world of scientific research is made to operate within the rules of the competitive, for-profit commercial world.

APPENDIX
List of participants

Dr. Ghassem Asrar
NASA Headquarters
Mail Code YS, 300 E Street, SW
Washington, DC 20546
United States
Tel: [1] (202) 358-2165
Fax: [1] (202) 358-3092
E-mail: gasrar@mail.hq.nasa.gov

Dr. Wim Bastiaanssen
International Water Management Institute (IWMI)
P.O. Box 2075
Colombo
Sri Lanka
Tel: [94] (1) 86 74 04
Fax: [94] (1) 86 68 54
E-mail: W.Bastiaanssen@cgiar.org

Dr. Alan S. Belward
Space Applications Institute (SAI)
EC Joint Research Centre, TP 440
I-21020 Ispra (VA)
Italy
Tel: [39] 03 32 78 92 98
Fax: [39] 03 32 78 90 73
E-mail: alan.belward@jrc.it

Dr. Bhaskar J. Choudhury
NASA Goddard Space Flight Center (GSFC)
Hydrological Sciences Branch
Mail Code 974
Greenbelt, MD 20771
United States
Tel: [1] (301) 614-5767
Fax: [1] (301) 614-5808
E-mail: bhaskar@te.gsfc.nasa.gov

Prof. Emilio Chuvieco
Department of Geography
Colegios, 2
ES-28801 Alcalá de Henares (Madrid)
Spain
Tel: [34] (1) 885-4438/4429
Fax: [34] (1) 885-4439
E-mail: ggecs@geogra.alcala.es

Mr. Alan Cross
European Commission, DG XII/D/4
200 rue de la Loi
B-1049 Brussels
Belgium
Tel: [32] (2) 296-4961
Fax: [32] (2) 296-0588
E-mail: alan.cross@dg12.cec.be

Prof. Jürgen Fischer
Free University of Berlin
Institute for Space Sciences
Fabeckstrasse 69
D-14195 Berlin 33
Germany
Tel: [49] (30) 838-6663
Fax: [49] (30) 832-8648
E-mail: fischer@zedat.fu-berlin.de

Dr. Anatoly Gitelson
Ben-Gurion University
Remote Sensing Laboratory
Sede-Boker Campus 84990
Israel
Tel: [972] (7) 659-6858
Fax: [972] (7) 659-6909
E-mail: gitelson@bgumail.bgu.ac.il

APPENDIX

Dr. Nadine Gobron
Space Applications Institute
EC Joint Research Centre, TP 440
I-21020 Ispra (VA)
Italy
Tel: [39] 03 32 78 63 38
Fax: [39] 03 32 78 90 73
E-mail: nadine.gobron@jrc.it

Dr. Yves Govaerts
EUMETSAT
Am Kavalleriesand 31
D-64295 Darmstadt
Germany
Tel: [49] (6151) 80 73 62
Fax: [49] (6151) 80 73 04
E-mail: govaerts@eumetsat.de

Dr. Ian Grant
CSIRO Division of Atmospheric Research
PMB 1
Aspendale VIC 3195
Australia
Tel: [61] (3) 92 39 46 68
Fax: [61] (3) 92 39 44 44
E-mail: ifg@dar.csiro.au

Prof. Martti Hallikainen
Helsinki University of Technology
Laboratory of Space Technology
Otakaari 5 A FIN-02015 HUT
Finland
Tel: [358] (9) 451-2371
Fax: [358] (9) 451-2898
E-mail: Martti.Hallikainen@hut.fi

Dr. Tuomas Häme
VTT
Automation
PO Box 13002
FIN-02044 VTT
Finland
Tel: [358] (9) 456-6282
Fax: [358] (9) 456-6475
E-mail: Tuomas.Hame@vtt.fi

Dr. Tamotsu Igarashi
National Space Development Agency of Japan
Roppongi First Bldg. 13 F
1-9-9, Roppongi, Minato-ku, Tokyo, 160
Japan
Tel: [81] (3) 32 24 70 53
Fax: [81] (3) 32 24 70 52
E-mail: igarashi@eorc.nasda.go.jp

Dr. Anne Jochum
Instituto de Desarrollo Regional
Universidad de Castilla-La Mancha
Avenida de Espana s/n
E-02071 Albacete
Spain
Tel: [34] (67) 59 92 00 ext 2629
Fax: [34] (67) 59 92 33
E-mail: ajochum@prov-ab.uclm.es

Dr. David Jupp
CSIRO
Earth Observation Centre , GPO Box 3023
Aspendale VIC 3195
Australia
Tel: [61] (6) 216-7203
Fax: [61] (6) 216-7222
E-mail: david.jupp@cossa.csiro.au

Prof. Risto Kuittinen
Finnish Geodetic Institute
Photogrammetry and Remote Sensing , POBox 15
FIN-02431 Masala
Finland
Tel: [358] (9) 29 55 53 05
Fax: [358] (9) 29 55 52 00
E-mail: Risto.Kuittinen@fgi.fi

Dr. Bernd P. Kunkel
Dornier Satellitensysteme GMBH
P.O.Box 80 11 69
D-81663 München
Germany
Tel: [49] (89) 60 72 41 61
Fax: [49] (89) 60 72 60 39
E-mail: bernd.kunkel@space.otn.dasa.de

Dr. Eric Lambin
Université Catholique de Louvain
Laboratoire de Télédétection et d'Analyse Régionale
3, Place Louis Pasteur
B-1348, Louvain-la Neuve
Belgium
Tel: [32] (10) 47 44 77
Fax: [32] (10) 47 28 77
E-mail: lambin@geog.ucl.ac.be

Heikki Laurila
Niklas Data Oy
Sinikalliontie 6
FIN-02630 Helsinki
Finland
Tel: [358] (9) 525-9020
Fax: [358] (9) 525-9022
E-mail: heikki.laurila@niklasdata.fi

Dr. Anssi Lohi
VTT Automation
P.O.Box 13002
FIN-02044 VTT
Finland
Tel: [358] (9) 456-4369
Fax: [358] (9) 456-6475
E-mail: anssi.lohi@vtt.fi

Dr. Wolfgang Lucht
Department of Global Change and Natural Systems
Potsdam Institute for Climate Impact Research
Telegrafenberg C4
Postfach 60 12 03
D-14412 Potsdam
Germany
Tel: [49] (331) 288-2533
Fax: [49] (331) 288-2600
E-mail: wlucht@pik-potsdam.de

Dr. John S. MacDonald
MacDonald Dettwiler
13800 Commerce Parkway
Richmond, B.C. V6V 2J3
Canada
Tel: [1] (604) 231-2223

Fax: [1] (604) 273-9830
E-mail: jsm@mda.ca

Dr. Massimo Menenti
Winand Staring Centre
P.O.Box 125
6700 Wageningen
The Netherlands
Tel: [31] (8370) 74-324
Fax: [31] (8370) 24-812
E-mail: menenti@sc.agro.nl

Dr. José Moreno
University of Valencia
Faculty of Physics, Department of Thermodynamics
46100 Burjassot
Valencia
Spain
Tel: [34] (6) 386-4350
Fax: [34] (6) 364-2345
E-mail: moreno@vm.ci.uv.es

Dr. Karri Muinonen
University of Helsinki
Observatory
P.O.Box 14
00014 Helsingin Yliopisto
Finland
Tel: [358] (9) 19 12 29 41
E-mail: Karri.Muinonen@Helsinki.fi

Dr. Ranga B. Myneni
Boston University
Department of Geography
Boston, MA 02215
United States
Tel: [1] (617) 353-5742
Fax: [1] (617) 353-8399
E-mail: rmyneni@bu.edu

Dr. Joel Noilhan
Météo-France
Groupe de Météorologie de Moyenne Echelle CNRM/GMME
42, av. Gustave Coriolis
F-31057 Toulouse Cedex
France

Tel: [33] (5) 61 07 94 74
Fax: [33] (5) 61 07 96 26
E-mail: noilhan@meteo.fr

Dr. Jouni Peltoniemi
Finnish Geodetic Institute
Photogrammetry and Remote Sensing
P.O.Box 15
FIN-02431 Masala
Finland
Tel: [358] (9) 29 55 52 12
Fax: [358] (9) 29 55 52 00
E-mail: Jouni.Peltoniemi@fgi.fi

Dr. Bernard Pinty
Space Applications Institute
EC Joint Research Centre TP 440
I-21020 Ispra (VA)
Italy
Tel: [39] 03 32 78 61 40
Fax: [39] 03 32 78 90 73
E-mail: bernard.pinty@jrc.it

Dr. Michael Rast
ESA/ESTEC
Earth Sciences Division
Postbus 299
NL-2200 AG Noordwijk
Keplerlaan 1
The Netherlands
Tel: [31] (71) 565-6565
Fax: [31] (71) 565-6040
E-mail: mrast@jw.estec.esa.nl

Dr. Mats Rosengren
Swedish Space Corporation
P.O.Box 4207
S-171 04 Solna
Sweden
Tel: [46] (8) 627-6200
Fax: [46] (8) 98 70 69
E-mail: mro@ssc.se

Dr. Gilbert Saint
Centre National d'Etudes Spatiales (CNES)
DP/MP/OT

BPI 2534, 18, Av. Edouard Belin
F-31401 Toulouse Cedex
France
Tel: [33] (5) 61 27 36 54
Fax: [33] (5) 61 27 41 72
E-mail: gilbert.saint@cnes.fr

Dr. Jesus San Miguel-Ayanz
Space Applications Institute
EC Joint Research Centre TP 950
I-21020 Ispra (VA)
Italy
Tel: [39] 03 32 78 61 38
Fax: [39] 03 32 78 55 00
E-mail: jesus.san-miguel@jrc.it

Dr. Michael Schaale
Free University Berlin
Institute for Space Sciences
Fabeckstrasse 69
D-14195 Berlin 33
Germany
Tel: [49] (30) 838-6663
Fax: [49] (30) 832-8648
E-mail: schaale@zedat.fu-berlin.de

Dr. Johannes Schmetz
EUMETSAT
Am Kavalleriesand 31
D-64295 Darmstadt
Germany
Tel: [49] (6151) 80 75 90 / 75 91
Fax: [49] (6151) 80 75 55
E-mail: schmetz@eumetsat.de

Dr. Madis Sulev
Tartu Observatory
EE2444 Toravere, Tartumaa
Estonia
Tel: [372] 741-0278
Fax: [372] 741-0205
E-mail: sulev@aai.ee

Dr. Kari Tilli
TEKES, Technology Development Centre
Space Technology

PL 69
00101 Helsinki
Finland
Tel: [358] (10) 521-5850
Fax: [358] (10) 521-5900
E-mail: Kari.Tilli@tekes.fi

Prof. Erkki Tomppo
METLA, Finnish Forest Research Institute
Unioninkatu 40 A
00170 Helsinki
Finland
Tel: [358] (9) 85 70 53 40
E-mail: erkki.tomppo@metla.fi

Dr. Jean Verdebout
Space Applications Institute
EC Joint Research Centre TP 440
I-21020 Ispra (VA)
Italy
Tel: [39] 03 32 78 50 34
Fax: [39] 03 32 78 54 61
E-mail: jean.verdebout@jrc.it

Dr. Michel M. Verstraete
Space Applications Institute
EC Joint Research Centre TP 440
I-21020 Ispra (VA)
Italy
Tel: [39] 03 32 78 55 07
Fax: [39] 03 32 78 98 30
E-mail: michel.verstraete@jrc.it

Dr. Barry K. Wyatt
Institute of Terrestrial Ecology
Abbots Ripton
Huntingdon, Cambridgeshire PE17 2LS
United Kingdom
Tel: [44] (1487) 77 33 81
Fax: [44] (1487) 77 34 67
E-mail: B.Wyatt@ite.ac.uk

Advances in Global Change Research

1. P. Martens and J. Rotmans (eds.): *Climate Change: An Integrated Perspective.* 1999
   ISBN 0-7923-5996-8
2. A. Gillespie and W.C.G. Burns (eds.): *Climate Change in the South Pacific: Impacts and Responses in Australia, New Zealand, and Small Island States.* 2000
   ISBN 0-7923-6077-X
3. J.L. Innes, M. Beniston and M.M. Verstraete (eds.): *Biomass Burning and Its Inter-Relationships with the Climate Systems.* 2000 ISBN 0-7923-6107-5
4. M.M. Verstraete, M. Menenti and J. Peltoniemi (eds.): *Observing Land from Space: Science, Customers and Technology.* 2000 ISBN 0-7923-6503-8

KLUWER ACADEMIC PUBLISHERS – DORDRECHT / BOSTON / LONDON